Influenza

Molecular Virology

Edited by

Qinghua Wang

Baylor College of Medicine
Houston, TX
USA

and

Yizhi Jane Tao

Department of Biochemistry and Cell Biology
Rice University
Houston, TX
USA

Caister Academic Press

Caister Academic Press
Norfolk, UK

www.caister.com

British Library Cataloguing-in-Publication Data
A catalogue record for this book is available from the British Library

ISBN: 978-1-904455-57-8

Cover image taken from Figure 5.1. Flu virus diagram by Yizhi Jane Tao and Robert Krug.

Printed and bound in Great Britain by the MPG Books Group

Contents

Contributors

Mark Bartlam
College of Life Sciences
Nankai University
Tianjin
China

bartlam@nankai.edu.cn

Zachary A. Bornholdt
Department of Immunology
The Scripps Research Institute
La Jolla, CA
USA

zachb@scripps.edu

Berenice Carrillo
Department of Molecular Virology and Microbiology
Baylor College of Medicine
Houston, TX
USA

bcarill@bcm.tmc.edu

James J. Chou
Department of BCMP
Harvard Medical School
Boston, MA
USA

james_chou@hms.harvard.edu

Erica D. Dawson
InDevR Inc.
Boulder, CO
USA

dawson@indeur.com

Michael W. Deem
Departments of Bioengineering and Physics & Astronomy
Rice University
Houston, TX
USA

mwdeem@rice.edu

Adolfo García-Sastre
Departments of Microbiology and Medicine
Global Health and Emerging Pathogens Institute
Mount Sinai School of Medicine
New York, NY
USA

adolfo.garcia_sastre@mssm.edu

Robert M. Krug
Institute for Cellular and Molecular Biology
University of Texas at Austin
Austin, TX
USA

rkrug@icmb.utexas.edu

Rei-Lin Kuo
Institute for Cellular and Molecular Biology
University of Texas at Austin
Austin, TX
USA

rlkuo@mail.utexas.edu

Yingfang Liu
National Laboratory of Biomacromolecules
Institute of Biophysics
Chinese Academy of Sciences
Beijing
China

liuy@ibp.ac.cn

Zhiyong Lou
Laboratory of Structural Biology
Tsinghua University
Beijing
China

louzy@xtal.tsinghua.edu.cn

Peter Palese
Departments of Microbiology and Medicine
Mount Sinai School of Medicine
New York, NY
USA

peter.palese@mssm.edu

Natalie Pica
Department of Microbiology
Mount Sinai School of Medicine
New York, NY
USA

natalie.pica@mssm.edu

Ramdas S. Pophale
Department of Bioengineering
Rice University
Houston, TX
USA

ramdas@rice.edu

B.V. Venkataram Prasad
Verna Marrs Mclean Department of Biochemistry and Molecular Biology
Baylor College of Medicine
Houston, TX
USA

vprasad@bcm.tmc.edu

Zihe Rao
Laboratory of Structural Biology
Life Sciences Building
Tsinghua University
Beijing
China

raozh@xtal.tsinghua.edu.cn

Kathy L. Rowlen
InDevR Inc.
Boulder, CO
USA

rowlen@indevr.com

Jason R. Schnell
Department of Biochemistry
University of Oxford
Oxford
UK

jason.schnell@bioch.ox.ac.uk

David A. Steinhauer
Department of Microbiology and Immunology
Emory University School of Medicine
Atlanta, GA
USA

steinhauer@microbio.emory.edu

Yizhi Jane Tao
Department of Biochemistry and Cell Biology
Rice University
Houston, TX
USA

ytao@rice.edu

Terrence M. Tumpey
Influenza Division
National Center for Immunization and Respiratory Diseases
Centers for Disease Control and Prevention
Atlanta, GA
USA

tft9@cdc.gov

Qinghua Wang
Department of Biochemistry and Molecular Biology
Baylor College of Medicine
Houston, TX
USA

qinghuaw@bcm.tmc.edu

Qiaozhen Ye
Department of Biochemistry and Cell Biology
Rice University
Houston, TX
USA

qiaozhen@rice.edu

Chen Zhao
Institute for Cellular and Molecular Biology
University of Texas at Austin
Austin, TX
USA

chenzhao@mail.utexas.edu

Hao Zhou
Department of Bioengineering
Rice University
Houston, TX
USA

mwdeem@rice.edu

Preface

The battle between humans and the influenza virus is never-ending. In the last century, there have been three major pandemics, the 1918 Spanish Flu, the 1957 Asian Flu and the 1968 Hong Kong Flu, which claimed approximately 50 million, 2 million and 1 million human lives, respectively. In addition, the annual epidemics from influenza A virus H1N1 and H3N2 subtypes and influenza B virus affect 5–20% of human populations, cause severe illness in 3–5 million patients and result in 250,000–500,000 deaths each year worldwide. It was estimated that the cumulative impacts of the annual influenza epidemics in the 20th century even exceeded those of the three major pandemics combined. Emerging in 1997, influenza A virus H5N1 subtype has a record high fatality rate of more than 50% among infected humans, much higher than the 2–20% rate of the most devastating 1918 Spanish Flu.

Over the past few years, driven largely by the grave concern for a new influenza pandemic, the influenza research community has made many scientific breakthroughs. These breakthroughs span broad areas in influenza virus research, such as structure and assembly of the virus, the virus–host interactions, host immune responses, new methods for rapid diagnostics, and novel strategies for effective vaccine and antiviral developments. Thus, a major goal of this book is to highlight some of the most exciting discoveries in recent years. It is our hope that by staying at the forefront of influenza research, this book will help to identify some of the most pressing scientific problems in influenza molecular virology.

As this book goes to press, a major outbreak of influenza A virus H1N1 subtype is spreading to more than 50 countries and causing more than 100 human deaths. Although the effects of this new H1N1 virus are rather mild, influenza virologists and health officials fear that further mutations in the virus could lead to a new and potentially more dangerous outbreak in subsequent months. With its haemagglutinin from North American swine influenza, the new H1N1 virus is easily transmissible among human populations. Thus, the worst case is a 1918-like pandemic emerging from reassortments between the more deadly H5N1 virus and the more transmissible H1N1 virus.

At a time in which humanity faces such serious threats from influenza virus, we hope that this book will stimulate much more vigorous research in the field. Eventually, the combined efforts of scientists working on various aspects of influenza virology will help improve our preparedness for future outbreaks of this dangerous virus.

We are deeply indebted to all authors for their contributions, which made this book possible, and to Annette Griffin for initiative and help along the way.

Qinghua Wang and Yizhi Jane Tao
Houston, Texas, USA
May 2009

The NS1 Protein of Influenza A Virus

1

Chen Zhao, Rei-Lin Kuo and Robert M. Krug

Abstract

The NS1 protein of influenza A viruses is a small (230–237 amino acids), multifunctional dimeric protein that participates in both protein–RNA and protein–protein interactions. It comprises two functional domains: an N-terminal (amino acids 1–73) RNA-binding domain and a C-terminal (amino acids 74–230/237) effector domain. Here we focus on several of the best-characterized functional interactions of the NS1 protein. A major role of the NS1 protein is to counter host cell antiviral responses. Thus, the RNA-binding domain binds double-stranded (ds) RNA, thereby inhibiting the dsRNA activation of the antiviral oligo A synthetase/RNase L pathway that is induced by interferon-α/β (IFN-α/β). A region of the effector domain binds the protein kinase PKR, thereby preventing its activation that would otherwise lead to the shutdown of both viral and host protein synthesis. Another region of the effector domain binds the 30-kDa subunit of the cleavage and polyadenylation specificity factor (CPSF30), a cellular protein required for the 3′ end processing of all cellular pre-mRNAs. As a consequence, the substantial amount of IFN-β pre-mRNA that is synthesized in virus-infected cells is not processed to form mature IFN-β mRNA, thereby suppressing the IFN response. The NS1 protein also has other functions that are not directly involved in countering host antiviral responses. The effector domain of the NS1 protein binds the P85β regulatory subunit of phosphoinositide 3-kinase (PI3K), resulting in the activation of PI3K and the Akt kinase, which in turn inhibits apoptosis. The C-termini of pathogenic influenza A viruses have a PDZ-binding motif that has been implicated in pathogenicity. The NS1 protein also interacts with the cellular nuclear export protein (TAP), and may have a role in the nuclear export of viral mRNAs. Finally, the NS1 protein functionally interacts with the viral polymerase complex in infected cells and probably has a role in the regulation of viral RNA synthesis.

Introduction

The NS1 protein, which is encoded by the smallest genomic RNA segment (NS segment) of influenza A viruses, is designated as non-structural because it is synthesized in large amounts in infected cells, but is not incorporated into virions (Lamb and Krug, 2001). It is a small, multifunctional dimeric protein that participates in both protein–RNA and protein–protein interactions (Hale *et al.*, 2008c). Here we focus on several of the best-characterized functional interactions of the NS1 protein, which are diagrammed in Fig. 1.1.

Structure of the NS1 protein

The NS1 protein comprises two functional domains: (i) an N-terminal (amino acids 1–73) RNA-binding domain (RBD) and (ii) a C-terminal effector domain (ED) (Fig. 1.1) (Qian *et al.*, 1994). Most NS1 proteins are 230–237 amino acids long (as shown in Fig. 1.1). However, a small fraction are shorter, including the 2009 H1N1 ('swine flu') NS1, which is 219 amino acids long.

The NS1 RBD binds double-stranded RNA (dsRNA) in a non-sequence-specific manner

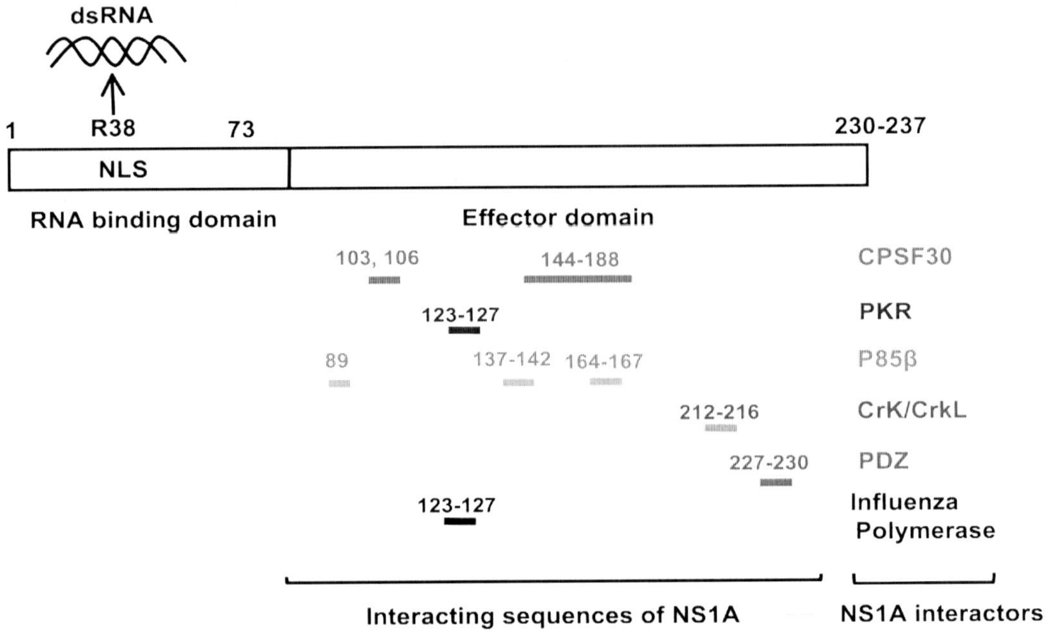

Figure 1.1 Schematic diagram of the NS1 protein and its sequences that interact with RNA and specific cellular/viral proteins. The NS1 amino acid residues involved in these interactions are denoted.

(Chien *et al.*, 2004; Hatada and Fukuda, 1992; Wang *et al.*, 1999). The solution NMR and X-ray crystal structures of the isolated RBD of the NS1A protein of influenza A/Udorn/72 (Ud) virus showed that it has a unique six-helical symmetric head-to-tail homodimer structure (Chien *et al.*, 1997; Liu *et al.*, 1997) (Fig. 1.2A). The symmetric subunits in this novel RNA-binding structure contain three long alpha helices (α1-α2-α3). The homodimer is stabilized by numerous hydrogen bonds, and by hydrophobic and electrostatic interactions. R38, which is the only amino acid that is absolutely required for dsRNA binding (Wang *et al.*, 1999), along with several other basic amino acids, are located in symmetry related, antiparallel helices (α2 and α2′) on one face of the protein (Chien *et al.*, 1997; Liu *et al.*, 1997). The dsRNA-binding surface is comprised of tracks of highly conserved basic and hydrophilic residues that are complementary to the phosphate backbone of A-form dsRNA (Fig. 1.2B). These tracks, which include R38 (and symmetry related R38′), are consistent with NMR chemical shift perturbation data that identified amino acids in Ud NS1(1–73) that interact with a 16-basepair dsRNA (Yin *et al.*, 2007). An important feature of the dsRNA-binding surface is a deep pocket consisting of numerous hydrophilic and hydrophobic residues, which is an inviting target for antiviral drug design. This binding mode was confirmed in a X-ray crystal structure of the NS1 RBD (amino acids 1–70) from influenza A/Puerto Rico/8/34 (PR8) in complex with a 19-basepair dsRNA (Cheng *et al.*, 2009). In this structure NS1 amino acids interact with backbone atoms and 2′-hydroxyls along the major groove of the RNA duplex (Fig. 1.2C). Numerous hydrophilic/charged amino acids on the binding face of NS1 contribute to dsRNA binding, particularly R38/R38′, which penetrates into the helix, forming hydrogen bonds with phosphates on both strands. In addition, the symmetry related R38/R38′ pair form hydrogen bonds with each other (Cheng *et al.*, 2009).

The X-ray crystal structure of the ED of several NS1 proteins showed that the ED is a homodimer, as is the case for the RBD (Bornholdt and Prasad, 2006; Hale *et al.*, 2008a; Xia *et al.*, 2009). Each ED monomer comprises seven β-strands and three α-helices (Fig. 1.3A). The β-strands form a twisted anti-parallel β-sheet around a long, central α-helix. The ~25 C-terminal

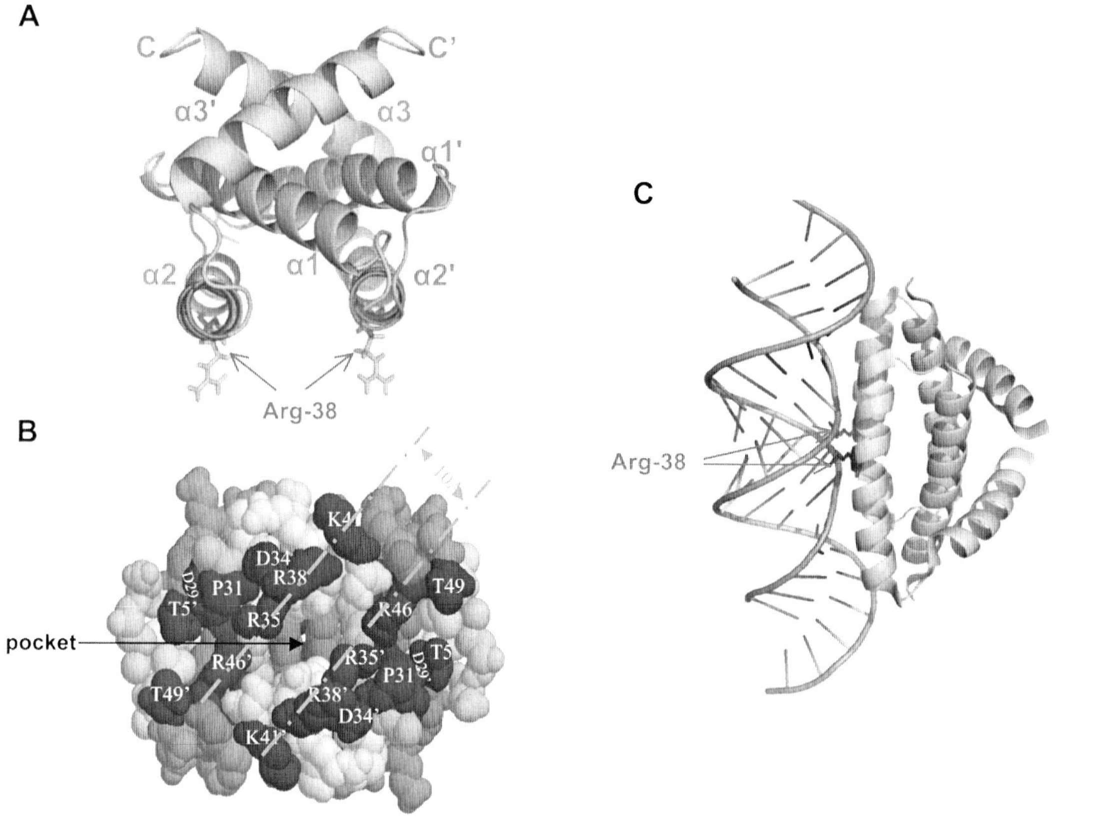

Figure 1.2 Structure of the NS1 RBD revealed by NMR and X-ray crystallography. (A) Ribbon stereo diagrams of the Ud NS1 RBD homodimer in free state. (B) Space filling model of the Ud NS1 RBD, showing tracks of conserved amino acids on the RNA-binding face of the protein. Amino acids are coloured according to their degree of conservation across the entire NS1 family, ranging from highest (black) to lowest (white). The separation across the basic edges of the conserved tracks corresponds to 10 Å, the width of the major groove in dsRNA. The pocket on the RNA-binding face is indicated. (C) Co-crystal structure of a NS1 RBD and a 19-bp dsRNA, confirming that Arg38 pairs as critical residues in dsRNA recognition.

amino acids of the ED were disordered in the crystal structure. Two different dimer interfaces have been proposed. One proposed interface is formed by hydrophobic interactions between the first N-terminal β-strand of each monomer (Fig. 1.3B) (Bornholdt and Prasad, 2006). The other dimer interface is formed by interactions of two long α-helices, with W at position 187 playing a crucial role in this dimerization (Fig. 1.3C) (Hale *et al.*, 2008a; Xia *et al.*, 2009). Because replacement of W with A at this position disrupted the dimer, as assayed by gel filtration, it is likely that the later dimerization interface exists in solution (Hale *et al.*, 2008a).

A crystal structure of the full-length NS1 protein of a pathogenic H5N1 virus (influenza A/Vietnam/1203/2004) has been published (Bornholdt and Prasad, 2008). To obtain this crystal structure, the investigators had to eliminate its dsRNA-binding activity by changing both R38 and K41 to A. Based on this crystal structure, the investigators proposed an intriguing, but speculative model for the way that multiple H5N1 NS1 proteins bind to long dsRNAs. They proposed that the RBD and ED monomeric subunits, instead of dimerizing to form a single dimeric NS1 protein, interact with their respective monomeric subunits from neighbouring NS1 proteins, thereby forming a chain of NS1 molecules with alternating RBD and ED dimers. They then speculated that this NS1 chain wraps around a long dsRNA to form a tubular structure. The ED dimerization interface in this model differs from both of the ED interfaces previously proposed

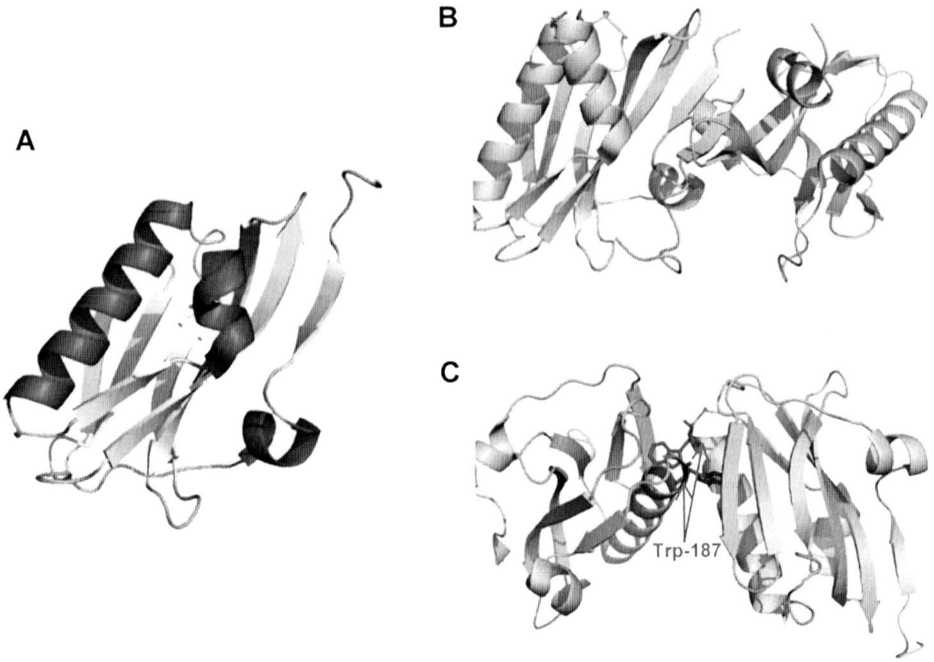

Figure 1.3 Crystal structures of the monomeric and dimeric NS1 ED from different influenza A viruses, demonstrating a common overall ED monomer structure but with different dimer interfaces. (A) Ribbon stereo diagram of the NS1 ED monomer. The three α-helices are dark coloured, and the β-sheets are light coloured. (B) Crystal structure of a NS1 ED dimer that locates the dimeric interface to the first N-terminal β-stand of each monomer. (C) Crystal structure of a NS1 ED dimer, showing a dimer surface provided by interactions of two long α-helices. The pair of W187 residues crucial for this interaction is indicated.

for the isolated ED, and the mode of dsRNA binding by the RBD predicted from this structure differs from that established for the isolated RBD described above.

The role of the NS1 protein in countering host antiviral responses

Cells respond to virus infection by mounting several antiviral defences. The major cellular response is the synthesis of interferon-α/β (IFN-α/β), which in turn activates the transcription of an array of genes encoding proteins that establish an antiviral state (Hale *et al.*, 2008b; Haller *et al.*, 2006; Randall and Goodbourn, 2008). Transcription of the IFN-β gene in infected cells requires the activation of two transcription factors, IRF-3 and NF-κB. Activation of these transcription factors in non-immune cells infected by influenza A virus is initiated by the cytoplasmic RIG-I protein, which recognizes viral RNA molecules containing a 5′ triphosphate (Pichlmair *et al.*, 2006). Binding of such a viral RNA molecule triggers a signalling pathway that activates IRF-3 and NF-κB. IRF-3 is activated by phosphorylation, leading to its dimerization and transport into the nucleus. NF-κB, which is retained in the cytoplasm by its association with IκB, is activated by the ubiquitin-mediated degradation of IκB (Hiscott, 2007). Another important component of the cellular antiviral system is the protein kinase PKR, which is constitutively expressed in mammalian cells and is further increased by IFN-α/β stimulation (Hovanessian, 1989; Meurs *et al.*, 1990). PKR is activated by binding either dsRNA or the cellular PACT protein, and then phosphorylates the α-subunit of the eIF2 translation initiation factor, resulting in the inhibition of cellular and viral protein synthesis and virus replication (Gale and Katze, 1998; Samuel, 1993).

The NS1 protein counters several of these host antiviral responses, mediated either by the dsRNA-binding activity of the RBD or by the binding of specific cellular proteins to regions of the ED.

Role of the dsRNA-binding activity of the RBD in countering the host antiviral response

Initially the functions of dsRNA-binding by the NS1 RBD were determined by overexpression of the NS1 protein in transient transfection experiments (Salvatore *et al.*, 2002; Talon *et al.*, 2000; Wang *et al.*, 2000). However, it was later established that the functions indicated by these assays do not correspond to the actual function of NS1 protein-mediated dsRNA binding in influenza A virus-infected cells. The actual function was established by determining the defect in replication of a recombinant influenza A/Udorn/72 (Ud) virus that encodes a NS1 protein in which R38 was changed to A, thereby eliminating dsRNA-binding activity (Min and Krug, 2006). This amino acid substitution also inactivates a nuclear localization signal (NLS) present in the RBD, but the Ud NS1 protein has another NLS (NLS2), a bipartite NLS that extends from amino acid 219 to the C-terminal 237 amino acid (Melen *et al.*, 2007). Consequently, in cells infected by the R38A mutant virus, the NS1 protein is localized in the nucleus (Min and Krug, 2006), as is the case for the wild-type (wt) Ud NS1 protein, so that the ~1000-fold attenuation of the R38A mutant virus could be attributed to the loss of the dsRNA-binding activity of the NS1 protein. In contrast, the NS1 protein of influenza A/WSN/33 (WSN) virus does not have a NLS2, and the NS1 protein of a WSN virus encoding a NS1 protein with R38A, K41A substitutions is not imported into the nucleus (Min and Krug, 2006). Consequently, its replication defect cannot be attributed to the loss of dsRNA-binding activity.

Analysis of the defect of the Ud R38A mutant virus showed that NS1 dsRNA-binding activity has no detectable role in inhibiting the production of IFN-β mRNA or inhibiting the activation of PKR (Li *et al.*, 2006; Min and Krug, 2006). Rather, these experiments demonstrated that the primary, if not the only, role of NS1 dsRNA-binding activity in infected cells is the inhibition of the IFN-α/β-induced oligo A synthetase (OAS)/RNase L pathway (Min and Krug, 2006). The 2′–5′ OAS enzyme is strongly induced by IFN-α/β, and once it is activated, 2′–5′ OAS polymerizes ATP into 2′–5′ oligo (A) chains, which in turn activate the latent RNase L that degrades mRNAs and ribosomal RNAs and as a result inhibits virus replication (Silverman, 2007). Because the dsRNA-binding activity of the NS1 protein is required for efficient virus replication (Min and Krug, 2006), its novel dsRNA-binding surface is a valid target for the development of antivirals directed against influenza A virus.

Role of the NS1 ED in inhibiting the activation of PKR

Insight into the mechanism by which the NS1 protein inhibits PKR activation came from *in vitro* binding assays, which showed that direct binding of the NS1 protein to the N-terminal 230-amino-acid region of PKR inhibited PKR activation (Li *et al.*, 2006). Further, these assays showed that the NS1 amino acids at positions 123–127 are required for PKR binding *in vitro*. Subsequently, recombinant Ud virus was generated, which expresses an NS1 protein in which amino acids 123/124 or 126/127 were changed to alanines (Min *et al.*, 2007). In cells infected by these recombinant viruses, PKR is activated, eIF2a is phosphorylated, and viral protein synthesis is dramatically inhibited. In mouse cells lacking PKR, inhibition of viral protein synthesis does not occur in cells infected with the 123/124 mutant virus, verifying that protein synthesis inhibition in cells infected by this mutant virus is caused solely by PKR activation (Min *et al.*, 2007). These results showed that the inhibition of PKR activation in influenza A virus-infected cells is mediated by the direct binding of PKR to the 123–127 region of the NS1 protein.

Role of the NS1 ED in inhibiting the production of IFN-β and other antiviral proteins

The Ud ED also has a binding site for the 30-kDa subunit of the cleavage and polyadenylation specificity factor (CPSF30) (Nemeroff *et al.*, 1998), a cellular factor required for the 3′ end processing of all cellular pre-mRNAs (Colgan and Manley, 1997). As a consequence of sequestration of CPSF30 by the NS1 protein, the substantial amount of IFN-β pre-mRNA that is synthesized in Ud virus-infected cells is not processed to form mature IFN-β mRNA, thereby suppressing the IFN response (Das *et al.*, 2008; Kim *et al.*, 2002; Noah *et al.*, 2003; Twu *et al.*, 2006, 2007). The production of other antiviral mRNAs should also be suppressed. The CPSF30 binding site is crucial for virus replication, as shown by the fact that a recombinant Ud virus expressing an NS1 protein lacking a binding site for CPSF30 is highly attenuated (Noah *et al.*, 2003). Further evidence for the importance of the CPSF30 binding site was obtained using a 61-amino-acid domain of CPSF30, comprising two of its zinc fingers (F2F3), which binds efficiently to the NS1 protein (Twu *et al.*, 2006). The expression of the F2F3 domain in cells leads to the inhibition of influenza A virus replication, coupled with increased production of IFN-β mRNA, presumably as a result of F2F3 occupying the CPSF30 binding site on the NS1 protein and hence blocking the binding of endogenous CPSF30 to this site (Twu *et al.*, 2006).

The binding interface between the Ud NS1 protein and CPSF30 was revealed by the X-ray crystal structure of the Ud ED in complex with the F2F3 domain of CPSF30 (Das *et al.*, 2008). This structure revealed that the ED undergoes a dramatic structural rearrangement from its original dimer structure to form a tetrameric complex comprised of two F2F3 units wrapped around two NS1 EDs that interact with each other head-to-head (Fig. 1.4A). This structure identifies a largely hydrophobic CPSF30-binding pocket on the NS1A protein comprised of eight amino acids that interact with aromatic side chains of the F3 zinc finger (Fig. 1.4B). Six of these amino acids are conserved in >98% of human influenza A viruses, including H5N1 viruses and the 1918 virus, strongly suggesting that this CPSF30 binding site is used by all human influenza A viruses to suppress the production of IFN-β mRNA. A recombinant Ud virus expressing a NS1 protein with a single amino acid change within this binding pocket (G184 changed to R) is attenuated and does not inhibit IFN-β pre-mRNA processing, demonstrating that this binding pocket is required for virus replication (Das *et al.*, 2008).

The crystal structure also shows that the interaction surface between NS1 and F2F3 in the tetrameric complex extends beyond this binding pocket (Fig. 1.5). Two hydrophobic NS1 amino acids outside the binding pocket, F103

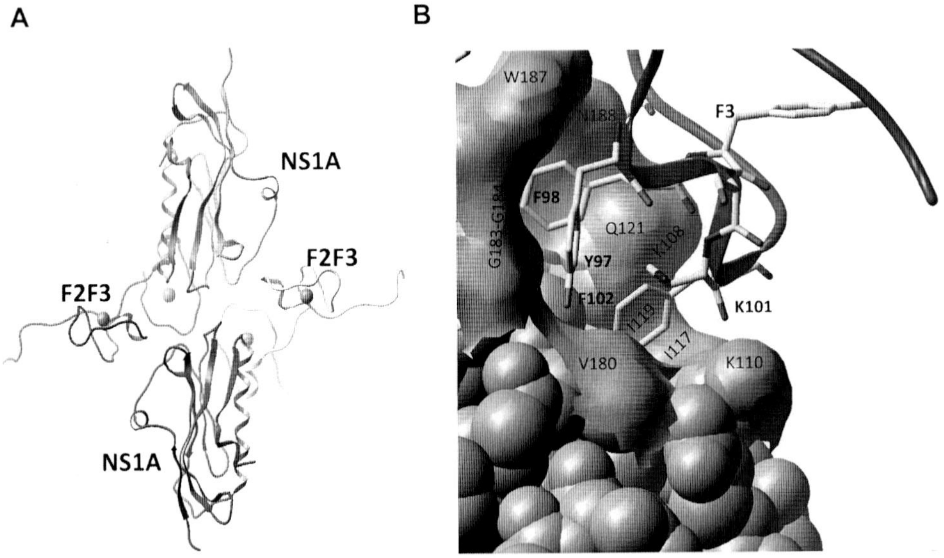

Figure 1.4 Crystal structure of the complex formed between the Ud NS1 ED and the F2F3 domain of CPSF30. (A) The NS1A–F2F3 complex is a tetramer assembled by two NS1A effector domains and two F2F3 domains. (B) A hydrophobic pocket of NS1 mediates interaction with the F3 Zinc finger of F2F3 molecule. Eight NS1 amino acids (K110, I117, I119, Q121, V180, G183, G184, and W187) interact with the aromatic side chains of three F2F3 amino acids: Y97, F98 and F102, labelled in bold.

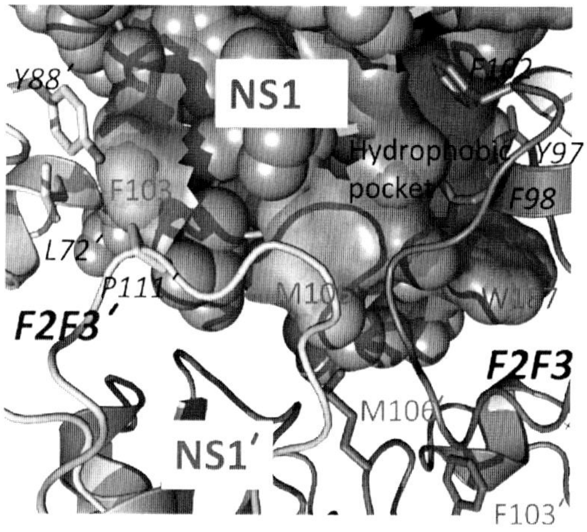

Figure 1.5 The stabilizing role of two NS1 amino acids, F103 and M106, in the formation of the NS1: F2F3 tetramer. The side chain of M106 from one NS1 molecule interacts with not only the side chain of M106′ of the other NS1 molecule (NS1′), but also amino acids of both F2F3 domains. The aromatic side chain of F103 of NS1 makes extensive contacts with three hydrophobic amino acids of F2F3′, L72′, Y88′, and P111′. The NS1 and F2F3 amino acids are labelled in grey and black, respectively.

and M106, stabilize the tetrameric complex by interacting with hydrophobic amino acids in F2F3 (Das *et al.*, 2008). In addition, M106 in one NS1 molecule interacts with M106 in the second NS1A molecule. These two amino acids, which are highly conserved (>99%) in the NS1 proteins of human influenza A viruses, are required for tight binding *in vitro* (Twu *et al.*, 2007).

However, some CPSF30 binding can occur *in vivo* when the NS1 protein contains two different hydrophobic amino acids at positions 103 and 106, specifically L instead of F at 103 and I instead of M at 106, as is the case for the NS1 protein of the 1997 pathogenic H5N1 influenza A/Hong Kong/483/97 (HK97) virus, as well as the other H5N1 viruses transmitted to humans in 1997 (Twu *et al.*, 2007). The HK97 NS1 protein binds CPSF30 to a significant, though not optimum, extent in infected cells because its cognate viral polymerase complex stabilizes the NS1–CPSF30 complex (discussed further below). Because of weakened CPSF30 binding, replication of the HK97 virus is attenuated in tissue culture cells. Thus, changing L103 to F and I106 to M in the HK97 NS1 protein results in a 20-fold increase in the rate of virus replication in MDCK cells, coupled with a ninefold decrease in the production of IFN-β mRNA, indicating that the wild-type wt HK97 virus is impaired in its ability to suppress of IFN-β and other antiviral effectors in the host cell (Twu *et al.*, 2007). In addition, these two amino acid substitutions in the H5N1 NS1 protein result in a dramatic 250-fold increase in virulence of the H5N1 virus in mice (L.-M. Chen, R. T. Davis, R.-L. Kuo, M. Malur, , R.M. Krug and R.O. Donis, submitted for publication). Thus, strong CPSF30 binding by the NS1 protein enhances H5N1 virulence, and conversely a H5N1 influenza virus can retain virulence, although reduced, even when it encodes a defective NS1 protein.

The NS1 protein and the RIG-I pathway

The UdNS1:F2F3 structure predicts that a hydrophilic residue at position 103 in the NS1 protein should strongly attenuate CPSF30 binding (Fig. 1.5), and indeed the influenza A/PR/8/34 (PR8) NS1 protein that contains S103 does not bind CPSF30 *in vitro* and in infected cells (Kochs *et al.*, 2007). Only four other seasonal viruses have encoded a NS1 protein with a hydrophilic amino acid (S) at position 103, namely, two viruses in addition to PR8 that were isolated in 1934–1936, one virus isolated in 1954 (A/Leningrad/54), and one virus isolated in 1976 (A/New Jersey/76) (Macken *et al.*, 2001). The absence of such viruses since 1976 shows that influenza A viruses encoding NS1 proteins with S103 are selected against during replication in humans.

CPSF30 binding by the PR8 NS1 protein is not required during virus infection because, unlike many other influenza A viruses, IRF-3 is not activated in PR8 virus-infected cells, and the resulting activation of IFN-β gene expression does not occur. There is disagreement concerning the molecular mechanism(s) that lead to the absence of IRF-3 activation in PR8 virus-infected cells. In one set of experiments cells were infected with a recombinant Ud virus that expresses the PR8 instead of the Ud NS1 protein. IRF-3 was activated, indicating that the PR8 NS1 protein itself does not suppress IRF-3 activation and that other mechanisms are responsible for the lack of IRF-3 activation in PR8 virus-infected cells (Das *et al.*, 2008). In contrast, the PR8 NS1 protein has been reported to form a complex with RIG-I that leads to the inhibition of IRF-3 activation (Mibayashi *et al.*, 2007; Pichlmair *et al.*, 2006). It is not known whether the interaction of PR8 NS1 with RIG-I is direct. Co-immunoprecipitation experiments indicate that a PR8 NS1 protein lacking dsRNA-binding activity (R38A, K41A mutant) binds RIG-I poorly, suggesting that the interaction of the PR8 NS1 protein with RIG-I is mediated by RNA (Pichlmair *et al.*, 2006). In addition, an interaction between bacterially expressed recombinant PR8 NS1 and RIG-I proteins was not detected (Mibayashi *et al.*, 2007). However, it is possible that other components of the RIG-I pathway may be required for an interaction with the PR8 NS1 protein.

The role of the NS1 protein in activating the PI3-kinase (PI3K) pathway

The phosphoinositide 3-kinases, PI3-kinases (PI3Ks), are a family of lipid kinases that catalyse the phosphorylation of the 3′-inositol position on membrane-associated phosphatidylinositols (Cantrell, 2001). A major product of this reaction is phosphatidylinositol-3,4,5-triphosphate (PIP3), which acts as second messenger molecule by interacting with a variety of cellular effectors to regulate their functions. One downstream event is the activation (phosphorylation) of Akt kinase, which in turn inhibits apoptosis and promotes cellular survival. Class IA PI3K kinases are comprised of heterodimers that contain various combinations of a regulatory subunit (P85) isoform and a catalytic subunit (P110) isoform. There are three P85 isoforms, P85α, P85β, P55γ; and three p110 isoforms, p110α, p110β, or p110δ. Both P85α and P85β are ubiquitously expressed, while P55γ shows more tissue-specific expression. The P85 subunit inhibits the kinase activity of the catalytic p110 subunit in quiescent cells (Bayascas and Alessi, 2005; Brazil *et al.*, 2004; Cantley, 2002).

Several groups showed that the NS1 protein binds the P85β regulatory subunit of PI3K (Ehrhardt *et al.*, 2007; Ehrhardt *et al.*, 2007; Hale *et al.*, 2008b; Hale *et al.*, 2006; Li *et al.*, 2008; Shin *et al.*, 2007a; Shin *et al.*, 2007b). Interestingly, binding of the P85α regulatory subunit was not detected. Initially, the 89–93 amino acid region of NS1 was proposed as the P85β binding site. This sequence, which is YLTDM in the PR8 NS1 protein, fits the consensus motif (YXXM) for binding to the SH2 domain of P85β (Songyang *et al.*, 1993). Consistent with this prediction, individual amino acid substitutions in this NS1 sequence, Y89 changed to F, or M93 changed to A, strongly decreased P85β binding (Hale *et al.*, 2006). However, later studies argued against the NS1 89–93 sequence being directly involved in P85β interaction. First, the crystal structure of the NS1 effector domain revealed that only Y89, but not M93, is exposed on the surface of the molecule (Bornholdt and Prasad, 2006; Hale *et al.*, 2008a), suggesting that the inhibition of P85β binding by the M93A substitution resulted from an alteration of the overall structure of the NS1 protein. Second, although the interaction between the YXXM motif and the P85β SH2 domain has been shown to require Y (tyrosine) phosphorylation, a mutant NS1 protein in which Y89 was replaced with E to mimic constitutive Y phosphorylation did not bind P85β (Hale *et al.*, 2008b). Two alternative P85β interaction regions of the NS1 protein were subsequently identified: amino acids 164–167, PSLP, which resembles known SH3 binding sites; and amino acids 137–142, NFDRLE. Alanine substitutions of specific amino acids in either of these two NS1 sites, namely P164AP167A and L141AE142A, strongly inhibit P85β binding (Li *et al.*, 2008; Shin *et al.*, 2007a).

As a consequence of the binding of P85β by the NS1 protein, PI3K is activated, resulting in the downstream activation (phosphorylation) of the Akt kinase (Hale *et al.*, 2006). The binding of the NS1 protein to P85β does not detach P85β from its complex with the P110 catalytic unit, but instead the binding of the NS1 protein to P85β most likely results in a trimeric complex in which specific contacts between P85β and P110 that are required for inhibition of P110 activity are disrupted (Hale *et al.*, 2008b).

Transient expression of the NS1 protein is sufficient to induce the phosphorylation of Akt in transfected cells (Hale *et al.*, 2006). Akt phosphorylation occurs in cells infected with influenza

A viruses expressing a wild-type NS1 protein, but not in cells infected with a recombinant virus expressing a mutant NS1 protein that is incapable of P85β binding. Such a mutant NS1 protein leads to attenuation of the Ud virus, but not the two mouse-adapted influenza A viruses, PR8 and WSN (Hale *et al.*, 2006). The basis for this difference between these virus strains has not been determined.

What is the role of the 212–216 amino acid region of the NS1 protein?

The amino acid sequence of the 212–216 amino acid region of avian influenza A viruses is PxΦPx+ (where x, Φ and + represent any amino acid, hydrophobic residues, and positively charged residues respectively), which is a class II SH3-binding motif (Heikkinen *et al.*, 2008). This motif is also found in the 1918 pandemic virus and H5N1 viruses, but not in any seasonal human influenza A viruses. In human influenza A viruses the P at 215 is replaced by a T, thereby eliminating the SH3-binding motif. One group of investigators used a phage display approach to identify SH3 domain-containing proteins that bind to the SH3-binding motif of avian NS1 proteins (Heikkinen *et al.*, 2008). They found that the major binding partner was Crk/CrkL, multifunctional adapter proteins involved in many cellular processes. Substitution of the P at 215 with other amino acids, including T, resulted in the loss of Crk/CrkL binding, indicating that the SH3-binding motif is required for this binding and that human NS1 proteins lack this binding activity (Heikkinen *et al.*, 2008).

Subsequently, another group of investigators showed that T215 in human influenza A virus NS1 proteins is phosphorylated during virus infection (Hale *et al.*, 2009). Replacement of T for A in a recombinant virus resulted in attenuation of virus replication, indicating that T215 phosphorylation is important for virus replication. The cause of attenuation was not identified, although it was shown that it was not due the loss of two known NS1 functions, specifically, antagonizing the host IFN response or activating Akt kinase. *In vitro* experiments showed that cyclin-dependent kinases (CDKs) and extracellular signal-regulated kinases (ERKs) phosphorylate a recombinant NS1 protein at T215 (Hale *et al.*, 2009).

One interpretation of these two sets of results is that the 212–216 region of avian and human NS1 proteins have different functions, one (avian) mediated by a SH3-binding motif, and the other (human) mediated by post-translation phosphorylation of T215.

Role of the c-terminal domain of the NS1 protein in virulence/pathogenesis

Analysis of the sequences of the NS1 proteins of a large number of human and avian influenza A viruses showed that the C-termini of human and avian NS1 proteins exhibited different signature sequences (Obenauer *et al.*, 2006). The C-terminal four amino acids of almost all NS1 proteins of human virus isolates are RSKV or RSEV, whereas the NS1 proteins of avian virus isolates contain ESEV or EPEV C-termini. The latter two sequences constitute a potential PDZ-binding motif (Songyang *et al.*, 1997). The NS1 proteins of pathogenic H5N1 viruses isolated from either avian species or humans have this PDZ-binding motif (Obenauer *et al.*, 2006). The NS1 protein of the pandemic 1918 virus has a different PDZ-binding motif, KSEV. The ESEV/EPEV domain was shown to bind to known PDZ domains in an array of human PDZ domains derived from proteins with diverse cellular functions ranging from cell polarity to T cell proliferation (Obenauer *et al.*, 2006). It was proposed that the binding to particular PDZ-containing proteins played an important role in the pathogenesis of H5N1 viruses and the 1918 virus in humans. Consistent with this hypothesis, a recombinant WSN virus expressing a NS1 protein in which its C-terminal amino acids were changed to an avian PDZ-binding domain was 7- to 10-fold more virulent than wild type WSN virus in mice (Jackson *et al.*, 2008). However, such a PDZ-binding domain is not required for high virulence of H5N1 viruses, because the NS1 protein of one of the most virulent H5N1 viruses, A/Vietnam/1203/03, expresses a NS1 protein that is 215 amino acids long and lacks the ESEV/EPEV domain (Govorkova *et al.*, 2005).

Role of the NS1 protein in the nuclear export of cellular and viral mRNAs

As discussed above, the NS1 protein of almost all (>99%) human influenza A viruses inhibit the nuclear export of cellular mRNAs by binding CPSF30, thereby inhibiting the 3′ end processing of cellular pre-mRNAs (Nemeroff *et al.*, 1998). These unprocessed cellular pre-mRNAs accumulate in the nucleus. When the NS1 binding site for CPSF30 was inactivated by mutation, the cellular pre-mRNAs were processed and the resulting mature cellular mRNAs were efficiently transported to the cytoplasm where they were translated (Das *et al.*, 2008; Noah *et al.*, 2003). Several lines of evidence show that this is the only mechanism by which NS1 inhibits the nuclear export of cellular mRNAs. For example, one group of investigators established and maintained a HEp-2 cell line that constitutively expresses substantial levels of full-length functional PR8 NS1 protein (Hale *et al.*, 2008b), which as discussed above is one of the very few NS1 proteins that do not bind CPSF30. They used this cell line to demonstrate that the NS1 protein binds P85β. This cell line grew extremely well, definitively proving that a NS1 protein that does not bind CPSF30 does not significantly impair the nuclear export of cellular mRNAs (Hale *et al.*, 2008b). Interestingly, these investigators were unable to establish cell lines expressing NS1 proteins of other influenza A virus strains that bind CPSF30.

The nuclear export of influenza virus mRNAs is not inhibited by the sequestering of CPSF30 by the NS1 protein because the virus does not use the cellular 3′ end processing machinery. Instead, poly(A) is added to viral mRNAs by the viral polymerase, which reiteratively copies a short stretch of U residues on virion RNA (vRNA) templates (Poon *et al.*, 2000; Poon *et al.*, 1999; Pritlove *et al.*, 1998). Viral mRNA synthesis occurs in the nucleus, and the viral mRNAs have to be exported from the nucleus. Three different types of viral mRNAs have to be exported. Six viral mRNAs lack introns; two viral mRNAs, encoding the M1 and NS1 proteins, contain introns, but are not spliced; and two viral mRNAs, encoding the M2 ion channel protein, and the NS2/NEP protein, are generated by inefficient splicing of the M1 and NS1 mRNAs, respectively (Krug *et al.*, 1989). A recent study used molecular beacon technology to visualize the subcellular distribution and export of influenza virus mRNAs in virus-infected cells, and showed that both the NS1 protein and the cellular TAP protein were associated with the three types of influenza virus mRNAs in the nucleus (Wang *et al.*, 2008). Another study also showed that the NS1 protein is associated with the TAP protein in infected cells (Satterly *et al.*, 200). The TAP protein mediates the nuclear export of most cellular mRNAs (Cullen, 2003). These results indicate that the NS1 protein, by binding both TAP and viral mRNAs, probably co-opts the TAP-mediated cellular mRNA export machinery for the selective transport of viral mRNAs. The ICP27 protein of herpes simplex virus has been shown to have a similar role in the selective nuclear export of viral mRNAs (Johnson and Sandri-Goldin, 2009).

Role of the NS1 protein in viral RNA synthesis

The influenza virus polymerase, which comprises three proteins (PB1, PB2 and PA), transcribes and replicates the eight vRNAs in infected cells (Krug *et al.*, 1989). Transcription is initiated with capped RNA primers excised from cellular pre-mRNAs by the intrinsic endonuclease of the viral polymerase (Krug *et al.*, 1979; Li *et al.*, 2001; Rao *et al.*, 2003). Viral RNA replication occurs in two steps: first a full-length copy of vRNA is made, termed complementary RNA (cRNA), and then this cRNA is copied to produce vRNA (Krug *et al.*, 1989). The synthesis of cRNAs and vRNAs is initiated without a primer, in contrast to the initiation of viral mRNA synthesis, and requires the viral nucleocapsid protein (NP) (Newcomb *et al.*, 2009). Several studies have provided strong evidence that the NS1 protein interacts with the viral polymerase complex in infected cells and has an important role in viral RNA synthesis (Falcon *et al.*, 2004; Marion *et al.*, 1997; Min *et al.*, 2007; Shimizu *et al.*, 1994). Thus, one study showed that vRNA synthesis was defective in cells infected with a recombinant virus expressing NS1 proteins with C-terminal deletions (Falcon *et al.*, 2004).

Another study showed that the NS1 protein plays a key role in the temporal control of viral RNA synthesis during infection. Viral RNA synthesis in influenza A virus-infected cells has been shown to be divided into an early and late phase. During the early phase, two vRNAs are selectively replicated and transcribed at a relatively low rate, whereas during the late phase all vRNAs are replicated and transcribed at a high rate (Shapiro *et al.*, 1987). In contrast, a recombinant Ud virus that expresses a NS1 protein in which only two amino acids (123 and 124) are changed to alanines deregulates this temporal control, resulting in the very early appearance of the late phase of rapid replication and transcription of all eight vRNAs (Min *et al.*, 2007). Interestingly, the same amino acid substitutions result in the activation of PKR activation, as described above, but the two phenotypes introduced by these amino acid substitutions are not coupled.

Further strong evidence for an interaction of the NS1 protein with the viral polymerase came from studies of the NS1–CPSF30 complex that forms in infected cells (Kuo and Krug, 2009). As already discussed, a NS1 protein that has non-consensus hydrophobic amino acids at positions 103 (L instead of F) and 106 (I instead of M) binds CPSF30 to a significant extent only when this NS1 protein is expressed in conjunction with cognate proteins of the viral polymerase complex (PB1, PB2, PA and NP). Expression of such a NS1 protein in infected cells in the presence of non-cognate viral polymerase complex proteins does not result in the binding of the NS1 protein to CPSF30 (Kuo and Krug, 2009). The requirement for cognate proteins clearly shows that there are specific interactions between the NS1 protein and cognate proteins of the viral polymerase complex that lead to the stabilization of the NS1–CPSF30 complex. In fact, the primary, if not the only, interaction probably occurs between the NS1 protein and its cognate PA and NP proteins.

The stabilization of the NS1–CPSF30 complex by a cognate viral polymerase complex suggested that the viral polymerase complex is an integral part of the NS1A–CPSF30 complex, which was then verified (Kuo and Krug, 2009). The macromolecular complex containing the viral polymerase complex and the NS1 protein probably includes other host cell proteins in addition to CPSF30. Because it had been shown previously that the binding of the NS1 protein to CPSF30 does not disrupt the interaction between CPSF30 and other CPSF subunits (Nemeroff *et al.*, 1998), it is likely that these other CPSF subunits are also part of this macromolecular complex. Other host factors may also be present that bind directly to the viral polymerase rather than to the NS1 protein. One such host factor is the large subunit of the cellular RNA polymerase II which has been shown to bind to the viral polymerase via its C-terminal domain (Engelhardt *et al.*, 2005). It is reasonable to propose that the macromolecular complex containing the host CPSF30 protein, the viral NS1 protein and the viral polymerase complex may have an important role in the regulation of viral RNA synthesis in infected cells.

Concluding remarks

We have discussed many, but not all, of the reported functions of the NS1 protein of influenza A viruses. For example, we have not discussed the possible interactions of the NS1 protein with cellular translation factors (Aragon *et al.*, 2000; Burgui *et al.*, 2003). It is remarkable that this small viral protein of 230–237 amino acids in length carries out such a multitude of functions. Clearly, a great deal of future research will be directed at elucidating many aspects of the already identified functions of the NS1 protein. In addition, it will not be surprising if additional functions of the NS1 protein are discovered in the future.

Acknowledgement

Research in the authors' laboratory is supported by grants from the National Institutes of Health USA.

References

Aragon, T., de la Luna, S., Novoa, I., Carrasco, L., Ortin, J., and Nieto, A. (2000). Eukaryotic translation initiation factor 4GI is a cellular target for NS1 protein, a translational activator of influenza virus. Mol. Cell. Biol. *20*, 6259–6268.

Bayascas, J.R., and Alessi, D.R. (2005). Regulation of Akt/PKB Ser473 phosphorylation. Mol. Cell *18*, 143–145.

Bornholdt, Z.A., and Prasad, B.V. (2006). X-ray structure of influenza virus NS1 effector domain. Nat. Struct. Mol. Biol. *13*, 559–560.

Bornholdt, Z.A., and Prasad, B.V. (2008). X-ray structure of NS1 from a highly pathogenic H5N1 influenza virus. Nature *456*, 985–988.

Brazil, D.P., Yang, Z.Z., and Hemmings, B.A. (2004). Advances in protein kinase B signalling: AKTion on multiple fronts. Trends Biochem. Sci. *29*, 233–242.

Burgui, I., Aragon, T., Ortin, J., and Nieto, A. (2003). PABP1 and eIF4GI associate with influenza virus

NS1 protein in viral mRNA translation initiation complexes. J. Gen. Virol. *84*, 3263–3274.

Cantley, L.C. (2002). The phosphoinositide 3-kinase pathway. Science *296*, 1655–1657.

Cantrell, D.A. (2001). Phosphoinositide 3-kinase signalling pathways. J. Cell Sci. *114*, 1439–1445.

Cheng, A., Wong, S.M., and Yuan, Y.A. (2009). Structural basis for dsRNA recognition by NS1 protein of influenza A virus. Cell Res. *19*, 187–195.

Chien, C.Y., Tejero, R., Huang, Y., Zimmerman, D.E., Rios, C.B., Krug, R.M., and Montelione, G.T. (1997). A novel RNA-binding motif in influenza A virus nonstructural protein 1. Nat. Struct. Biol. *4*, 891–895.

Chien, C.Y., Xu, Y., Xiao, R., Aramini, J.M., Sahasrabudhe, P.V., Krug, R.M., and Montelione, G.T. (2004). Biophysical characterization of the complex between double-stranded RNA and the N-terminal domain of the NS1 protein from influenza A virus: evidence for a novel RNA-binding mode. Biochemistry *43*, 1950–1962.

Colgan, D.F., and Manley, J.L. (1997). Mechanism and regulation of mRNA polyadenylation. Genes Dev. *11*, 2755–2766.

Cullen, B.R. (2003). Nuclear RNA export. J. Cell Sci. *116*, 587–597.

Das, K., Ma, L.C., Xiao, R., Radvansky, B., Aramini, J., Zhao, L., Marklund, J., Kuo, R.L., Twu, K.Y., Arnold, E., *et al.* (2008). Structural basis for suppression of a host antiviral response by influenza A virus. Proc. Natl. Acad. Sci. U.S.A. *105*, 13093–13098.

Ehrhardt, C., Wolff, T., and Ludwig, S. (2007a). Activation of phosphatidylinositol 3-kinase signaling by the nonstructural NS1 protein is not conserved among type A and B influenza viruses. J. Virol. *81*, 12097–12100.

Ehrhardt, C., Wolff, T., Pleschka, S., Planz, O., Beermann, W., Bode, J.G., Schmolke, M., and Ludwig, S. (2007b). Influenza A virus NS1 protein activates the PI3K/Akt pathway to mediate antiapoptotic signaling responses. J. Virol. *81*, 3058–3067.

Engelhardt, O.G., Smith, M., and Fodor, E. (2005). Association of the influenza A virus RNA-dependent RNA polymerase with cellular RNA polymerase II. J. Virol. *79*, 5812–5818.

Falcon, A.M., Marion, R.M., Zurcher, T., Gomez, P., Portela, A., Nieto, A., and Ortin, J. (2004). Defective RNA replication and late gene expression in temperature-sensitive influenza viruses expressing deleted forms of the NS1 protein. J. Virol. *78*, 3880–3888.

Gale, M., Jr., and Katze, M.G. (1998). Molecular mechanisms of interferon resistance mediated by viral-directed inhibition of PKR, the interferon-induced protein kinase. Pharmacol. Ther. *78*, 29–46.

Govorkova, E.A., Rehg, J.E., Krauss, S., Yen, H.L., Guan, Y., Peiris, M., Nguyen, T.D., Hanh, T.H., Puthavathana, P., Long, H.T., *et al.* (2005). Lethality to ferrets of H5N1 influenza viruses isolated from humans and poultry in 2004. J. Virol. *79*, 2191–2198.

Hale, B.G., Barclay, W.S., Randall, R.E., and Russell, R.J. (2008a). Structure of an avian influenza A virus NS1 protein effector domain. Virology *378*, 1–5.

Hale, B.G., Batty, I.H., Downes, C.P., and Randall, R.E. (2008b). Binding of influenza A virus NS1 protein to the inter-SH2 domain of p85 suggests a novel mechanism for phosphoinositide 3-kinase activation. J. Biol. Chem. *283*, 1372–1380.

Hale, B.G., Jackson, D., Chen, Y.H., Lamb, R.A., and Randall, R.E. (2006). Influenza A virus NS1 protein binds p85beta and activates phosphatidylinositol-3-kinase signaling. Proc. Natl. Acad. Sci. U.S.A. *103*, 14194–14199.

Hale, B.G., Knebel, A., Botting, C.H., Galloway, C.S., Precious, B.L., Jackson, D., Elliott, R.M., and Randall, R.E. (2009). CDK/ERK-mediated phosphorylation of the human influenza A virus NS1 protein at threonine-215. Virology *383*, 6–11.

Hale, B.G., Randall, R.E., Ortin, J., and Jackson, D. (2008c). The multifunctional NS1 protein of influenza A viruses. J. Gen. Virol. *89*, 2359–2376.

Haller, O., Kochs, G., and Weber, F. (2006). The interferon response circuit: induction and suppression by pathogenic viruses. Virology *344*, 119–130.

Hatada, E., and Fukuda, R. (1992). Binding of influenza A virus NS1 protein to dsRNA *in vitro*. J. Gen. Virol. *73*, 3325–3329.

Heikkinen, L.S., Kazlauskas, A., Melen, K., Wagner, R., Ziegler, T., Julkunen, I., and Saksela, K. (2008). Avian and 1918 Spanish influenza a virus NS1 proteins bind to Crk/CrkL Src homology 3 domains to activate host cell signaling. J. Biol. Chem. *283*, 5719–5727.

Hiscott, J. (2007). Triggering the innate antiviral response through IRF-3 activation. J. Biol. Chem. *282*, 15325–15329.

Hovanessian, A.G. (1989). The double stranded RNA-activated protein kinase induced by interferon: dsRNA-PK. J. Interferon. Res. *9*, 641–647.

Jackson, D., Hossain, M.J., Hickman, D., Perez, D.R., and Lamb, R.A. (2008). A new influenza virus virulence determinant: the NS1 protein four C-terminal residues modulate pathogenicity. Proc. Natl. Acad. Sci. U.S.A. *105*, 4381–4386.

Johnson, L.A., and Sandri-Goldin, R.M. (2009). Efficient nuclear export of herpes simplex virus 1 transcripts requires both RNA binding by ICP27 and ICP27 interaction with TAP/NXF1. J. Virol. *83*, 1184–1192.

Kim, M.J., Latham, A.G., and Krug, R.M. (2002). Human influenza viruses activate an interferon-independent transcription of cellular antiviral genes: outcome with influenza A virus is unique. Proc. Natl. Acad. Sci. U.S.A. *99*, 10096–10101.

Kochs, G., Garcia-Sastre, A., and Martinez-Sobrido, L. (2007). Multiple anti-interferon actions of the influenza A virus NS1 protein. J. Virol. *81*, 7011–7021.

Krug, R.M., Alonso-Caplen, F.V., Julkunen, I., and Katze, M. (1989). Expression and replication of the influenza virus genome. The influenza viruses, 89–152.

Krug, R.M., Broni, B.A., and Bouloy, M. (1979). Are the 5′ ends of influenza viral mRNAs synthesized *in vivo* donated by host mRNAs? Cell *18*, 329–334.

Kuo, R.L., and Krug, R.M. (2009). Influenza a virus polymerase is an integral component of the CPSF30-NS1A protein complex in infected cells. J. Virol. *83*, 1611–1616.

Lamb, R.A., and Krug, R.M. (2001). Orthomyxoviridae: the viruses and their replication. Field's Virology, 4th edn, 1487–1532.

Li, M.L., Rao, P., and Krug, R.M. (2001). The active sites of the influenza cap-dependent endonuclease are on different polymerase subunits. EMBO J. *20*, 2078–2086.

Li, S., Min, J.Y., Krug, R.M., and Sen, G.C. (2006). Binding of the influenza A virus NS1 protein to PKR mediates the inhibition of its activation by either PACT or double-stranded RNA. Virology *349*, 13–21.

Li, Y., Anderson, D.H., Liu, Q., and Zhou, Y. (2008). Mechanism of influenza A virus NS1 protein interaction with the p85beta, but not the p85alpha, subunit of phosphatidylinositol 3-kinase (PI3K) and up-regulation of PI3K activity. J. Biol. Chem. *283*, 23397–23409.

Liu, J., Lynch, P.A., Chien, C.Y., Montelione, G.T., Krug, R.M., and Berman, H.M. (1997). Crystal structure of the unique RNA-binding domain of the influenza virus NS1 protein. Nat. Struct. Biol. *4*, 896–899.

Macken, C., Lu, H., Goodman, J., and Boykin, L. (2001). The value of a database in surveillance and vaccine selection. In Options for the Control of Influenza IV. A.D.M.E. Osterhaus, N. Cox and A.W. Hampson, eds (Amsterdam: Elsevier Science), pp. 103–106.

Marion, R.M., Zurcher, T., de la Luna, S., and Ortin, J. (1997). Influenza virus NS1 protein interacts with viral transcription–replication complexes *in vivo*. J. Gen. Virol. *78*, 2447–2451.

Melen, K., Kinnunen, L., Fagerlund, R., Ikonen, N., Twu, K.Y., Krug, R.M., and Julkunen, I. (2007). Nuclear and nucleolar targeting of influenza A virus NS1 protein: striking differences between different virus subtypes. J. Virol. *81*, 5995–6006.

Meurs, E., Chong, K., Galabru, J., Thomas, N.S., Kerr, I.M., Williams, B.R., and Hovanessian, A.G. (1990). Molecular cloning and characterization of the human double-stranded RNA-activated protein kinase induced by interferon. Cell *62*, 379–390.

Mibayashi, M., Martinez-Sobrido, L., Loo, Y.M., Cardenas, W.B., Gale, M., Jr., and Garcia-Sastre, A. (2007). Inhibition of retinoic acid-inducible gene I-mediated induction of beta interferon by the NS1 protein of influenza A virus. J. Virol. *81*, 514–524.

Min, J.Y., and Krug, R.M. (2006). The primary function of RNA binding by the influenza A virus NS1 protein in infected cells: Inhibiting the 2′–5′ oligo (A) synthetase/RNase L pathway. Proc. Natl. Acad. Sci. U.S.A. *103*, 7100–7105.

Min, J.Y., Li, S., Sen, G.C., and Krug, R.M. (2007). A site on the influenza A virus NS1 protein mediates both inhibition of PKR activation and temporal regulation of viral RNA synthesis. Virology *363*, 236–243.

Nemeroff, M.E., Barabino, S.M., Li, Y., Keller, W., and Krug, R.M. (1998). Influenza virus NS1 protein interacts with the cellular 30 kDa subunit of CPSF and inhibits 3′end formation of cellular pre-mRNAs. Mol. Cell *1*, 991–1000.

Newcomb, L.L., Kuo, R.L., Ye, Q., Jiang, Y., Tao, Y.J., and Krug, R.M. (2009). Interaction of the influenza a virus nucleocapsid protein with the viral RNA polymerase potentiates unprimed viral RNA replication. J. Virol. *83*, 29–36.

Noah, D.L., Twu, K.Y., and Krug, R.M. (2003). Cellular antiviral responses against influenza A virus are countered at the posttranscriptional level by the viral NS1A protein via its binding to a cellular protein required for the 3′ end processing of cellular pre-mRNAS. Virology *307*, 386–395.

Obenauer, J.C., Denson, J., Mehta, P.K., Su, X., Mukatira, S., Finkelstein, D.B., Xu, X., Wang, J., Ma, J., Fan, Y., *et al.* (2006). Large-scale sequence analysis of avian influenza isolates. Science *311*, 1576–1580.

Pichlmair, A., Schulz, O., Tan, C.P., Naslund, T.I., Liljestrom, P., Weber, F., and Reis e Sousa, C. (2006). RIG-I-mediated antiviral responses to single-stranded RNA bearing 5′-phosphates. Science *314*, 997–1001.

Poon, L.L., Fodor, E., and Brownlee, G.G. (2000). Polyuridylated mRNA synthesized by a recombinant influenza virus is defective in nuclear export. J. Virol. *74*, 418–427.

Poon, L.L., Pritlove, D.C., Fodor, E., and Brownlee, G.G. (1999). Direct evidence that the poly(A) tail of influenza A virus mRNA is synthesized by reiterative copying of a U track in the virion RNA template. J. Virol. *73*, 3473–3476.

Pritlove, D.C., Poon, L.L., Fodor, E., Sharps, J., and Brownlee, G.G. (1998). Polyadenylation of influenza virus mRNA transcribed *in vitro* from model virion RNA templates: requirement for 5′ conserved sequences. J. Virol. *72*, 1280–1286.

Qian, X.Y., Alonso-Caplen, F., and Krug, R.M. (1994). Two functional domains of the influenza virus NS1 protein are required for regulation of nuclear export of mRNA. J. Virol. *68*, 2433–2441.

Randall, R.E., and Goodbourn, S. (2008). Interferons and viruses: an interplay between induction, signalling, antiviral responses and virus countermeasures. J. Gen. Virol. *89*, 1–47.

Rao, P., Yuan, W., and Krug, R.M. (2003). Crucial role of CA cleavage sites in the cap-snatching mechanism for initiating viral mRNA synthesis. EMBO J. *22*, 1188–1198.

Salvatore, M., Basler, C.F., Parisien, J.P., Horvath, C.M., Bourmakina, S., Zheng, H., Muster, T., Palese, P., and Garcia-Sastre, A. (2002). Effects of influenza A virus NS1 protein on protein expression: the NS1 protein enhances translation and is not required for shutoff of host protein synthesis. J. Virol. *76*, 1206–1212.

Samuel, C.E. (1993). The eIF-2 alpha protein kinases, regulators of translation in eukaryotes from yeasts to humans. J. Biol. Chem. *268*, 7603–7606.

Shapiro, G.I., Gurney, T., Jr., and Krug, R.M. (1987). Influenza virus gene expression: control mechanisms at early and late times of infection and nuclear-cytoplasmic transport of virus-specific RNAs. J. Virol. *61*, 764–773.

Shimizu, K., Handa, H., Nakada, S., and Nagata, K. (1994). Regulation of influenza virus RNA polymerase activity by cellular and viral factors. Nucleic Acids Res. 22, 5047–5053.

Shin, Y.K., Li, Y., Liu, Q., Anderson, D.H., Babiuk, L.A., and Zhou, Y. (2007a). SH3 binding motif 1 in influenza A virus NS1 protein is essential for PI3K/Akt signaling pathway activation. J. Virol. *81*, 12730–12739.

Shin, Y.K., Liu, Q., Tikoo, S.K., Babiuk, L.A., and Zhou, Y. (2007b). Influenza A virus NS1 protein activates the phosphatidylinositol 3-kinase (PI3K)/Akt pathway by direct interaction with the p85 subunit of PI3K. J. Gen. Virol. *88*, 13–18.

Silverman, R.H. (2007). Viral encounters with 2′,5′-oligoadenylate synthetase and RNase L during the interferon antiviral response. J. Virol. *81*, 12720–12729.

Songyang, Z., Fanning, A.S., Fu, C., Xu, J., Marfatia, S.M., Chishti, A.H., Crompton, A., Chan, A.C., Anderson, J.M., and Cantley, L.C. (1997). Recognition of unique carboxyl-terminal motifs by distinct PDZ domains. Science *275*, 73–77.

Songyang, Z., Shoelson, S.E., Chaudhuri, M., Gish, G., Pawson, T., Haser, W.G., King, F., Roberts, T., Ratnofsky, S., Lechleider, R.J., and *et al.* (1993). SH2 domains recognize specific phosphopeptide sequences. Cell *72*, 767–778.

Talon, J., Horvath, C.M., Polley, R., Basler, C.F., Muster, T., Palese, P., and Garcia-Sastre, A. (2000). Activation of interferon regulatory factor 3 is inhibited by the influenza A virus NS1 protein. J. Virol. *74*, 7989–7996.

Twu, K.Y., Kuo, R.L., Marklund, J., and Krug, R.M. (2007). The H5N1 influenza virus NS genes selected after 1998 enhance virus replication in mammalian cells. J. Virol. *81*, 8112–8121.

Twu, K.Y., Noah, D.L., Rao, P., Kuo, R.L., and Krug, R.M. (2006). The CPSF30 binding site on the NS1A protein of influenza A virus is a potential antiviral target. J. Virol. *80*, 3957–3965.

Wang, W., Cui, Z.Q., Han, H., Zhang, Z.P., Wei, H.P., Zhou, Y.F., Chen, Z., and Zhang, X.E. (2008). Imaging and characterizing influenza A virus mRNA transport in living cells. Nucleic Acids Res. *36*, 4913–4928.

Wang, W., Riedel, K., Lynch, P., Chien, C.Y., Montelione, G.T., and Krug, R.M. (1999). RNA binding by the novel helical domain of the influenza virus NS1 protein requires its dimer structure and a small number of specific basic amino acids. Rna *5*, 195–205.

Wang, X., Li, M., Zheng, H., Muster, T., Palese, P., Beg, A.A., and Garcia-Sastre, A. (2000). Influenza A virus NS1 protein prevents activation of NF-kappaB and induction of alpha/beta interferon. J. Virol. *74*, 11566–11573.

Xia, S., Monzingo, A.F., and Robertus, J.D. (2009). Structure of NS1A effector domain from the influenza A/Udorn/72 virus. Acta Crystallogr D Biol Crystallogr *65*, 11–17.

Yin, C., Khan, J.A., Swapna, G.V., Ertekin, A., Krug, R.M., Tong, L., and Montelione, G.T. (2007). Conserved surface features form the double-stranded RNA binding site of non-structural protein 1 (NS1) from influenza A and B viruses. J. Biol. Chem. *282*, 20584–20592.

Structural Insights into the Function of Influenza NS1

2

Zachary A. Bornholdt, Berenice Carrillo and B.V. Venkataram Prasad

Abstract

The non-structural protein 1 (NS1) of influenza virus is a potent antagonist of the cellular antiviral interferon (IFN) response. It is a multifunctional protein with two domains, a dsRNA binding domain (RBD) and an effector domain (ED) which interacts with various cellular proteins. Although, initially sequestration of dsRNA was considered the primary mechanism for countering IFN, subsequent studies have shown that the interactions of ED with various cellular proteins are probably involved. NS1 is shown to be a virulence determinant, especially in the highly pathogenic H5N1 viruses that are currently a threat for another influenza pandemic. Among various influenza virus strains, NS1 is relatively well conserved with major differences occurring in the linker region and the C-terminus, where several NS1 proteins contain truncations. How these differences contribute to virulence remains unknown but these differences seem to have an effect on NS1 function that may be strain specific. In recent years, substantial progress has been made towards understanding of the structural aspects of this two-domain protein. Here we review this progress and discuss the structural basis of various activities of NS1.

Introduction

Influenza A viruses are important pathogens that cause seasonal epidemics and have been responsible for three pandemics in the last century (Simonsen *et al.*, 1997). These viruses capable of infecting humans, animals, and birds are members of the Orthomyxoviridae family and contain eight negative sense ssRNA segments encoding eleven viral proteins (Palese and Shaw, 2007). Based on the antigenic and genetic diversity of the two surface glycoproteins, haemagglutinin (HA) and neuraminidase (NA), which are the targets for protective immune response, these viruses are classified into several subtypes, and each subtype consists of several strains (Palese and Shaw, 2007; Webster *et al.*, 1992). More recently, the emergence of the highly pathogenic avian H5N1 subtype has raised significant concern because of its epizootic and panzootic nature, and its association with lethal human infections (Abdel-Ghafar *et al.*, 2008; Beigel *et al.*, 2005). The virulence of influenza A viruses is multifactorial and several viral proteins contribute to the virulence. Increased virulence in H5N1 strains is specially attributed to the non-structural protein NS1 encoded by segment 8 of the viral genome (Egorov *et al.*, 1998; Garcia-Sastre *et al.*, 1998; Kochs *et al.*, 2007b).

NS1 is a multifunctional protein implicated in inhibiting key cellular processes, which are responsible for activating the host antiviral interferon response through multiple mechanisms to facilitate efficient viral replication during infection (Kochs *et al.*, 2007a; Lipatov *et al.*, 2005; Seo *et al.*, 2002, 2004). The inhibition of these cellular processes involves the ability of NS1 to interact with a variety of RNA species including dsRNA and several cellular proteins (Hale *et al.*, 2008b; Kochs *et al.*, 2007a). Although strain-dependent, NS1 functions include limiting IFN-β induction at the pre-transcriptional level (Kochs *et al.*, 2007a; Talon *et al.*, 2000; Wang *et al.*, 2000); sequestering of components of the cellular pre-mRNA processing machinery (Chen *et al.*, 1999; Nemeroff *et al.*, 1998; Noah *et al.*, 2003); inhibiting the antiviral effects of 2′–5′-oligo (A) synthetase (OAS)/Rnase L (Min and Krug, 2006) and protein kinase R (PKR) (Li *et al.*, 2006; Lu *et al.*, 1995; Tan and Katze, 1998); suppressing RIG-I signal transduction (Gack *et al.*, 2009; Guo *et al.*, 2007; Mibayashi *et al.*, 2007), activating phosphoinositide 3-kinase (PI3K) (Hale *et al.*, 2006; Shin *et al.*, 2007); inhibiting host gene expression by interacting with components of nuclear transport (Melen *et al.*, 2007), blocking mRNA export (Fortes *et al.*, 1994); and binding to PDZ ligand proteins (Jackson *et al.*, 2008; Obenauer *et al.*, 2006). In addition to its function as an antagonist of innate immunity by blocking the type I interferon response, NS1 is also implicated in suppressing adaptive immunity by attenuating human dendritic cell maturation and the capacity of dendritic cells to induce T-cell responses (Fernandez-Sesma *et al.*, 2006).

NS1 (~26 kDa) is typically about 230 amino acid residues in size, although the precise length varies between various influenza virus subtypes and strains (Ludwig *et al.*, 1991; Palese and Shaw, 2007; Suarez and Perdue, 1998). It contains two distinct domains, an N-terminal RNA-binding domain (RBD) and a C-terminal effector domain (ED) connected by a linker region (Fig. 2.1). Residues 1–72 of NS1 constitute the RBD, residues 73–84 constitute the linker region, and residues 85–220 comprise the ED (Bornholdt and Prasad, 2006; Chien *et al.*, 1997). The known binding sites for dsRNA and various cellular proteins in these two domains are summarized in Fig. 2.1. Although NS1 sequences from influenza A viruses are largely conserved, there are specific regions in the NS1 sequence that show considerable variations. First is the linker region, which in the case of NS1 from the majority of the recent H5N1 strains (~90%) contains a five amino acid deletion (residues 80–84) that is implicated in high virulence in the H5N1 influenza strains (Lipatov *et al.*, 2005; Long *et al.*, 2008). Close to this region is the point mutation D92E which in some H5N1 strains confers resistance to the interferon induced antiviral response and enhances virulence (Lipatov *et al.*, 2005; Seo *et al.*, 2002, 2004). Both of these mutations occur in the region that could potentially influence the relative orientations of RBD and ED in the NS1 structure. Another variable region is at the C-terminus. NS1 from some strains (H1N1, H2N2 and H3N2), which kept circulating until the 1980s, exhibit a seven amino acid extension, while in the current avian strains, NS1 is typically 230 residue long (Melen *et al.*, 2007; Palese and Shaw, 2007) (Fig. 2.1). Some avian strains exhibit C-terminal truncations of as many as 15–25 residues. For example, NS1 from a highly pathogenic H5N1 strain (A/Vietnam/1203/2004) consists of only 215 residues. In the majority of avian strains, the last four amino acid residues of NS1 contain a consensus PDZ binding motif (Obenauer *et al.*, 2006). Based on the phylogenic analysis, NS1 sequences are grouped into two alleles – A and B (Ludwig *et al.*, 1991). While allele A consists of NS1 from both human and avian strains, allele B exclusively consists of NS1 from avian strains. An analysis of the concatenated genomes from avian strains shows that, in addition to HA and NA, NS1 contributes the most to genome variability (Obenauer *et al.*, 2006). While immune pressure could account for the genetic variability of HA and NA, what drives the variability of NS1 is not clear.

Until recently, our understanding of the NS1 structure came mainly from the crystallographic analyses of individual domains and their complexes with ligands. Although these studies have provided detailed characterization of each domain, a more complete understanding of whether these two domains interact with one another, whether they influence the structural characteristics of one another, how they function in concert in binding to various ligands and whether the highly variable linker region has any role in the inter-domain interactions has been lacking. The focus of structural studies on the individual domains has been prevalent because the full-length NS1 readily precipitates at concentrations necessary for crystallographic studies. Recently, Bornholdt and Prasad (2008) developed a strategy to counter this problem by introducing two point mutations and successfully determined the structure of the full-length NS1 from a highly pathogenic avian H5N1 strain. This review focuses on our current understanding of the structural characteristics of

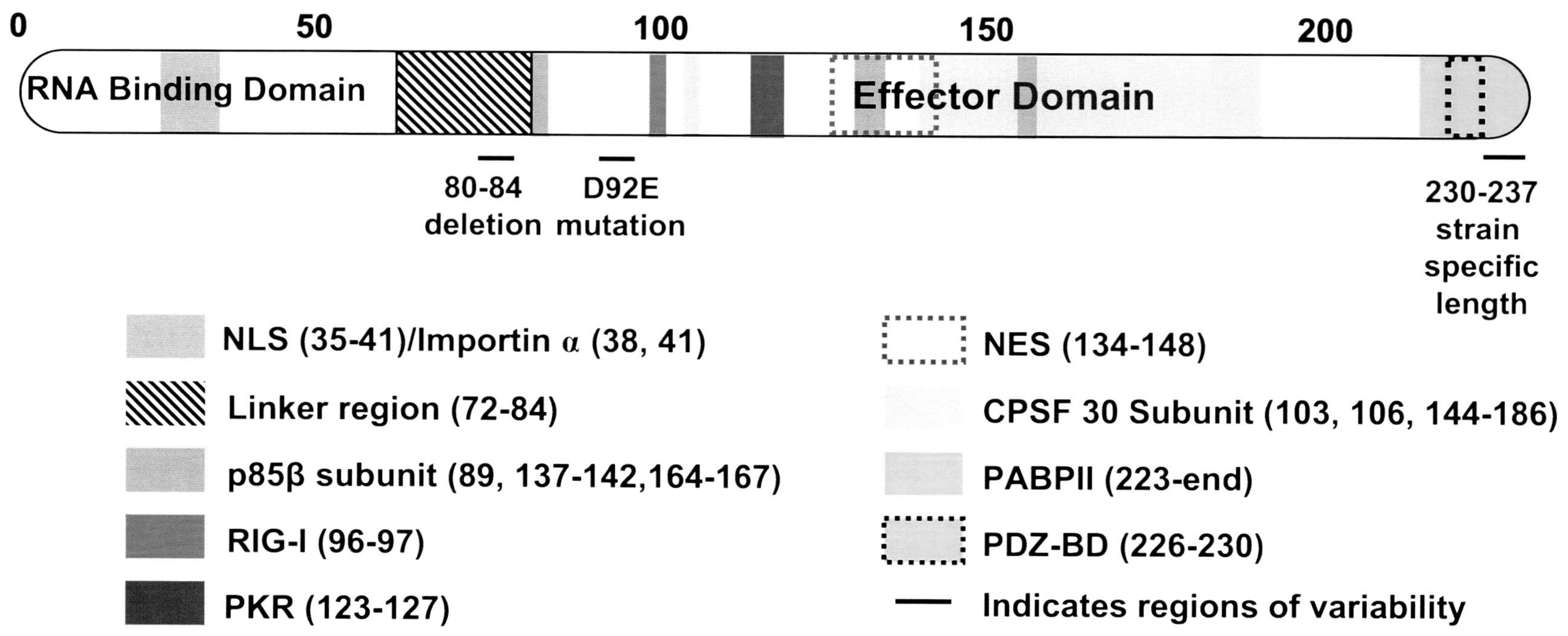

Figure 2.1 Schematic representation of NS1 and its interacting partners. NS1 is 215–237 amino acids in length depending on the strain. The N-terminus contains the RBD composed by residues 1–73, followed by a short linker region (black diagonally hatched box) that connects it to the ED, which encompasses residues 85–230. NS1 RBD contains a nuclear localization sequence that is coincident with the binding site for α-importin (grey). In addition, NS1 contains binding sites for: p85β (violet), RIG-I (blue), PKR (red), CPSF (yellow), PABPII (aqua), and PDZ-containing proteins (aqua and black dashed box). The ED also contains a nuclear export signal (dashed red box). The black bar indicates regions of significant variability in the NS1 sequence: deletion of amino acids 80–84 in the H5N1 strains, and C-terminal variations.

NS1 both at the level of individual domains and in the context of the full-length protein and their implications in NS1 function.

The RNA-binding domain of NS1

The RBD consists of the first 72 N-terminal amino acids of NS1 and contains residues critical for RNA binding. NS1 has been reported to bind to several types of RNA including mRNAs containing poly(A) (Qiu and Krug, 1994) and U6 snRNA (Qiu *et al.*, 1995). However, the ability of NS1 to inhibit the immune response is mostly linked to the dsRNA binding activity (Lu *et al.*, 1995). The dsRNA binding is not sequence specific (Hatada and Fukuda, 1992) and requires dimerization of the RBD (Wang *et al.*, 1999). It is suggested that NS1 inhibits the activation of the IFN-inducing PKR (Diebold *et al.*, 2003; Lu *et al.*, 1995) and OAS/RNase L (Min and Krug, 2006) pathways OAS/RNase L pathways (Min and Krug, 2006) through the sequestration of dsRNA that is produced during infection. During a viral infection, dsRNA of at least 30 bp is required to activate these pathways, with maximum activation occurring with a dsRNA greater than 80bp (Wang and Carmichael, 2004). Thus, NS1 must be able to sequester varying lengths of dsRNA and completely prevent exposure of the dsRNA to any interactions with cellular dsRNA binding proteins involved in activating the type I IFN response in the cell. How NS1 sequesters variable lengths of dsRNA is an interesting question (see Full-length NS1 forms a tubular structure, p. 24).

The RBD also contains a nuclear localization signal (NLS) (residues 35–41) that is highly conserved among influenza A virus strains (Greenspan *et al.*, 1988). The nuclear localization of NS1 is mediated by α-importin binding (Melen *et al.*, 2007). Interestingly, the α-importin binding site in NS1 is the same for dsRNA binding with the highly conserved Arg38 being critical for both binding activities (Melen *et al.*, 2007; Wang *et al.*, 1999). Despite this overlapping binding sites, the RBD is shown to be capable of binding dsRNA and α-importin non-competitively, suggesting that the RBD is dynamic in function and possibly capable of importing bound dsRNA to the nucleus (Melen *et al.*, 2007).

RBD structure

The first crystal structure of the RBD was determined from the A/Udorn/72 H3N2 strain (Liu *et al.*, 1997). Consistent with previous biochemical studies which had demonstrated that RBD dimerization was required for its ability to bind dsRNA (Chien *et al.*, 1997; Qian *et al.*, 1995), in the crystal structure, the RBD subunits related by crystallographic twofold symmetry engage in dimeric interactions. The RBD homodimer consists of six α-helices, three from each subunit (Fig. 2.2). The basic amino acid residues from these helices line a shallow concave surface suitable for interaction with the negatively charged RNA backbone. The overall fold of the NS1 RBD homodimer is unique among dsRNA binding proteins where the dominant secondary structures observed are β-sheets. Some examples include the ribonucleoprotein domain, the hnRNP-K homologous domain, the cold shock protein, and the dsRNA binding domain from *E. coli* RNase III, whose structures have been determined (reviewed in (Nagai, 1996)). Subsequently, the crystal structures of other NS1 RBDs from multiple strains of influenza have been described and demonstrate a strong conservation of overall dimeric structure (Bornholdt and Prasad, 2008; Cheng *et al.*, 2009).

RBD–dsRNA interactions

Until recently, it was not known how NS1 interacted with dsRNA, although models based on unliganded RBD crystal structures and NMR chemical shift perturbation upon complex formation had been proposed (Chien *et al.*, 1997; Liu *et al.*, 1997) (Fig. 2.2). These studies generally described the involvement of various residues including the highly conserved Thr5, Pro31, Asp34, Arg35, Arg38, Lys41, Gly45, Arg46, and Thr49 in dsRNA binding. However, these models did not unambiguously address whether the NS1 binding was through the minor or the major groove of dsRNA or whether both potential binding sites were interconvertably used. This past year, the structure of the NS1 RBD bound to a 21-nucleotide siRNA was reported and provided the structural details of how NS1 RBD interacts with dsRNA (Cheng *et al.*, 2009). This structural analysis unambiguously shows that the RBD recognizes the major groove of dsRNA. The dsRNA molecule sits atop the shallow concave

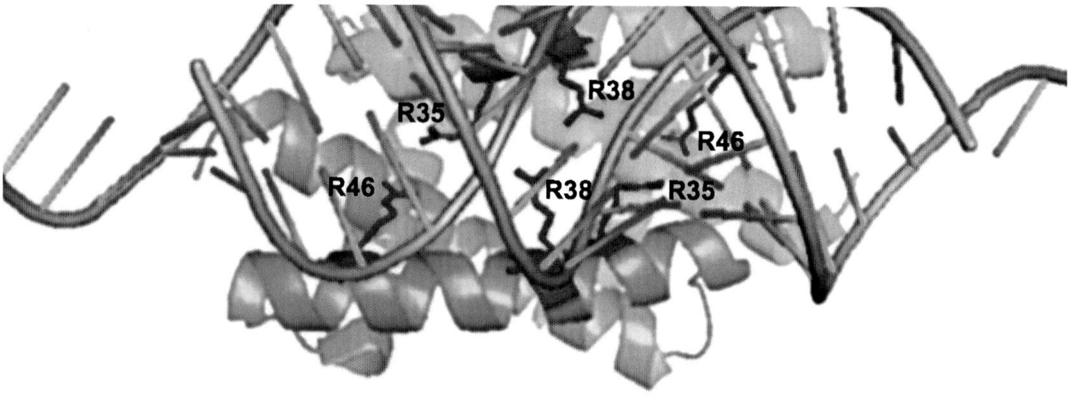

Figure 2.2 The NS1 RBD structure and its interactions with dsRNA. The NS1 RBD bound to dsRNA (orange) (PDB ID: 2ZKO, (Cheng *et al.*, 2009)) is shown here with each monomer coloured in marine and lime green. The Arg38-Arg 38 pair as well as the Arg35-Arg46 hydrogen bonding pair is depicted with sticks in red.

surface without inducing any drastic alterations in the overall structure of the RBD dimer. The complex formation is through hydrogen bonds and electrostatic interactions between the amino acid residues lining the shallow concave surface of the RBD dimer and the sugar phosphate backbone of the dsRNA. Involvement of only sugar phosphate backbone of the RNA is in line with the sequence-independent recognition of dsRNA by NS1. Many of the RBD residues implicated in dsRNA binding by the previous studies participate in the interactions with dsRNA. While the side chains of basic residues Arg38 and Arg35 and hydroxyl residues Ser42 and Thr49 make direct hydrogen bonding with the RNA backbone, other amino acids such as Thr5, Asp29, Asp34, Asp39, and Lys41 interact with dsRNA through water-mediated hydrogen bonding.

Among the residues that are involved in direct interactions with dsRNA, Arg38, Arg35 and Arg46 are critical for binding. The dsRNA binding is completely abolished in R38A single mutant and in the double mutant R35A/R46A, while mutations of other residues such as Ser42 and Thr49, which are also involved in direct interactions with dsRNA, reduce the dsRNA binding affinity by 10-fold (Cheng *et al.*, 2009). Both in the unliganded and liganded RBD structures, Arg35 and Arg46 from opposing subunits of the dimer, are paired through hydrogen bonding interactions. Similarly, Arg38 from opposing subunits also form a hydrogen bonding pair but only in the liganded structure representing the only conformational change induced by dsRNA binding. These observations suggest that while Arg35–Arg46 interactions are important for the stability of the dimer, the formation of the Arg38 pair, which penetrates into bound dsRNA, is critical for the dsRNA binding (Cheng *et al.*, 2009).

Interestingly, directly below the shallow concave surface where dsRNA binds, the RBD has a well-conserved pocket that is suggested as a potential site for drug targeting (Cheng *et al.*, 2009). The Arg pair induced by dsRNA binding forms a lid over this pocket. In the crystal structure of the RBD–dsRNA complex, this pocket is occupied by a glycerol molecule suggesting the possibility of designing molecules that bind to this pocket and disrupt the Arg38 pairing (Cheng *et al.*, 2009). Disruption of Arg38 pairing could affect dsRNA binding affinity and consequently result in attenuation of virulence.

The effector domain of NS1

The ED is the larger of the two NS1 domains consisting of residues 86–230 (residues can vary by strain) and within lethal H5N1 strains it is the ED which contains mutations that confer a virulent and/or cytokine resistant phenotype in recombinant viruses (Lipatov *et al.*, 2005; Seo *et al.*, 2002,2004). The ED contains binding sites for various cellular proteins including protein kinase R (PKR) (Li *et al.*, 2006; Lu *et al.*, 1995; Tan and Katze, 1998), the 30-kDa subunit of cleavage and polyadenylation specificity factor 30 (CPSF) (Nemeroff *et al.*, 1998; Noah *et al.*, 2003), poly-A-binding protein II (PABPII) (Chen *et al.*, 1999), p85β (Ehrhardt *et al.*, 2006; Hale *et al.*, 2006; Shin *et al.*, 2007), and some PDZ domain ligand proteins (Jackson *et al.*, 2008; Obenauer *et al.*, 2006) (Fig. 2.1). Alternative to the possibility of dsRNA sequestration involving RBD as a mechanism to inhibit PKR activation, direct binding involving ED is suggested for PKR inactivation (Li *et al.*, 2006; Tan and Katze, 1998). In some strains, NS1 binding to CPSF30 and binding to PABPII through ED modifies cellular pre-mRNA processing including 3′ end processing of pre-mRNAs to prevent the induction of interferons and interferon-induced genes in infected cells (Kochs *et al.*, 2007a). The sequestration of p85β by the NS1 ED is poorly understood and the resulting p85β binding site knockout phenotypes are varied among different strains (Hale *et al.*, 2006; Shin *et al.*, 2007). Recently, E96/E97 (as per numbering in the PR8 NS1 sequence) of the ED is suggested to mediate interactions with TRIM-25 in suppressing RIG-I mediated signal transduction (Gack *et al.*, 2009). It remains to be further explored as to what is the functional importance of ED interactions with PDZ proteins particularly in the avian strains. Interactions with PDZ proteins could potentially be important for enhancing pathogenicity of the virus by modulating the functions of these proteins in cell polarity, transport, and localization.

ED subunit structure

The first structure of the NS1 ED was determined from the A/Puerto Rico/8/34 H1N1 strain (PR8) (Bornholdt and Prasad, 2006). The ED structure consists of seven β-strands and three α-helices. The β-strands form an anti-parallel twisted β-sheet with the exception of the final strand, which is oriented in a parallel fashion. The six strands of the twisted β-sheet surround a central long α-helix, which is held in place by a series of hydrophobic interactions between the large β-sheet and the central α-helix (Fig. 2.3A). A DALI (Holm and Sander, 1996) search with NS1 ED indicated no homologous structures. A similar fold (not found by DALI) exists which is known as the 'hot dog' fold (Leesong *et al.*, 1996). However, in this fold the strands of the β-sheet run perpendicular to the central α-helix rather than parallel to the central α-helix as in the NS1 ED. Thus, the NS1 ED exhibits a novel fold which is described as an α-helix β-crescent fold, as the β-sheet forms a crescent-like shape about the central long α-helix (Bornholdt and Prasad, 2006). Following the determination of the PR8 NS1 ED structure, the structure of the NS1 ED from various other strains have been determined. Although the subunit fold of the NS1 ED is highly conserved in all of these strains, the dimeric and oligomeric interactions show significant variations.

ED dimeric interactions

In the crystal structure of the PR8 NS1 ED, the crystallographic asymmetric unit consists of two NS1 molecules interacting closely with one another through paring of their β-sheets with a substantial buried surface area of 1600 $Å^2$ (Fig. 2.3B). This non-crystallographic dimer interacts with neighbouring dimeric units related by crystallographic twofold axes through interactions involving the main α-helix and a conserved Trp187 that inserts into the hydrophobic pocket of the opposing subunit. Thus, from this crystal structure, two types of dimeric interaction can be envisaged, one within the asymmetric unit involving paring of the β-sheets (β-sheet dimer), and the other across the crystallographic twofold axis involving inter-subunit α-helix interactions (helix–helix dimer) (Fig. 2.3C). The buried surface area in the helix–helix dimer is substantially smaller than that in the β-sheet dimer. Based on these observations, Bornholdt and Prasad (2006) argued that the dimer observed within in the asymmetric unit would be biologically relevant. However, the crystallographic structure of the A/Duck/Albany/76/H12N5 NS1 ED was recently determined (Hale *et al.*, 2008a). In this structure, as well, there are two molecules in the asymmetric unit; however, the dimer formation involves helix–helix interactions (Fig. 2.3C), similar to the crystallographic dimer in the PR8 ED, prompting the authors to propose this type of dimer as biologically relevant. However, it should be mentioned here that, phylogenetically, avian H12N5 NS1 belongs to a separate group termed allele B, whereas the NS1 from PR8 strain belongs to the allele A group (Hale *et al.*, 2008b). Whether the observed differences in the dimeric interactions simply

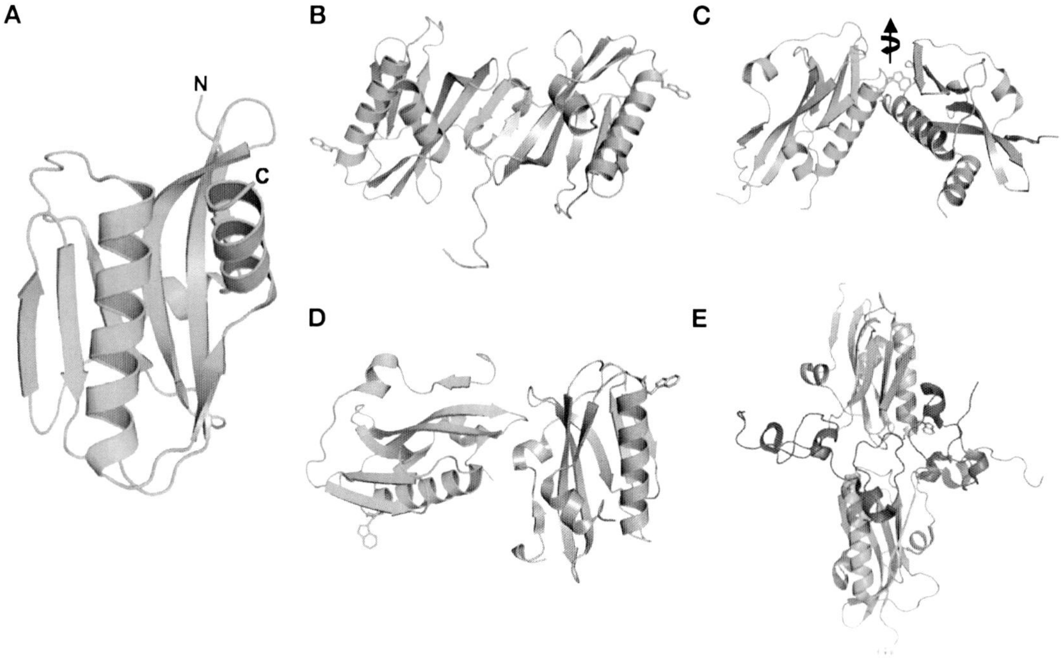

Figure 2.3 The structural conformations of the NS1 ED. (A) The A/PR/8/34 H1N1 ED subunit (PDB ID: 2GX9) is shown in aqua and consists of seven β-strands and three α-helices exhibiting a novel fold which is described as an α-helix– β-crescent fold. The N- and C-termini are marked. (B) The non-crystallographic dimer of the A/PR/8/34 H1N1 ED, with each monomer colored in aqua and bright orange, involves the pairing of β-sheets (β-sheet dimer). (C) Crystallographic dimer of the A/PR/8/34 H1N1 ED involves inter-subunit α-helix interactions (helix-helix dimer). The 2-fold rotation axis is indicated. Similar dimeric interactions are observed in the crystal structure of H12N5 ED between the subunits in the crystallographic asymmetric unit (PDB ID: 3D6R), and in the crystal structure of A/Udorn/72 H3N2 ED (PDB ID: 3EE8) between the subunits related by the crystallographic 2-fold axis. (D) Dimeric interactions between the subunits in the crystallographic asymmetric unit of A/Udorn/72 H3N2 ED (PDB ID: 3EE8). (E) The A/Udorn/72 H3N2 ED in complex with F2F3 domain of CPSF forming a tetramer composed of two NS1 ED subunits interacting head-to-head with each other, and two F2F3 molecules (magenta) that wrap around the ED subunits (PDB ID: 2RHK). All the dimers (B–E) contain one monomer coloured in aqua and another one coloured in bright orange. Trp187, important for CPSF binding, is depicted as sticks in all ED dimers.

represent allele-specific differences remains an open question.

More recently, the structure of the A/Udorn/72 NS1 ED (Udorn) was determined (Xia *et al.*, 2009). Interestingly, in this structure, although the interactions between the two subunits within the crystallographic asymmetric unit are quite different (Fig. 2.3D, PDB ID:3EE8), in the crystal packing a similar helix–helix dimer (Fig. 2.3C) related by the crystallographic two-fold symmetry is observed (Xia *et al.*, 2009). In the helix–helix dimer, the Trp187 residue that is shown to be important for CPSF binding is in the dimeric interface and is buried inside a hydrophobic pocket. If the helix–helix dimer were to be the functional state, then Trp187 would be inaccessible for interactions with CPSF. In the dimer that is stabilized by pairing of the β-sheets, as found in the asymmetric unit of the PR8 NS1 crystal structure, not only is the buried surface area substantially larger but also the residues implicated in the CPSF binding, including Trp187, remain accessible.

ED interactions with CPSF

Recently the X-ray structure of the Udorn NS1 ED in complex with the F2F3 domain of CPSF has been determined (Das *et al.*, 2008). In this crystal structure, ED exhibits entirely different quaternary interactions. The complex forms a tetramer composed of two NS1 ED subunits interacting head-to-head with each other, and two F2F3 molecules that wrap around the ED subunits (Fig. 2.3E). This head-to-head ED dimer is stable only in the presence of the F2F3 peptide. The ED interaction with F2F3 is mediated by an eight amino acid hydrophobic pocket (composed of Lys110, Ile117, Ile119, Gln121, Val180, Gly183, Gly184, and Trp187) in the NS1 ED as well as Phe103 and Met106 located outside the pocket, which are also critical for the formation of the tetrameric complex. Although many of these residues are highly conserved in NS1 including those from human influenza virus strains, there are some variations in the positions 103 and 106. The evolutionary origin of these substitutions and their implications in the NS1 functions is presently speculative at best. Some of the substitutions as found for example in H5N1 strains, Phe to Leu at 103 and/or Met to Ile at 106, weaken CPSF binding, or in some cases, as in PR8 NS1 containing Ser and Ile at 103 and 106 respectively, completely abolish CPSF binding (Das *et al.*, 2008).

Taken together, these structures of NS1 ED present a confusing picture as to which dimeric state is favoured, particularly in the context of the full-length protein. Whether the various dimeric states observed in the crystal structures of ED just reflect strain-dependent variations, or whether in the absence of RBD the ED has the flexibility to adapt alternative dimeric states, remain as open questions. Indeed in support of the former possibility, two regions, one at the dimeric interface of the β-sheet dimer and another close to it, exhibit sequence variations. While the majority of NS1 (>90%) have the LTD (residues 90–92 in PR8 NS1) sequence in the first region, both Udorn NS1 and avian H12N5 NS1 have ITD, and in the second region, instead of SET (residues 195–197 in PR8 NS1), Udorn NS1 has SKT and H12N5 has SKN (consensus is TET or SET in this region). In the NS1 from a highly pathogenic avian H5N1 strain, for which the crystallographic structure of the full-length protein is determined as discussed below, both the LTD and TET sequence motifs are conserved, and the ED exhibits a β-sheet dimer albeit with a twist between the opposing subunits in the dimer. In this context, although the crystal structure of the Udorn NS1 ED–F2F3 complex correctly depicts the binding site for CPSF in NS1, whether the observed quaternary association and the relative orientations of the ED and F2F3 subunits in the tetrameric complex are possible in the full-length NS1 needs further investigation.

Structure of the full-length NS1

To date, the only full-length structure of NS1 that has been determined is from an H5N1 strain (A/Vietnam/1203/2004) (Bornholdt and Prasad, 2008). As expected from the individual crystal structures of RBD and ED from non-H5N1 strains, the overall polypeptide folds of both the RBD and ED are conserved and both of these domains individually participate in dimeric interactions. A remarkable revelation in this structure, however, is that these dimeric interactions instead of resulting in a single dimeric unit, result in the formation of a chain of NS1 molecules with alternating RBD and ED dimers (Fig. 2.4A). In this chain formation, the two domains of each NS1 molecule separately interact with their respective domains from the neighbouring NS1 molecules, related by crystallographic 2-fold axes.

The RBD dimer in the full-length structure, except for minor differences, recapitulates the features observed in the other structures of the isolated RBD. The H5N1 sequence exhibits substitutions in nine amino acid position when compared, for instance, to the Udorn RBD. Except for F22V, other substitutions, including the two mutations, R38A and K41A, which were introduced to make the protein behave well for crystallographic studies, do not have any effect on the overall structure of RBD dimer. F22, located at the C-terminus of the first α-helix, causes a slight movement of the loop that connects this α-helix and the second main α-helix. Overall conservation of the dimeric structure of the RBD irrespective of strain differences, as an independent entity as well as within the full-length structure, is consistent with the observation that RBD dimerization is critical for dsRNA binding activity of the NS1 (Wang *et al.*, 1999).

In contrast, the ED dimer in the full-length H5N1 NS1 structure exhibits noticeable differences (Fig. 2.4B). Although similar to the β-sheet dimer discussed earlier, the two subunits in ED dimer of the full-length NS1 are considerably more twisted with respect to each other as the two

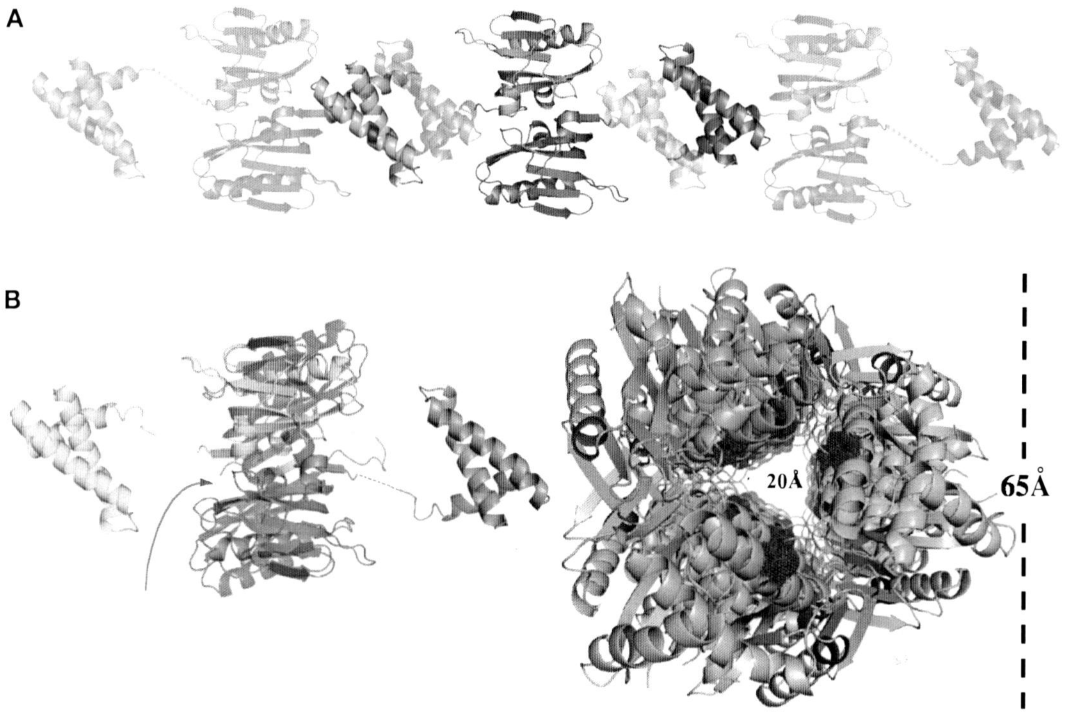

Figure 2.4 Full-length NS1 structure and higher-order oligomerization of NS1. (A) In the crystal structure of the full-length A/Vietnam/1203/2004 H5N1 NS1, RBD and ED dimeric interactions, instead of resulting in a single dimeric unit, result in the formation of a chain of NS1 molecules with alternating RBD and ED dimers. The monomers of are coloured in chartreuse and aqua, with the critical dsRNA binding residue Arg38 coloured in dark blue. The 2-fold symmetry related NS1 dimers displayed in a semi-transparent fashion. (B) The A/PR8/34 H1N1 NS1 ED dimer (in transparent red) is superimposed over the ED of the full-length A/Vietnam/1203/2004 H5N1 NS1 (PDB ID: 3F5T) protein to demonstrate the twisting motion (shown by the green and yellow arrows) in the ED dimer in the full-length protein. The monomers of the A/Vietnam/1203/2004 H5N1 NS1 protein dimer are coloured in yellow and green, with arrows showing the motion of their respectively coloured domains. (C) Three NS1 chains, related by the crystallographic 31 screw axis, come together to form a tube-like structure with a ~20-Å-wide tunnel. Each chain is coloured differently here in aqua, orange, and chartreuse. The residues involved in dsRNA binding point towards the central tunnel. The position of the Arg 38 residue critical for dsRNA binding is shown as an example (represented by the blue mesh). The residues (depicted in red) critical to binding CPSF F2F3 (PDB ID: 3F5T) are located on the outer surface of the tubular structure.

largest β-strands, residues 115–120 and 126–137 twist towards the centre of the ED monomer and the most N-terminal β-sheet of the ED move away from the dimeric interface. Consequently, the ED dimeric interface in the full-length NS1 consists of a series of electrostatic interactions in contrast to the dimeric interactions observed in the β-sheet dimer in the PR8 NS1 ED, which consists of primarily hydrophobic interactions. Considering that the residues involved in the dimeric interactions in the H5N1 ED dimer are well conserved compared to that in the β-sheet dimer of the PR8 NS1, a strong possibility is that the conformational difference in the H5N1 NS1 ED dimer is due to the presence of the RBD and the formation of the RBD dimer. It is also likely that the length of the linker between the RBD and ED may influence the degree of the ED dimer twisting. In the NS1 from many of the H5N1 strains, the linker region has a five-residue deletion. Thus, the stress induced within the H5N1 NS1 structure upon RBD dimerization has pulled the H5N1 NS1 ED N-termini towards their corresponding RBDs, where perhaps within the PR8 NS1 the additional five residues contained in the linker may limit the degree of the ED twisting.

Full-length NS1 forms a tubular structure

A fascinating observation from the crystal structure of the full-length NS1, which may have functional implications as discussed separately, is the tight association of the three NS1 chains, related by the crystallographic 31 screw, to form a tubular structure of ~70 Å in diameter (Fig. 2.4C and Fig. 2.5). The association of the NS1 chains is facilitated by interactions between the ED dimer of one chain and an RBD from each of the other chains, with the RBDs of the two chains creating a saddle like depression for the ED dimer to fit. Interactions along the second main α-helix of the RBD namely Leu33, Asp34, Leu36, Arg37, and probably Lys41along with the ED facilitate the formation of the tubular NS1 structure. None of these residues is involved in the dsRNA binding with the exception of Asp34 and Lys41, which interact with dsRNA through water-mediated hydrogen bonding (Cheng *et al.*, 2009).

Possible mechanism of dsRNA sequestration by NS1

Is there any functional relevance to the tubular structure observed in the context of the crystal packing? Although the structure of RBD in complex with dsRNA, as discussed earlier in a separate section, has provided molecular details of RBD interaction with dsRNA, it does not provide a mechanism for how NS1 sequesters dsRNA to prevent activation of the type I IFN response. Several observations point to the possibility that dsRNA sequestration by NS1 may involve tubular organization. The tubular structure formed by the associations of three NS1 chains has a ~20 Å tunnel that tantalizingly seems appropriate to sequester dsRNA (Fig. 2.4C). Is the orientation of the RBD dimer in the tubular structure appropriate for interacting with the dsRNA? In each NS1 chain, the RBD dimer is turned at a 45° angle relative to the ED, aligning the shallow concave surface in the RBD dimer implicated in dsRNA binding to run the length of the NS1 chain. This predominantly basic surface of the RBD, in level with a polar saddle formed along the top of the ED dimer, creates a polar surface running the length of the NS1 chain (Fig. 2.3A). This polar surface created by each of the three NS1 chains face inward lining the tunnel in the NS1 tubular structure (Fig. 2.4C). The Arg38 residue that is critical for dsRNA binding projects into the tunnel. Thus, the diameter of the tunnel inside the tubular structure, the inner polar lining,

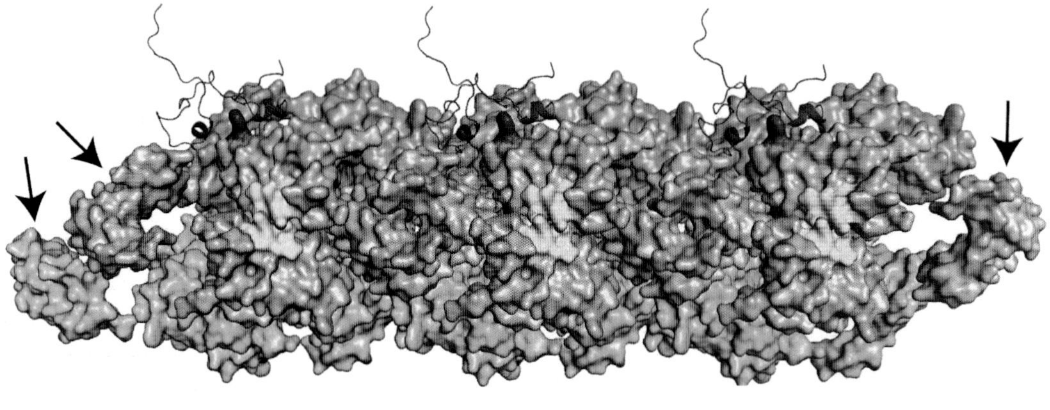

Figure 2.5 Some of the protein ligand binding sites are accessible on the outer surface of the tubular structure. A surface representation of the H5N1 NS1 tubular structure with the individual chains coloured in aqua, chartreuse, and orange. The binding site for CPSF F2F3 peptide (magenta) is accessible on the surface of the NS1 tubular structure, as shown by superimposing a monomer A/Udorn/72 ED bound to the CSPF F2F3 peptide (PDB ID: 2RHK) to the A/Vietnam/1203/2004 H5N1 NS1 ED (PDB ID: 3F5T) of one for the three chains (orange). The critical Trp187 residue is coloured red. Arrows point to 'free RBDs' at the end of the tubular structures, and yellow patches represent the C-term of that would project outwards. The C-terminal tail consists of binding sites for PABII and PDZ domain proteins.

and the appropriately located Arg38 residue, all point to the possibility that formation of the tubular structure indeed could be the mechanism for sequestering dsRNA.

In support of such a possibility, Bornholdt and Prasad (2008) present cryo-EM images of native full-length PR8 NS1 in the presence of dsRNA. These images show formation of smooth tubular structures of similar width to that observed in the crystal structure of the full-length H5N1 NS1. It appears that the tubular formation by NS1 is contingent upon the presence of dsRNA, as observed in the cryo-EM images, or R38A and K41A mutations, as observed in the crystal structure of the full-length H5N1 NS1. As mentioned earlier, the native full-length NS1 from several of the influenza strains precipitates readily. However, in the presence of dsRNA, or with R38A and K41A mutations NS1 remains in solution. It is possible that these charged residues in the native protein need to be neutralized by the bound dsRNA or mutated to uncharged residues for the native protein to remain in solution.

The formation of the tubular structure by the mutated full-length H5N1 NS1 protein is concentration dependent. At lower concentrations, as shown by size exclusion chromatography, mutated NS1 is a dimer, and with increasing concentrations, it forms higher order oligomers as indicated by dynamic light scattering (Bornholdt and Prasad, 2008). During crystallization, the concentration of NS1 is necessarily high to promote the formation of tubular organization as observed in the crystal structure of the full-length NS1. A plausible model then is that NS1 cooperatively oligomerizes in the presence of dsRNA to form a tubular structure capable of sequestering any length of dsRNA within the central tunnel. The inherent flexibility in the linker between the ED and RBD domains in such a model would allow for any structural variations in the dsRNA, including changes in the helical pitch. Thus, this model provides an attractive mechanism for the way NS1 can sequester varying lengths of dsRNA produced during infection and prevent exposure of the dsRNA to any interactions with cellular dsRNA binding proteins involved in activating the type I IFN response in the cell. In addition to dsRNA, NS1 is known to bind poly(A) (Qiu and Krug, 1994) and U6 snRNA (Qiu *et al.*, 1995). The proposed tubular structure is also possible for NS1 interactions with these nucleic acids.

NS1 interactions with cellular protein ligands

As mentioned earlier, in addition to dsRNA sequestration, other mechanisms involving NS1 interactions with various cellular proteins are probably in play to counter the IFN response. Presently it is not clear whether these mechanisms are independent of each other, but some of the studies do indicate strain-dependent preference for these mechanisms. Formation of a tubular structure is unlikely in the absence of dsRNA or at low concentrations as mentioned above. Several observations suggest the possibility that dsRNA sequestration by cooperative oligomerization of NS1 could be a predominant mechanism and provide an attractive scaffold to interact with various cellular proteins. First, NS1 is expressed at high levels in infected cells (Krug and Etkind, 1973; Palese and Shaw, 2007). Second, recombinant protein even at moderate concentrations readily precipitates but remains soluble in the presence of dsRNA. Third, RBD is highly conserved in both sequence and structure. However, it has to be mentioned that sequestration of dsRNA as an obligatory event is not yet shown. If it indeed is the case, the residues implicated in binding several of the cellular proteins such as CPSF30 (103, 106, 184–188), p85β (89), and PDZ binding proteins and PABII (C-terminal tail) remain accessible on the outer surface of the NS1 tubular structure (Fig. 2.5). The tubular organization also provides an attractive explanation for the perplexing data indicating that despite overlapping binding sites, NS1 simultaneously interacts with both dsRNA and α-importin (Melen *et al.*, 2007). The cooperative interactions leading to NS1 tubular formation in the presence of dsRNA would allow for a 'free' RBD at either end of the tubular structure, which is not involved in the interactions with dsRNA, to bind α-importin (Bornholdt and Prasad, 2008) (Fig. 2.5). Although cooperative oligomerization of NS1 into tubular structure provides an attractive mechanism to sequester dsRNA and simultaneously interact with some cellular proteins, in the absence of dsRNA, however, NS1 could indeed exhibit other conformational states appropriate for its interactions with various other cellular proteins. Further structural and biochemical studies are required to address questions such as whether NS1 forms such tubular structures *in vivo*, whether cooperative oligomerization of NS1 is strain independent, how the length of the linker region affects the interdomain interactions, and what other quaternary interactions are possible with NS1 in the absence and presence of various NS1 binding ligands.

Summary and future directions

From various biochemical studies, it is becoming evident that NS1 is a highly evolved protein capable of antagonizing the antiviral host immune response by using multiple mechanisms. Such a strategy could indeed confer upon the virus the ability to adapt to various host environments, support efficient virus replication and increase the zoonotic potential. Structural studies on the individual domains of NS1 show that while the RBD exhibits highly conserved dimeric interactions, the ED exhibits considerable variations in quaternary interactions. In the context of the full-length NS1, whether there is a unifying common structural scaffold independent of strains that could support the multifunctional properties of NS1 remains yet to be established. The crystallographic structure of the full-length NS1 from an avian strain and cryo-EM imaging of NS1 from a non-avian (PR8) strain in complex with dsRNA bring in a new perspective in suggesting the possibility of such a common scaffold. A hypothesis that emerges from these structural studies of full-length NS1 is that cooperative oligomerization of NS1 into a tubular structure provides a plausible mechanism for how NS1 sequesters varying lengths of dsRNA to counter cellular antiviral dsRNA response pathways, while also allowing interactions with other cellular ligands during an infection. In such a model, a highly variable linker region could play an important role in modulating the ligand binding properties of NS1 by influencing the relative orientations of the two domains within the context of the tubular organization. An alternative hypothesis is that the ability of the ED to exist in different quaternary states allows NS1 to engage in a different or particular antagonistic mechanism. Clearly, additional crystallographic structures of the full-length NS1 from various strains, with shorter and longer linkers and NS1 complexed with protein/dsRNA ligands, are necessary to further understand the structural basis for how NS1 counters interferon pathways using different mechanisms. Such structural studies will also be invaluable in targeting NS1 for antiviral drug design (see, for example, Basu *et al.* 2009; Darapaneni *et al.*, 2009) or vaccine development (see Chambers *et al.*, 2009) to counter influenza A virus infection, which is a major global health concern.

Acknowledgement

We acknowledge support from NIH (AI36040) and Robert Welch Foundation.

References

Abdel-Ghafar, A.N., Chotpitayasunondh, T., Gao, Z., Hayden, F.G., Nguyen, D.H., de Jong, M.D., Naghdaliyev, A., Peiris, J.S., Shindo, N., Soeroso, S., *et al.* (2008). Update on avian influenza A (H5N1) virus infection in humans. N. Engl. J. Med. *358*, 261–273.

Basu, D., Walkiewicz, M.P., Frieman, M., Baric, R.S., Auble, D.T., and Engel, D.A. (2009). Novel influenza virus NS1 antagonists block replication and restore innate immune function. J. Virol. *83*, 1881–1891.

Beigel, J.H., Farrar, J., Han, A.M., Hayden, F.G., Hyer, R., de Jong, M.D., Lochindarat, S., Nguyen, T.K., Nguyen, T.H., Tran, T.H., *et al.* (2005). Avian influenza A (H5N1) infection in humans. N. Engl. J. Med. *353*, 1374–1385.

Bornholdt, Z.A., and Prasad, B.V. (2006). X-ray structure of influenza virus NS1 effector domain. Nat. Struct. Mol. Biol. *13*, 559–560.

Bornholdt, Z.A., and Prasad, B.V. (2008). X-ray structure of NS1 from a highly pathogenic H5N1 influenza virus. Nature *456*, 985–988.

Chambers, T.M., Quinlivan, M., Sturgill, T., Cullinane, A., Horohov, D.W., Zamarin, D., Arkins, S., Garcia-Sastre, A., and Palese, P. (2009). Influenza A viruses with truncated NS1 as modified live virus vaccines: pilot studies of safety and efficacy in horses. Equine. Vet. J. *41*, 87–92.

Chen, Z., Li, Y., and Krug, R.M. (1999). Influenza A virus NS1 protein targets poly(A)-binding protein II of the cellular 3′-end processing machinery. EMBO J. *18*, 2273–2283.

Cheng, A., Wong, S.M., and Yuan, Y.A. (2009). Structural basis for dsRNA recognition by NS1 protein of influenza A virus. Cell Res. *19*, 187–195.

Chien, C.Y., Tejero, R., Huang, Y., Zimmerman, D.E., Rios, C.B., Krug, R.M., and Montelione, G.T. (1997). A novel RNA-binding motif in influenza A virus nonstructural protein 1. Nat. Struct. Biol. *4*, 891–895.

Darapaneni, V., Prabhaker, V.K., and Kukol, A. (2009). Large-scale analysis of Influenza A virus sequences reveals potential drug-target sites of NS proteins. J. Gen. Virol. (More details?)

Das, K., Ma, L.C., Xiao, R., Radvansky, B., Aramini, J., Zhao, L., Marklund, J., Kuo, R.L., Twu, K.Y., Arnold, E., *et al.* (2008). Structural basis for suppression of a host antiviral response by influenza A virus. Proc. Natl. Acad. Sci. U.S.A. *105*, 13093–13098.

Diebold, S.S., Montoya, M., Unger, H., Alexopoulou, L., Roy, P., Haswell, L.E., Al-Shamkhani, A., Flavell, R., Borrow, P., and Reis e Sousa, C. (2003). Viral infection switches non-plasmacytoid dendritic cells into high interferon producers. Nature *424*, 324–328.

Egorov, A., Brandt, S., Sereinig, S., Romanova, J., Ferko, B., Katinger, D., Grassauer, A., Alexandrova, G., Katinger, H., and Muster, T. (1998). Transfectant influenza A viruses with long deletions in the NS1 protein grow efficiently in Vero cells. J. Virol. *72*, 6437–6441.

Ehrhardt, C., Marjuki, H., Wolff, T., Nurnberg, B., Planz, O., Pleschka, S., and Ludwig, S. (2006). Bivalent role of the phosphatidylinositol-3-kinase (PI3K) during influenza virus infection and host cell defence. Cell. Microbiol. *8*, 1336–1348.

Fernandez-Sesma, A., Marukian, S., Ebersole, B.J., Kaminski, D., Park, M.S., Yuen, T., Sealfon, S.C., Garcia-Sastre, A., and Moran, T.M. (2006). Influenza virus evades innate and adaptive immunity via the NS1 protein. J. Virol. *80*, 6295–6304.

Fortes, P., Beloso, A., and Ortin, J. (1994). Influenza virus NS1 protein inhibits pre-mRNA splicing and blocks mRNA nucleocytoplasmic transport. EMBO J. *13*, 704–712.

Gack, M.U., Albrecht, R.A., Urano, T., Inn, K.S., Huang, I.C., Carnero, E., Farzan, M., Inoue, S., Jung, J.U., and Garcia-Sastre, A. (2009). Influenza A virus NS1 targets the ubiquitin ligase TRIM25 to evade recognition by the host viral RNA sensor RIG-I. Cell Host Microbe *5*, 439–449.

Garcia-Sastre, A., Egorov, A., Matassov, D., Brandt, S., Levy, D.E., Durbin, J.E., Palese, P., and Muster, T. (1998). Influenza A virus lacking the NS1 gene replicates in interferon-deficient systems. Virology *252*, 324–330.

Greenspan, D., Palese, P., and Krystal, M. (1988). Two nuclear location signals in the influenza virus NS1 nonstructural protein. J. Virol. *62*, 3020–3026.

Guo, Z., Chen, L.M., Zeng, H., Gomez, J.A., Plowden, J., Fujita, T., Katz, J.M., Donis, R.O., and Sambhara, S. (2007). NS1 protein of influenza A virus inhibits the function of intracytoplasmic pathogen sensor, RIG-I. Am. J. Respir. Cell. Mol. Biol. *36*, 263–269.

Hale, B.G., Barclay, W.S., Randall, R.E., and Russell, R.J. (2008a). Structure of an avian influenza A virus NS1 protein effector domain. Virology *378*, 1–5.

Hale, B.G., Jackson, D., Chen, Y.H., Lamb, R.A., and Randall, R.E. (2006). Influenza A virus NS1 protein binds p85beta and activates phosphatidylinositol-3-kinase signaling. Proc. Natl. Acad. Sci. U.S.A. *103*, 14194–14199.

Hale, B.G., Randall, R.E., Ortin, J., and Jackson, D. (2008b). The multifunctional NS1 protein of influenza A viruses. J. Gen. Virol. *89*, 2359–2376.

Hatada, E., and Fukuda, R. (1992). Binding of influenza A virus NS1 protein to dsRNA *in vitro*. J. Gen. Virol. *73*, 3325–3329.

Holm, L., and Sander, C. (1996). Mapping the protein universe. Science *273*, 595–603.

Jackson, D., Hossain, M.J., Hickman, D., Perez, D.R., and Lamb, R.A. (2008). A new influenza virus virulence determinant: the NS1 protein four C-terminal residues modulate pathogenicity. Proc. Natl. Acad. Sci. U.S.A. *105*, 4381–4386.

Kochs, G., Garcia-Sastre, A., and Martinez-Sobrido, L. (2007a). Multiple anti-interferon actions of the influenza A virus NS1 protein. J. Virol. *81*, 7011–7021.

Kochs, G., Koerner, I., Thiel, L., Kothlow, S., Kaspers, B., Ruggli, N., Summerfield, A., Pavlovic, J., Stech, J., and Staeheli, P. (2007b). Properties of H7N7 influenza A virus strain SC35M lacking interferon antagonist NS1 in mice and chickens. J. Gen. Virol. *88*, 1403–1409.

Krug, R.M., and Etkind, P.R. (1973). Cytoplasmic and nuclear virus-specific proteins in influenza virus-infected MDCK cells. Virology *56*, 334–348.

Leesong, M., Henderson, B.S., Gillig, J.R., Schwab, J.M., and Smith, J.L. (1996). Structure of a dehydratase-isomerase from the bacterial pathway for biosynthesis of unsaturated fatty acids: two catalytic activities in one active site. Structure *4*, 253–264.

Li, S., Min, J.Y., Krug, R.M., and Sen, G.C. (2006). Binding of the influenza A virus NS1 protein to PKR mediates the inhibition of its activation by either PACT or double-stranded RNA. Virology *349*, 13–21.

Lipatov, A.S., Andreansky, S., Webby, R.J., Hulse, D.J., Rehg, J.E., Krauss, S., Perez, D.R., Doherty, P.C., Webster, R.G., and Sangster, M.Y. (2005). Pathogenesis of Hong Kong H5N1 influenza virus NS gene reassortants in mice: the role of cytokines and B- and T-cell responses. J. Gen. Virol. *86*, 1121–1130.

Liu, J., Lynch, P.A., Chien, C.Y., Montelione, G.T., Krug, R.M., and Berman, H.M. (1997). Crystal structure of the unique RNA-binding domain of the influenza virus NS1 protein. Nat. Struct. Biol. *4*, 896–899.

Long, J.X., Peng, D.X., Liu, Y.L., Wu, Y.T., and Liu, X.F. (2008). Virulence of H5N1 avian influenza virus enhanced by a 15-nucleotide deletion in the viral nonstructural gene. Virus Genes *36*, 471–478.

Lu, Y., Wambach, M., Katze, M.G., and Krug, R.M. (1995). Binding of the influenza virus NS1 protein to double-stranded RNA inhibits the activation of the protein kinase that phosphorylates the eIF-2 translation initiation factor. Virology *214*, 222–228.

Ludwig, S., Schultz, U., Mandler, J., Fitch, W.M., and Scholtissek, C. (1991). Phylogenetic relationship of the nonstructural (NS) genes of influenza A viruses. Virology *183*, 566–577.

Melen, K., Kinnunen, L., Fagerlund, R., Ikonen, N., Twu, K.Y., Krug, R.M., and Julkunen, I. (2007). Nuclear and nucleolar targeting of influenza A virus NS1 protein: striking differences between different virus subtypes. J. Virol. *81*, 5995–6006.

Mibayashi, M., Martinez-Sobrido, L., Loo, Y.M., Cardenas, W.B., Gale, M., Jr., and Garcia-Sastre, A. (2007). Inhibition of retinoic acid-inducible gene I-mediated induction of beta interferon by the NS1 protein of influenza A virus. J. Virol. *81*, 514–524.

Min, J.Y., and Krug, R.M. (2006). The primary function of RNA binding by the influenza A virus NS1 protein in infected cells: Inhibiting the 2′–5′ oligo (A) synthetase/RNase L pathway. Proc. Natl. Acad. Sci. U.S.A. *103*, 7100–7105.

Nagai, K. (1996). RNA–protein complexes. Curr. Opin. Struct. Biol. *6*, 53–61.

Nemeroff, M.E., Barabino, S.M., Li, Y., Keller, W., and Krug, R.M. (1998). Influenza virus NS1 protein interacts with the cellular 30 kDa subunit of CPSF and inhibits 3′ end formation of cellular pre-mRNAs. Mol. Cell *1*, 991–1000.

Noah, D.L., Twu, K.Y., and Krug, R.M. (2003). Cellular antiviral responses against influenza A virus are countered at the posttranscriptional level by the viral NS1A protein via its binding to a cellular protein required for the 3′ end processing of cellular pre-mRNAS. Virology *307*, 386–395.

Obenauer, J.C., Denson, J., Mehta, P.K., Su, X., Mukatira, S., Finkelstein, D.B., Xu, X., Wang, J., Ma, J., Fan, Y., *et al.* (2006). Large-scale sequence analysis of avian influenza isolates. Science *311*, 1576–1580.

Palese, P., and Shaw, M.L. (2007). Orthomyxoviridae: the viruses and their replication.. In In Fields Virology, D.M. Knipe, and P.M. Howley, eds. (Philadelphia, Lippincott Williams & Wilkins), pp. 1647–1689.

Qian, X.Y., Chien, C.Y., Lu, Y., Montelione, G.T., and Krug, R.M. (1995). An amino-terminal polypeptide fragment of the influenza virus NS1 protein possesses specific RNA-binding activity and largely helical backbone structure. RNA *1*, 948–956.

Qiu, Y., and Krug, R.M. (1994). The influenza virus NS1 protein is a poly(A)-binding protein that inhibits nuclear export of mRNAs containing poly(A). J. Virol. *68*, 2425–2432.

Qiu, Y., Nemeroff, M., and Krug, R.M. (1995). The influenza virus NS1 protein binds to a specific region in human U6 snRNA and inhibits U6–U2 and U6–U4 snRNA interactions during splicing. RNA *1*, 304–316.

Seo, S.H., Hoffmann, E., and Webster, R.G. (2002). Lethal H5N1 influenza viruses escape host anti-viral cytokine responses. Nat. Med. *8*, 950–954.

Seo, S.H., Hoffmann, E., and Webster, R.G. (2004). The NS1 gene of H5N1 influenza viruses circumvents the host anti-viral cytokine responses. Virus. Res. *103*, 107–113.

Shin, Y.K., Liu, Q., Tikoo, S.K., Babiuk, L.A., and Zhou, Y. (2007). Influenza A virus NS1 protein activates the phosphatidylinositol 3-kinase (PI3K)/Akt pathway by direct interaction with the p85 subunit of PI3K. J. Gen. Virol. *88*, 13–18.

Simonsen, L., Clarke, M.J., Williamson, G.D., Stroup, D.F., Arden, N.H., and Schonberger, L.B. (1997). The impact of influenza epidemics on mortality: introducing a severity index. Am. J. Public. Health. *87*, 1944–1950.

Suarez, D.L., and Perdue, M.L. (1998). Multiple alignment comparison of the non-structural genes of influenza A viruses. Virus. Res. *54*, 59–69.

Talon, J., Horvath, C.M., Polley, R., Basler, C.F., Muster, T., Palese, P., and Garcia-Sastre, A. (2000). Activation of interferon regulatory factor 3 is inhibited by the influenza A virus NS1 protein. J. Virol. *74*, 7989–7996.

Tan, S.L., and Katze, M.G. (1998). Biochemical and genetic evidence for complex formation between the influenza A virus NS1 protein and the interferon-induced PKR protein kinase. J. Interferon. Cytokine. Res. *18*, 757–766.

Wang, Q., and Carmichael, G.G. (2004). Effects of length and location on the cellular response to double-stranded RNA. Microbiol. Mol. Biol. Rev. *68*, 432–452.

Wang, W., Riedel, K., Lynch, P., Chien, C.Y., Montelione, G.T., and Krug, R.M. (1999). RNA binding by the novel helical domain of the influenza virus NS1 protein requires its dimer structure and a small number of specific basic amino acids. RNA *5*, 195–205.

Wang, X., Li, M., Zheng, H., Muster, T., Palese, P., Beg, A.A., and Garcia-Sastre, A. (2000). Influenza A virus NS1 protein prevents activation of NF-kappaB and induction of alpha/beta interferon. J. Virol. *74*, 11566–11573.

Webster, R.G., Bean, W.J., Gorman, O.T., Chambers, T.M., and Kawaoka, Y. (1992). Evolution and ecology of influenza A viruses. Microbiol. Rev. *56*, 152–179.

Xia, S., Monzingo, A.F., and Robertus, J.D. (2009). Structure of NS1A effector domain from the influenza A/Udorn/72 virus. Acta Crystallogr. D Biol. Crystallogr. *65*, 11–17.

Influenza Type B Virus Haemagglutinin: Antigenicity, Receptor Binding and Membrane Fusion

3

Qinghua Wang

Abstract

Influenza B virus infection remains a significant cause of morbidity and mortality worldwide, with particularly severe outcomes in the young and the elderly. However, compared to influenza A virus, influenza B virus was very poorly studied. Most recently, a series of research have been published that drastically deepened our understanding of influenza B virus. In this review, I will summarize the recent advances on influenza B virus haemagglutinin from aspects of antigenicity, receptor binding and membrane fusion. In addition, I will discuss some of the remaining challenges in influenza B virology.

Introduction

Influenza virus is a negative-stranded RNA virus belonging to the Orthomyxoviridae family. Influenza virus-caused infection remains a significant cause of morbidity and mortality worldwide. In the past century, there were three major pandemics – 1918, 1957 and 1968 – among which the 1918 pandemic was the most dreadful one, causing approximately 50 million human deaths. However, it was estimated that the annual epidemics caused by influenza A virus H1N1 and H3N2 subtypes and influenza B virus over the past 100 years have had an even greater cumulative impact than the three pandemics combined (Schweiger *et al.*, 2002; Webster, 1992, 1998).

The first influenza B virus strain, B/Lee/40, was isolated in 1940 (Krystal *et al.*, 1982). Over the years, influenza B virus has diverged into two major lineages, the Yamagata lineage and the Victoria lineage (Kanegae *et al.*, 1990; Kinnunen *et al.*, 1992; Rota *et al.*, 1990; Shaw *et al.*, 2002; Yamashita *et al.*, 1988). The evolution of influenza B virus is characterized by the co-circulation of antigenically distinct strains (Kanegae *et al.*, 1990; Nakagawa *et al.*, 2000) and the isolation of antigenically similar viruses many years apart (Bao-Lan *et al.*, 1983; Chakraverty, 1972; Oxford *et al.*, 1984; Rota *et al.*, 1992, 1990; Yamashita *et al.*, 1988).

Haemagglutinin (HA) is one of the two major glycoproteins on the surface of influenza B virus. It is the major antigen that elicits immune responses in hosts, and is responsible for attaching influenza virus to cell-surface sialic acid receptors and for releasing the virus into cell cytosol via fusion of host endosomal membrane and viral envelope membrane. Capitalizing on our recent structural studies and sequence analysis of influenza B virus HA (Shen *et al.*, 2009; Wang *et al.*, 2008; Wang *et al.*, 2007), this review will focus on the recent advancement in the field from aspects of antigenicity, receptor binding and membrane fusion. Some of the remaining questions in influenza B virology will also be discussed.

Three types of influenza viruses

There are three types of influenza viruses, A, B and C, based on the lack of serological cross-reactivity between their internal proteins. They all have a major surface glycoprotein, HA for influenza A and B viruses, and haemagglutinin-esterase-fusion (HEF) protein for influenza C virus (Air and Compans, 1983; Cox and Subbarao, 2000; Palese and Young, 1982; Skehel and Wiley, 2000; Stuart-Harris *et al.*, 1985; Webster *et al.*,

1982; Wiley and Skehel, 1987). Both influenza A and B virus HA glycoproteins recognize the most abundant sialic acid, N-acetylneuraminic acid, while influenza C virus HEF binds to 9-O-acetylated sialic acid (Rogers *et al.*, 1986; Rosenthal *et al.*, 1998; Skehel and Wiley, 2000; Vlasak *et al.*, 1987; Wiley and Skehel, 1987; Zhang *et al.*, 1999).

Influenza A virus infects a wide range of host species including humans, birds, horses and swine. Wild aquatic birds are the natural reservoirs for influenza A virus, where a total of 16 different subtypes of HA, H1–H16 (Fouchier *et al.*, 2005; WHO-MEMORANDUM, 1980), has been found. Among the 16 HA subtypes, only influenza viruses with H1, H2, and H3 HA have established in humans, causing the seasonal influenza epidemics and responsible for all known human influenza pandemics, including the pandemics in 1918 (H1N1), 1957 (H2N2) and 1968 (H3N2) in the last century (Murphy and Webster, 2001). In recent years, avian H5N1 influenza A virus has caused considerable concerns of potential pandemics due to its high fatality rate (more than 50%) in humans (Taubenberger and Morens, 2008). Two types of antigenic variation have been observed for influenza A virus: antigenic shift with substantial pandemic potential resulting from genetic reassortment among different influenza A viruses that are co-infecting the same hosts, and antigenic drift with epidemic potential as a result of the accumulation of amino-acid mutations in HA that lead to the emergence of antigenically distinct strains.

Influenza B virus has a much limited host range, infecting humans almost exclusively and seals to a lesser degree (Osterhaus *et al.*, 2000). Due to the host range limitation, influenza B virus in general does not impose pandemic threats. Different from influenza A virus HA, influenza B virus HA has no subtypes, with an average of 0.3% amino acid mutation rate per year (1.5 residues), which is about 3–5 times slower than influenza A virus HA (Air *et al.*, 1990; Bootman and Robertson, 1988; Chakraverty, 1972; Chen and Holmes, 2008; Cox and Bender, 1995; Hay *et al.*, 2001; Kanegae *et al.*, 1990; Krystal *et al.*, 1982, 1983; Lubeck *et al.*, 1980; Pechirra *et al.*, 2005; Schild *et al.*, 1973; Verhoeyen *et al.*, 1983; Yamashita *et al.*, 1988).

Similar to influenza B virus, influenza C virus also infects mostly humans with mild symptoms, usually among the young and the elderly (Air and Compans, 1983; Katagiri *et al.*, 1983; Matsuzaki *et al.*, 2006; Stuart-Harris *et al.*, 1985), has a slower amino-acid mutation rate, and is characterized by the co-circulation of many antigenically distinct strains (Buonagurio *et al.*, 1985; Buonagurio *et al.*, 1986).

The Yamagata and Victoria lineages of influenza B virus

Influenza B/Lee/40 strain was the first influenza B virus that had been identified (Krystal *et al.*, 1982). Currently, there are two influenza B virus lineages, the Yamagata lineage represented by the B/Yamagata/16/88 strain and the Victoria lineage represented by the B/Victoria/1/87 strain (Kanegae *et al.*, 1990; Kinnunen *et al.*, 1992; Rota *et al.*, 1990; Shaw *et al.*, 2002; Yamashita *et al.*, 1988). The divergence of these two lineages can be dated back to the late 1970s (Chen *et al.*, 2007; Shen *et al.*, 2009). Over the years, the Yamagata lineage HA was further divided into four sublineages, while the Victoria lineage HA was split into two sublineages (Nerome *et al.*, 1998; Shen *et al.*, 2009).

Crystal structures of influenza B virus HA

Similar to influenza A virus HA, influenza B virus HA is synthesized as a single polypeptide chain, which is then cleaved into two disulphide-bonded fragments, HA1 and HA2, during transport to the cellular surface. Influenza B virus HA shares about 20% sequence identity in HA1 and ~30% in HA2 with X:31 HA, the best understood HA glycoprotein of human influenza A virus H3 subtype. The HA1 polypeptide chain of influenza B virus HA is longer than that of influenza A virus HA, and contains many insertions and deletions (Matrosovich *et al.*, 1993).

We have most recently determined the crystal structures of influenza virus B/Hong Kong/8/73 (B/HK) HA in various states (Wang *et al.*, 2008; Wang *et al.*, 2007). Despite the low sequence identity, B/HK HA shares an overall similar fold (Wang *et al.*, 2008) to influenza A virus HA glycoproteins (Gamblin *et al.*, 2004; Ha *et al.*, 2002, 2003; Russell *et al.*, 2004; Stevens *et al.*, 2006; Stevens *et al.*, 2004; Wilson *et al.*, 1981; Yamada *et al.*, 2006). It has an elongated fusion domain (F), a globular receptor-binding domain (R), and a vestigial esterase domain (E′) (Fig. 3.1A). Each subunit of B/HK HA has six intra-subunit disulphide bonds and one inter-subunit disulphide bond (Fig. 3.1A). Five disulphide bonds have counterparts in X:31 HA: $Cys4_1$–$Cys137_2$, $Cys60_1$–$Cys72_1$, $Cys94_1$–$Cys143_1$, $Cys292_1$–$Cys318_1$ and $Cys144_2$–$Cys148_2$ (Krystal *et al.*, 1983), while two disulphide bonds, $Cys54_1$–$Cys57_1$ and $Cys178_1$–$Cys272_1$, are unique for influenza B virus HA. On each HA1/HA2 subunit, there are a total of ten potential glycosylation sites, seven on HA1 (25, 59, 123, 145, 163, 301, 330) and three on HA2 (145, 171, and 184). The gain or loss of the glycosylation at HA1 194 (Wang *et al.*, 2007) was frequently found in monoclonal antibody (mAb)-escape mutants (Berton *et al.*, 1984; Berton and Webster, 1985), in egg-adapted variants (Gambaryan *et al.*, 1999; Oxford *et al.*, 1990, 1991; Robertson *et al.*, 1990; Robertson *et al.*, 1985; Saito *et al.*, 2004; Schild *et al.*, 1983), or in field isolates (Ikonen *et al.*, 2005; Nakagawa *et al.*, 2000, 2004). However, in the B/HK HA used for our crystallographic study, the residue at HA1 194 is Asp, thus no glycosylation was found at this site.

Antigenicity of influenza B virus HA

Antigenic structure

The slower mutation rate of influenza B virus HA, in comparison to that of influenza A virus HA, has allowed the identification of approximately two dozens single amino-acid substitutions that have been confirmed to cause antigenicity changes (Fig. 3.1B) in site-specific mutants, mAb-selected escape variants, or field isolates (Berton *et al.*, 1984; Berton and Webster, 1985; Bootman and Robertson, 1988; Hovanec and Air, 1984; Krystal *et al.*, 1982, 1983; Lubeck *et al.*, 1980; Rota *et al.*, 1990, 1992; Verhoeyen *et al.*, 1983; Webster and Berton, 1981). Based on the new structure of B/HK HA, these mutations can be roughly divided into four groups, the 120-loop, HA1 116–137; the 150-loop, HA1 141–150; the 160-loop, HA1 162–167; and the 190-helix, HA1 194–202, and their surrounding regions (Fig. 3.1B).

The 120-loop (Fig. 3.1B, in cyan) is the most frequently mutated region in field isolates (Verhoeyen *et al.*, 1983). Virus variants of a Victoria lineage strain with specific substitutions at residues HA1 129 and 137, produced using reverse genetics, were found to cause altered antigenicity (Lugovtsev *et al.*, 2007). Similarly, HA1 75, 77, 116, 118, and 122 were found to present an epitope that is unique to the Victoria lineage strains (Nakagawa *et al.*, 2006). Collectively, these residues constitute an epitope that partially overlaps with site E in influenza A virus HA H3 subtype (X:31 HA) (Wiley and Skehel, 1987) and sites Sa and Cb in influenza A virus HA H1 subtype (Caton *et al.*, 1982; Gerhard *et al.*, 1981).

The 150-loop (Fig. 3.1B, in green) is a long protruding loop, with its tip (Thr147) pointing away from the main body of the structure by approximately 9 Å. For multiple times, laboratory-selected escape variants were shown to harbour single substitutions in this region: $Gly141_1$→Val (Hovanec and Air, 1984; Webster and Berton, 1981), $Gly141_1$→Arg, $Thr147_1$→Ile, $Ser148_1$→Gly, $Arg149_1$→Gly (Nakagawa *et al.*, 2003), and $Asn150_1$→Ser (Berton *et al.*, 1984; Berton and Webster, 1985). In strains isolated from fields, a single substitution of $Arg149_1$→Lys (Nakagawa *et al.*, 2001a) was shown to block reactions with Arg-strain specific antisera, and the Lys-strains rapidly became predominant in subsequent epidemic seasons (Abed *et al.*, 2003). Moreover, influenza B virus grown in embryonated eggs was found to possess altered antigenic properties due to single amino-acid substitutions at $Gly141_1$ (Lugovtsev *et al.*, 2005, 2007; Oxford *et al.*, 1990, 1991). In recent isolates, the 150-loop region became the neutralizing epitope unique for the Yamagata lineage strains (Nakagawa *et al.*, 2003). The 150-loop partially overlaps with site A in X:31 HA (Wiley *et al.*, 1981) and site Ca2 in H1 HA (Caton *et al.*, 1982; Gerhard *et al.*, 1981).

The 160-loop (Fig. 3.1B, in blue) is the only region in influenza B virus HA where insertions and deletions have been repeatedly detected in field isolates from different epidemic seasons (Fig. 3.1C) (McCullers *et al.*, 1999; Nerome *et al.*, 1998), consistent with its protruding nature. Known changes with altered antigenicity include single deletions at HA1 162‴ or 167 (Fig. 3.1C for numbering) (Nakagawa *et al.*, 2001b), a single

A

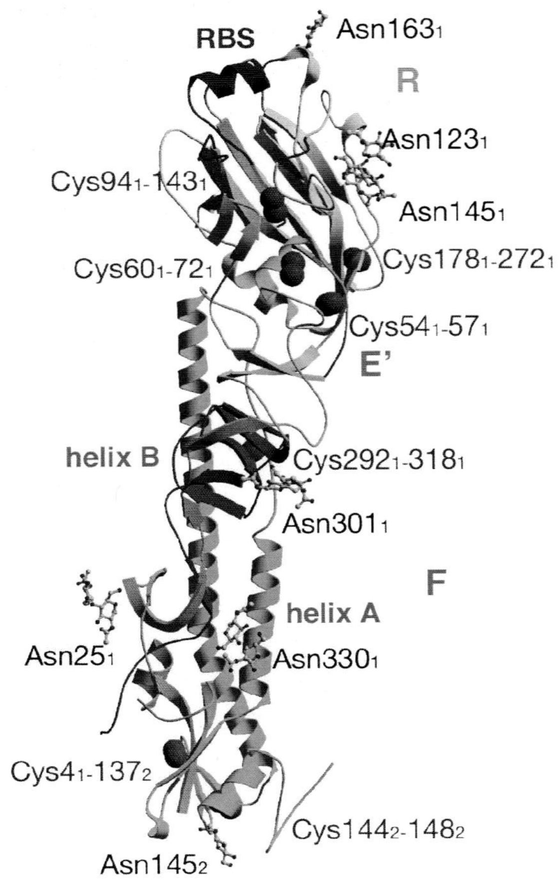

B

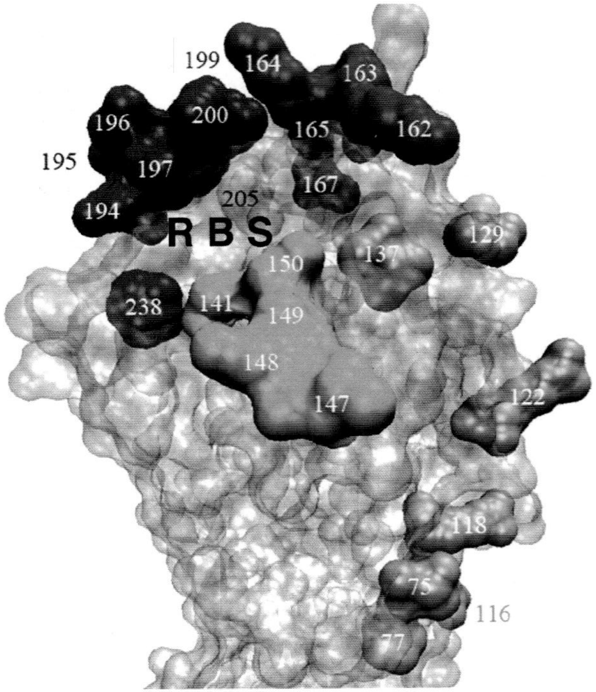

C

	161 163
	. .
B/HK/73	PK...NKTA
B/Lee/40	PK.DNNKTA
B/GL/54	PK...NKTA
B/MD/59	PKNK.NKTA
B/SI/64	PKNK.NKTA
B/SI/79	PK.D.NKTA
B/OR/80	PK.D.NKTA
B/ID/86	PKN.NNKTA
B/GA/86	PKNNNNKTA
B/VI/87	PKNDNNKTA
B/YM/88	PR.D.NKTA

insertion of Asn following HA1 163 (Webster and Berton, 1981), and single amino-acid substitutions such as Lys162$_1$→Ile (Hovanec and Air, 1984), Lys164$_1$→Ile or Asn (Berton *et al.*, 1984; Berton and Webster, 1985); Asp162$_1''$→Tyr; Asn162$_1'''$→Ser or Thr (Nakagawa *et al.*, 2001a); and Asp164$_1$→Glu or Asn165$_1$→Lys (Nakagawa *et al.*, 2005). Given the fact that B/HK HA has the shortest 160-loop among all known sequences, further extension of the 160-loop in other influenza B virus HA will make it a particularly strong epitope (Fig. 3.1C). In recent isolates, the 160-loop became unique for the Victoria lineage strains (Nakagawa *et al.*, 2001b). The 160-loop partially overlaps with site B in X:31 HA (Wiley *et al.*, 1981) and site Sa in H1 HA (Caton *et al.*, 1982; Gerhard *et al.*, 1981).

All residues at the external face of the 190-helix have important antigenic roles (Fig. 3.1B, in red). Single mutations in this region were repeatedly found in mAb-escape variants: Glu195$_1$→Lys, Gln197$_1$→Lys, Val199$_1$→Ala, Leu or Glu (Berton *et al.*, 1984; Berton and Webster, 1985), Lys200$_1$→Asn, Arg or Thr (Berton *et al.*, 1984; Berton and Webster, 1985; Gerhard *et al.*, 1981; Hovanec and Air, 1984; Nakagawa *et al.*, 2001a), Ser205$_1$→Leu or Pro (Berton *et al.*, 1984; Berton and Webster, 1985). Moreover, single mutations Thr196$_1$→Pro and Gln197$_1$→Ile completely abolished the binding of an mAb raised against influenza B/HK virus (Rivera *et al.*, 1995).

The hot spot on the 190-helix is HA1 194–196, a potential glycosylation site. A single mutation of Ala196$_1$→Thr, which potentially creates a new glycosylation site at HA1 194–196, rendered the virus epidemic (Nakagawa *et al.*, 2004). On the other hand, the loss of the glycosylation site at this region, in mAb-escape mutants (Asn194$_1$→Asp or Lys) (Berton *et al.*, 1984; Berton and Webster, 1985), in egg-adapted variants (Gambaryan *et al.*, 1999; Oxford *et al.*, 1990, 1991; Robertson *et al.*, 1985, 1990; Saito *et al.*, 2004; Schild *et al.*, 1983), or in field isolates (Ikonen *et al.*, 2005; Nakagawa *et al.*, 2000), was found to cause antigenic variation. Furthermore, a single difference at residue HA1 194, Asn versus Ser, caused distinct differences between two co-circulating Yamagata lineage strains in antigenic reactivity (Muyanga *et al.*, 2001). Moreover, the Asp194$_1$→Tyr substitution on egg-adapted influenza B virus was found to substantially alter its antigenic properties, stressing its importance in antigenicity independent of the glycans that are attached to it (Lugovtsev *et al.*, 2007). The 190-helix partially overlaps with site B in X:31 HA (Wiley *et al.*, 1981) and site Sb in H1 HA (Caton *et al.*, 1982; Gerhard *et al.*, 1981).

Residue HA1 238 is in proximity to the 190-helix and partially overlaps site D in X:31 HA (Wiley *et al.*, 1981) and site Ca1 in H1 HA (Caton *et al.*, 1982; Gerhard *et al.*, 1981). Substitutions at HA1 238, Pro238$_1$→Ser (Nakagawa *et al.*, 2001a), Gln, or Thr (Berton *et al.*, 1984; Berton and Webster, 1985), were observed in mAb-escape variants of influenza B virus (Berton *et al.*, 1984; Berton and Webster, 1985; Nakagawa *et al.*, 2001a). However, given the critical role of HA1 238 in receptor binding (Wang *et al.*, 2007),

---

Figure 3.1 (opposite) Structures of B/HK HA. (A) Structure of a B/HK HA subunit. The elongated fusion domain (F) is composed of HA2 (green) and the N- (pink) and C-terminal (blue) segments of HA1. The membrane-distal domain contains a receptor-binding subdomain (R, yellow) and vestigial esterase subdomain (E', cyan). The two major helices on HA2, helix A and B, are also labelled. Disulphide bonds on each subunit are shown in CPK as purple, and the sugar residues are in ball-and-stick. The receptor-binding site (RBS) is highlight in red. The subscripts, X1 and X2, of each residue indicate that the residue X is on HA1 and HA2, respectively. (B) Antigenic structure of B/HK HA. Mutations in four regions, the 120-loop (cyan), the 150-loop (green), the 160-loop (blue) and the 190-helix (red), have been found to cause antigenicity variation. The RBS is labelled. All labelled residues are on HA1 and thus no subscripts are indicated. (C) Sequence alignment of influenza B virus HA glycoproteins in the region of HA1 161–166 to highlight the insertion or deletion in this region. '.' indicates deletions in the sequence. By using the sequence of B/HK HA as a reference, the extra residues between HA1 162 and 163 present in other strains are treated as insertions and numbered accordingly. For example, B/Victoria/1/87 (B/VI/87) has a three-residue insertion numbered as Asn162$_1'$, Asp162$_1''$ and Asn162$_1'''$. Thus, all the sequences mentioned in the main text are based on B/HK HA numbering. B/MD/59, B/Maryland/59; B/SI/64, B/Singapore/64; B/SI/79, B/Singapore/222/79; B/OR/80, B/Oregon/5/80; B/ID/86, B/Idaho/1/86; B/GA/86, B/Georgia/86; B/YM/88, B/Yamagata/1/88. Adapted from Wang *et al.*, (2007, 2008).

it is not clear how substitutions at HA1 238 are accommodated without compromising receptor binding.

Antigenic differences between the Yamagata and Victoria lineage HA of influenza B virus

Ever since their divergence in the late 1970s, the Yamagata and Victoria lineages of influenza B viruses have developed fundamental differences in antigenicity. Among the four major epitopes that we identified on influenza B virus HA – the 120-loop, the 150-loop, the 160-loop and the 190-helix (Wang *et al.*, 2008), the 150-loop is a unique epitope specific for the Yamagata lineage (Nakagawa *et al.*, 2001a, 2003), while the 160-loop is the unique epitope specific for the Victoria lineage (Nakagawa *et al.*, 2001b, 2005). Moreover, our most recent sequence analysis of influenza B virus HA has indicated that although both Yamagata lineage and Victoria lineage viruses share the 120-loop as a major epitope, they tend to use different sites for evading antibody recognition (Shen *et al.*, 2009). Consequently, influenza B viruses of one lineage rarely cross-react with antisera induced by viruses of the other lineage. This lack of cross-reactivity posts a severe problem in choosing which strain to include in annual influenza vaccine. The currently recommended trivalent vaccine contains three vaccine strains, an influenza A H3N2 strain, an influenza A H1N1 strain, and an influenza B strain. However, selecting the right influenza B virus strain for vaccine development each year has been proven very difficult. For instance, influenza B virus strains in the vaccine have missed the dominant influenza B virus lineage in circulation in five out of nine seasons in the USA since the 2000–2001 season. When the mismatches happened, the lack of cross-protection between the two lineages led to a high frequency of illness, particularly in young children, and to deaths and severe outcomes in all age groups. It was for this reason that recently it has been proposed that influenza B virus strains from both Yamagata and Victoria lineages be included in the vaccine, thus turning the conventional trivalent vaccine into a quadrivalent vaccine. However, the quadrivalent vaccine proposal was met with many concerns, in particular regarding the more complex protocol for manufacturers, higher cost, lower overall production, and uncertainties on increased side effects and any impaired immune responses to the different strains. Apparently, various new and alternative strategies need to be evaluated for resolving this dilemma.

How many antigenic site(s) on influenza B virus HA?

It remains an open question whether the four major epitopes of influenza B virus HA, the 120-loop, the 150-loop, the 160-loop and the 190-helix, constitute one or more antigenic sites (Berton *et al.*, 1984; Berton and Webster, 1985; Hovanec and Air, 1984; Webster and Berton, 1981). In an operational and topological mapping with 17 mAbs against influenza B/OR/5/80 strain (Berton and Webster, 1985), Berton and Webster have demonstrated that the 150-loop, the 160-loop and the 190-helix belong to a single, continuous antigenic site on influenza B virus HA. Similarly, an mAb10B8 was found to induce antigenic variants by single amino-acid substitutions in the 160-loop or the 190-helix of recent Victoria lineage strains, suggesting the overlapping nature of the 160-loop/190-helix region (Nakagawa *et al.*, 2001b). However, escape mutants selected by an mAb 9B2, the first such confirmed to recognize the 120-loop of HA of the Victoria lineage strains, remained reactive to the mAbs 10B8 and 10D7 that recognize the 160-loop/190-helix region (Nakagawa *et al.*, 2006), indicating that the 120-loop may belong to an antigenic site different from that of the 160-loop/190-helix region. The lack of cross-reactivity between antibodies directed towards the 120-loop and the 160-loop/190-helix region is reminiscent of the study by Berton and Webster (Berton and Webster, 1985) in which two mAbs 55/1 and 94/2 against influenza B/OR/5/80 strain did not cross-react with any epitopes belonging to the 150-loop/160-loop/190-helix region. Unfortunately, the exact epitopes of these two mAbs were not identified due to their rather low haemagglutination inhibition and neutralization activities on intact virus. Nevertheless, both mAbs showed significant binding to detergent-disrupted virus antigen by enzyme-linked immunosorbent assay and both immunoprecipitated the B/OR/5/80 HA from infected Madin–Darby canine kidney (MDCK) cells (Berton and Webster, 1985). These results suggested that the influenza B virus HA epitopes to which the mAbs 55/1 and 94/2 bind are not easily accessible on intact virus. The limited accessibility of the epitopes of the mAbs 55/1 and 94/2 is consistent with the location of the 120-loop that is proximal to the base of the membrane-distal domain (Fig. 3.1B), similar to what was noted for antibodies targeting the same region of influenza A virus HA (Barbey-Martin *et al.*, 2002; Bizebard *et al.*, 2001; Bizebard *et al.*, 1995; Fleury *et al.*, 1999; Gigant *et al.*, 2000; Knossow *et al.*, 2002). Thus, these two studies collectively suggest that epitopes of the mAbs 9B2 (Nakagawa *et al.*, 2006), 55/1 and 94/2 (Berton and Webster, 1985) be located at the same or a similar region of the 120-loop on influenza B virus HA, and that the 120-loop is probably a second antigenic site.

Although the epidemiological significance of the 120-loop is not extensively tested, residues in the region of the 120-loop (HA1 73, 75, 129, 177 and 295) were found to be involved in positive selective pressure with high probability (Fig. 3.2) (Nunes *et al.*, 2008; Shen *et al.*, 2009). Given their rather distant locations from the receptor-binding site, the selections are mostly likely to avoid antibody neutralization (rather than modulating receptor binding), emphasizing the importance of the 120-loop in influenza B virus HA antigenicity. Indeed, field isolates of the Yamagata lineage differing by only two residues at HA1 58 and 126, both at the 120-loop, had significantly different reactivity to post-infection ferret antisera (Nakagawa *et al.*, 2002). Moreover, the mAb 9B2 recognizing the 120-loop had haemagglutination inhibition activity against all the 2002/2003 Victoria lineage isolates and all other representative Victoria lineage strains, suggesting the conserved nature of this epitope (Nakagawa *et al.*, 2006).

One unique feature of influenza B virus is that antigenic variants selected with mAbs, often possessing single amino-acid substitutions, can be frequently differentiated from the parental virus with polyclonal ferret sera (Webster and Berton, 1981). This feature may be due to the smaller number of non-overlapping antigenic sites of influenza B virus HA, of one or two antigenic sites as discussed above, when compared with those of influenza A virus HA glycoproteins of various subtypes (Gerhard *et al.*, 1981; Kaverin *et al.*, 20002, 2004, 2007; Okamatsu *et al.*, 2008; Raymond *et al.*, 1986; Skehel and Wiley, 2000; Wiley and Skehel, 1987; Wiley *et al.*, 1981). As a result, single amino acid changes in influenza B virus HA may be sufficient to alter the antigenicity of the virus, and render most of the influenza B virus variants epidemiologically potent (Nakagawa *et al.*, 2001a, 2002; Rota *et al.*, 1990). This may partially explain the observed slower amino acid mutation rate (Chakraverty, 1972; Krystal *et al.*, 1983; Schild *et al.*, 1973; Verhoeyen *et al.*, 1983) and lower selection pressure exerted by host antibodies (Chen and Holmes, 2008; Nunes *et al.*, 2008; Pechirra *et al.*, 2005; Shen *et al.*, 2009), as observed for influenza B virus compared with influenza A virus. However, it remains unclear why it is difficult to obtain escape influenza B virus mutants when selected by mAbs (Berton *et al.*, 1984; Berton and Webster, 1985; Webster and Berton, 1981). Probably, other factors, such as the tolerability of influenza B virus HA to amino-acid changes, may come into play in this case. Further studies are needed to address these issues.

Reassortment and insertion–deletion in the evolution of influenza B virus HA

With a genome composed of segmented negative stranded RNA, reassortment among different viruses infecting the same host cells is an important means for antigenic variation of influenza A and C viruses (Lindstrom *et al.*, 1998; Peng *et al.*, 1994, 1996). For instance, reassortment among avian and human influenza A viruses was responsible for the major pandemics in 1957 and in 1968 in the last century (Webster *et al.*, 1992), and reassortment among human influenza viruses was implicated in recent epidemics in Japan (Lindstrom *et al.*, 1999). Although influenza B virus does not have subtypes, the two lineages – the Victoria and the Yamagata lineage viruses – co-circulate and reassortment between them has been observed (Chen and Holmes, 2008; Hiromoto *et al.*, 2000; Jian *et al.*, 2008; Li *et al.*, 2008; Lindstrom *et al.*, 1999; Luo *et al.*, 1999; McCullers *et al.*, 1999; Puzelli *et al.*, 2004), where the reassortants seem to carry some selective advantages (Chi *et al.*, 2003; Daum *et al.*, 2006; McCullers *et al.*, 1999;

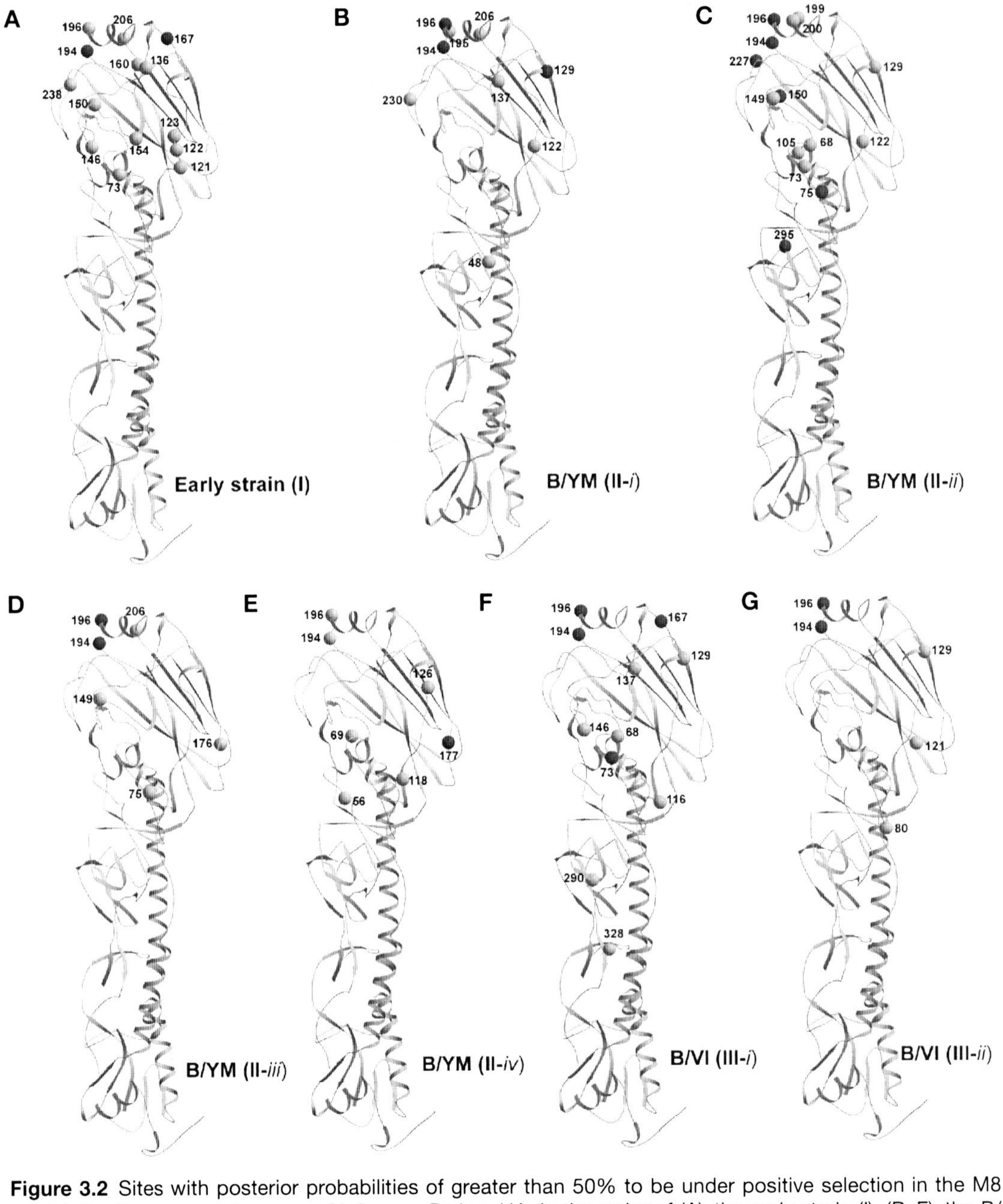

Figure 3.2 Sites with posterior probabilities of greater than 50% to be under positive selection in the M8 models for the seven subgroups of influenza B virus HA, in the order of (A) the early strain (I), (B–E) the B/Yamagata lineage (B/YM II-i-II-iv) and (F and G) B/Victoria lineage (B/VI III-i and III-ii). Each site is shown as a ball centred at its Cα atom in the structure (Protein Data Bank code 3BT6) (Wang *et al.*, 2008). Sites with greater than 95% posterior probability to be under positive selection are shown in dark colour and the rest are in light colour. All labelled residues are on HA1 and thus no subscripts are indicated. Adapted from Shen *et al.* (2009).

Shaw *et al.*, 2002; Tsai *et al.*, 2006; Xu *et al.*, 2004; Yamashita *et al.*, 1988).

In marked contrast to influenza A virus HA glycoproteins, amino-acid insertion and deletion on influenza B virus HA have been frequently observed in the region of HA1 162–163 (Fig. 3.1C) on the 160-loop (McCullers *et al.*, 1999; Nerome *et al.*, 1998). In relation to B/Lee/40 HA, the Victoria lineage HA has one amino acid insertion at HA1 162′, while the Yamagata lineage HA has one or two amino acid deletions in the same region. This difference between the Victoria- and Yamagata lineage HA glycoproteins caused fundamental differences in antigenicity: the 160-loop became a specific epitope unique for the Victoria lineage influenza B virus (Nakagawa *et al.*, 2001b; Wang *et al.*, 2008). Altering the patterns of insertions and deletions seems to play an important role in antigenic variation of influenza B virus, including HA and neuraminidase (McCullers *et al.*, 1999).

Host immune selection on antigenic drift of influenza B virus HA

Several studies have recently appeared in the literatures exploring the evolutionary dynamics of influenza B virus (Chen and Holmes, 2008; Nunes *et al.*, 2008; Pechirra *et al.*, 2005; Shen *et al.*, 2009). Antigenic drift via accumulation of amino-acid changes selected by host immune system can be most directly evaluated using computational programs such as CODEML in PAML (phylogenetic analysis by maximum likelihood) (Yang, 2007) that detect positive selection on protein-coding sequences by allowing different selective pressure on different amino-acid positions (Yang, 2007; Yang *et al.*, 2000). A larger than 1.0 rate ratio between the non-synonymous (amino-acid altering) substitutions over synonymous (silent) substitutions at a given position indicates positive selection pressure at the protein level. Similar studies have been previously performed on influenza A virus HA where positively selected codons have been identified to correlate well with the potential of influenza A virus strains to become predominant in future epidemics (Bush *et al.*, 1999a,b; Fitch *et al.*, 1997; Yang *et al.*, 2000).

By grouping a total of 271 influenza B virus HA sequences into seven subgroups: the early strains isolated between 1940 and 1970 (I), the Yamagata lineage strains since 1972 (II) including four (II-i–II-iv) sublineages and the Victoria lineage strains since 1975 (III) including two (III-i and III-ii) sublineages, our group found (Shen *et al.*, 2009) that all the identified positively selected sites in the seven subgroups were located on the four major antigenic epitopes (Fig. 3.2) – the 120-loop, the 150-loop, the 160-loop and the 190-helix (Wang *et al.*, 2008) – supporting the important roles of antibody selection in the molecular evolution of influenza B virus HA.

Among the four major epitopes, residue HA1 150 in the 150-loop was under positive selection with 96% posterior probability in the Yamagata II-ii sublineage, consistent with the recent study that the 150-loop region appeared to be the neutralizing epitope specific for the Yamagata lineage strains (Nakagawa *et al.*, 2001a, 2003). The 160-loop is the only region in influenza B virus HA where insertions and deletions were detected (Berton *et al.*, 1984; Berton and Webster, 1985; McCullers *et al.*, 1999; Nakagawa *et al.*, 2001a,b, 2005; Nerome *et al.*, 1998; Webster and Berton, 1981). Residue HA1 167 on the 160-loop was found to be selected positively in the early strains (I) and in the Victoria III-i sublineage with 95% and 96% posterior probability, respectively, which agrees well with the discovery that the 160-loop in recent isolates became specific for the Victoria lineage strains (Nakagawa *et al.*, 2001b; Nakagawa *et al.*, 2005). Most interestingly, HA1 194 and 196 on the 190-helix were constantly identified to be under positive selective pressure, with greater than 99% probability in 11 out of 14 cumulative cases (combining both sites in seven groups), and over 85% in three other cases. Amino acid substitutions in the region of HA1 194–196 frequently result in the addition or removal of a glycosylation site. Thus, as observed in influenza A virus HA (Caton *et al.*, 1982; Daniels *et al.*, 1983; Schulze, 1997; Skehel *et al.*, 1984; Skehel and Wiley, 2000), influenza B virus HA glycoproteins also utilize the addition or removal of glycosylation as a mechanism for antigenicity alteration.

In addition, the following sites on the 120-loop were found to be positively selected: HA1 129 in the Yamagata II-i sublineage, HA1 75 and 295 in the Yamagata II-ii sublineage, HA1 177 in the Yamagata II-iv sublineage, and HA1 73 in the Victoria III-i sublineage. Although the 120-loop region appeared to be one of the most frequently mutated regions in field isolates (Verhoeyen *et al.*, 1983), its role in antigenicity of influenza B virus HA was not recognized until most recently (Lugovtsev *et al.*, 2007; Nakagawa *et al.*, 2006; Nunes *et al.*, 2008; Shen *et al.*, 2009; Wang *et al.*, 2008). One possibility for such a delay in recognition is that the 120-loop is proximal to the viral envelope membrane, making the access by antibodies more difficult, as observed for influenza A virus HA (Barbey-Martin *et al.*, 2002; Bizebard *et al.*, 1995, 2001; Fleury *et al.*, 1999; Gigant *et al.*, 2000; Knossow *et al.*, 2002). Thus, the discovery of strong positive selection in the 120-loop region further supports its significance in antigenicity of influenza B virus HA (Nunes *et al.*, 2008; Shen *et al.*, 2009).

Interplay of influenza B virus with influenza A virus

Chen and Holmes (2008) investigated the prevalence patterns of influenza A and B viruses over a period of ~30 years, and found that there are population bottlenecks in influenza B virus concurring with a high prevalence of influenza A virus, and influenza B virus generally becomes prevalent following a peak of influenza A virus (Hay *et al.*, 2001). Thus, in competing for the same population of susceptible humans, infection by one type of virus would reduce the probability of infection or replication by another type of virus (Mikheeva and Ghendon, 1982).

Receptor binding of influenza B virus HA

Receptor-binding site

As observed for influenza A virus HA, the receptor-binding site of influenza B virus HA is located at the top of the membrane-distal domain (Fig. 3.1A), formed by the 190-helix (HA1 193–202), the 240-loop (HA1 237–242), and the 140-loop (HA1 136–143) (Fig. 3.3A). Four residues from the HA1 chain, Phe95$_1$, Trp158$_1$, His191$_1$, and Tyr202$_1$, all absolutely conserved among all known sequences of influenza B virus HA, constitute the base of the receptor-binding site (Air *et al.*, 1990; Krystal *et al.*, 1982, 1983; Lindstrom *et al.*, 1999; Muyanga *et al.*, 2001; Nakagawa *et al.*, 2002; Rota *et al.*, 1992; Shaw *et al.*, 2002; Xu *et al.*, 1994, 1996). π-stacking interactions among these four residues and a hydrogen bond between His191$_1$ and Tyr202$_1$ stabilize the receptor-binding site. In addition, four more hydrogen bonds between Asp193$_1$ and Ser240$_1$ (Fig. 3.3A) are found in the receptor-binding site, which makes the left entrance of the binding site narrower by approximately 0.3 Å than that of X:31 HA (between Glu190$_1$ and Ser228$_1$).

One unique feature of the receptor-binding site of influenza B virus HA is a phenylalanine residue at position 95, in marked contrast to a tyrosine residue for all influenza A virus HA at the equivalent position (Fig. 3.3B). This difference, together with the one-residue deletion following His191$_1$ and a substitution of a proline residue by Ser192$_1$, could be responsible for the different shape of the receptor-binding site and the different orientation of Trp158$_1$ in influenza B virus HA (Fig. 3.3B) (Matrosovich *et al.*, 1993). The structural difference at this part of the receptor-binding site can be appreciated by a shift of 3.6 Å between the Cα atoms of Asp193$_1$ in B/HK HA and its equivalent residue Ser186$_1$ in X:31 HA (Fig. 3.3B).

In the regions of the 190-helix and the 140-loop, the receptor-binding sites of B/HK HA and X:31 HA are very similar (Matrosovich *et al.*, 1993). The 140-loop region contributes the conserved main-chain and side-chain atoms for the interactions with the sialic acid moiety (Sia-1) of the receptors. These atoms include the main-chain carbonyl of Thr139$_1$, the side-chain hydroxyl of Ser140$_1$, and the main-chain amide and carbonyl of Gly141$_1$ (equivalent to Gly135$_1$, Ser136$_1$, and Asn137$_1$ in X:31 HA, respectively). However, despite the conserved interactions, residue Ser140$_1$ of B/HK HA is located approximately 1.3 Å further away from the 240-loop than its equivalent residue, Ser136$_1$, in X:31 HA (Fig. 3.3B).

The 240-loop forms the left edge of the B/HK HA receptor-binding site. It harbours residues Pro238$_1$ and Ser240$_1$, which are at equivalent positions to residues HA1 226 and 228 in X:31 HA, respectively. Both residues HA1 226 and 228 of X:31 HA were implicated as major determinants for receptor specificity (Skehel and Wiley, 2000; Wiley and Skehel, 1987). Owing to a four-residue insertion (HA1 233–236) proceeding Leu237$_1$ in influenza B/HK HA, this part of the receptor-binding site is markedly different from that of X:31 HA (Fig. 3.3B), with much looser interactions between the 240-loop and its proceeding loop of HA1 230–236. As a result, the side chain of Leu237$_1$ in B/HK HA is flipped over to be located similarly to Trp222$_1$ (Fig. 3.3B). The highly conserved residue Ser240$_1$ is located approximately 3.6 Å higher than its equivalent

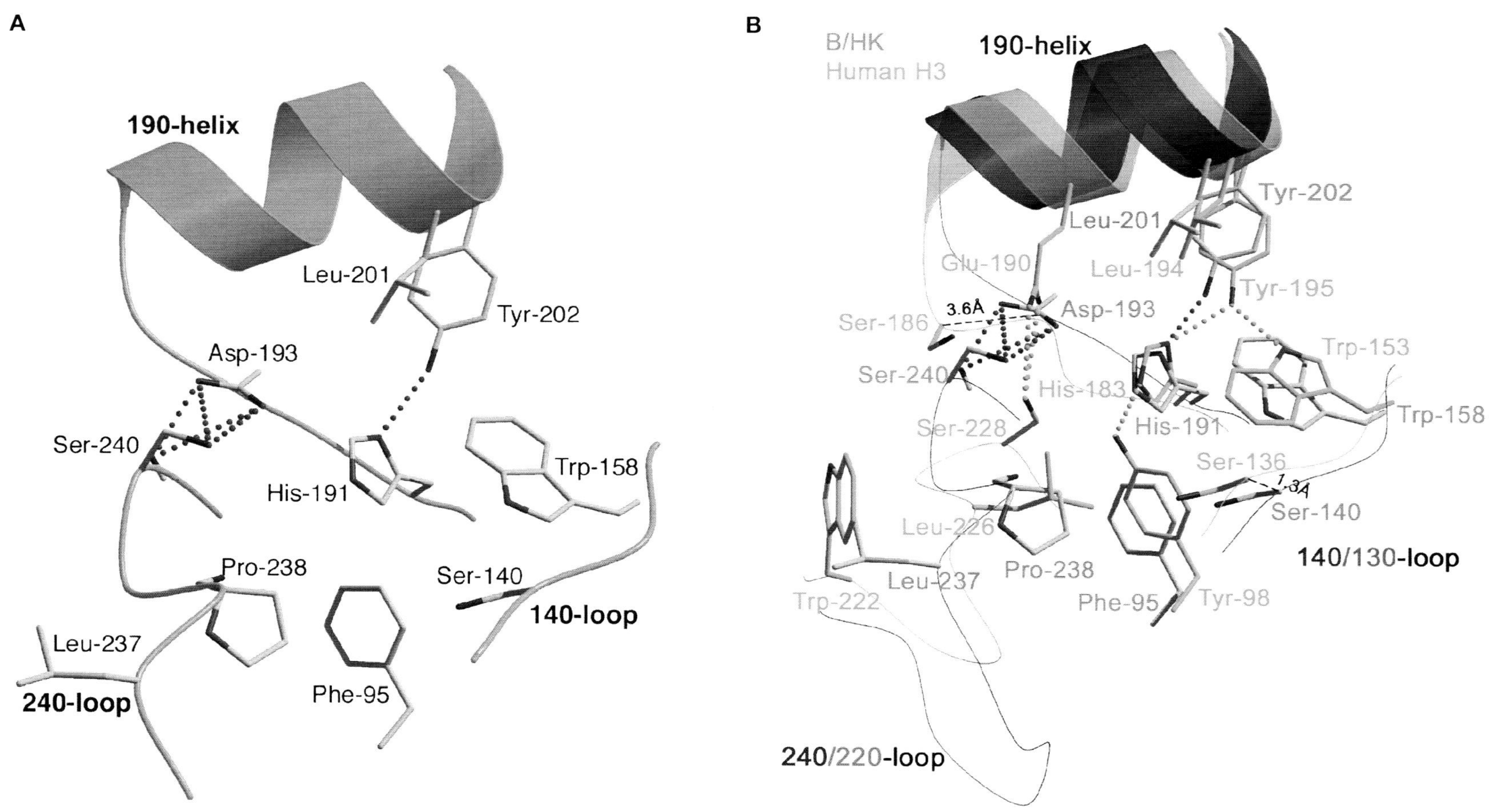

Figure 3.3 Structure and receptor-binding site of influenza B/HK HA. (A) The receptor-binding site of B/HK HA. Important residues forming the receptor-binding site are explicitly drawn, with hydrogen-bonding interactions among them shown as dashed lines. (B) Comparison of the receptor-binding sites of B/HK (red) and human H3 (X:31) (grey) HA glycoproteins. Their corresponding residues are shown in yellow and cyan, respectively. Two large structural shifts between them, at residues $Ser140_1$ and $Asp193_1$ respectively, are indicated by black dashed lines with the distances labelled. The hydrogen-bonding interactions are shown as dashed lines. All labelled residues are on HA1 and thus no subscripts are indicated. Adapted from Wang *et al.* (2007).

residue $Ser228_1$ in X:31 HA. Although the side chain of $Pro238_1$ is shorter than that of $Leu226_1$ in X:31 HA, the main chains of the 240-loop in B/HK HA is shifted towards the 140-loop so that the side-chain atoms responsible for interacting with the receptors are similarly located. Together with the ~1.3 Å shift of the 140-loop in B/HK HA, the lower opening of the B/HK HA binding site (between $Ser140_1$ and $Pro238_1$) is approximately 1.4 Å wider than that of X:31 HA (between $Ser136_1$ and $Leu226_1$).

Interactions with human receptor analogue

In order to study the interactions of human receptors with influenza B/HK HA, the human receptor analogue, α(2,6)-linked sialo-pentasaccharide LSTc, lactoseries tetrasaccharide c (Fig. 3.4B), was used for crystallographic study (Eisen *et al.*, 1997). Between the Sia-1 moiety and B/HK HA, there are a total of eight hydrogen bonds, six of which are contributed by the 140-loop, including the main-chain amide and carbonyl of $Gly141_1$, the side-chain hydroxyl of $Ser140_1$, and the main-chain carbonyl of $Thr139_1$, and the remaining two hydrogen bonds are between the 9-hydroxyl group of the Sia-1 and the side chains of residues $Asp193_1$ and $Ser240_1$ (Fig. 3.5A). Except in only one case for $Gly141_1$ (Xu *et al.*, 1996; Xu *et al.*, 1994), all residues that interact with the Sia-1 moiety are absolutely conserved among different field isolates of influenza B viruses (Air *et al.*, 1990; Krystal *et al.*, 1982; Krystal *et al.*, 1983; Lindstrom *et al.*, 1999; Muyanga *et al.*, 2001; Nakagawa *et al.*, 2002; Rota *et al.*, 1992; Shaw *et al.*, 2002; Xu *et al.*, 1996; Xu *et al.*, 1994), highlighting their importance in receptor binding for influenza B virus infection.

Fig. 3.5B shows the interactions of human receptor analogue LSTc with X:31 HA (Weis *et al.*, 1988). Different from influenza B/HK HA, five hydrogen bonds are found between the Sia-1 and the 130-loop of X:31 HA, among which a new hydrogen bond between the side chain of $Asn137_1$ and the Sia-1 of LSTc is at an optimal distance of 2.7 Å. In addition, the glycerol moiety of the Sia-1 makes a total of five hydrogen bonds, contributed by the hydroxyl of $Tyr98_1$ and the side chains of $Glu190_1$ and $Ser228_1$ (Fig. 3.5B). Therefore, the Sia-1 moiety of human receptor analogue LSTc overall makes more extensive interactions with the binding site of X:31 HA than with B/HK HA. The fewer interactions between the Sia-1 moiety and B/HK HA provide a molecular explanation for the experimental observation that influenza B viruses in general have lower binding affinities for Neu5Ac (representing only the Sia-1 moiety) than influenza A virus H1 and H3 subtypes (Matrosovich *et al.*, 1993).

In comparing the complex structures of X:31 HA-LSTc with B/HK HA-LSTc, we noticed that the interactions made by $Tyr98_1$ seem to be the main force that pulls the ring of the Sia-1 down deeper into the pocket in influenza A virus HA (Fig. 3.5C). In contrast, in B/HK HA where a $Phe95_1$ is present, the glycerol moiety of Sia-1 sits higher in the receptor-binding site and exclusively interacts with the side chains of $Asp193_1$ and

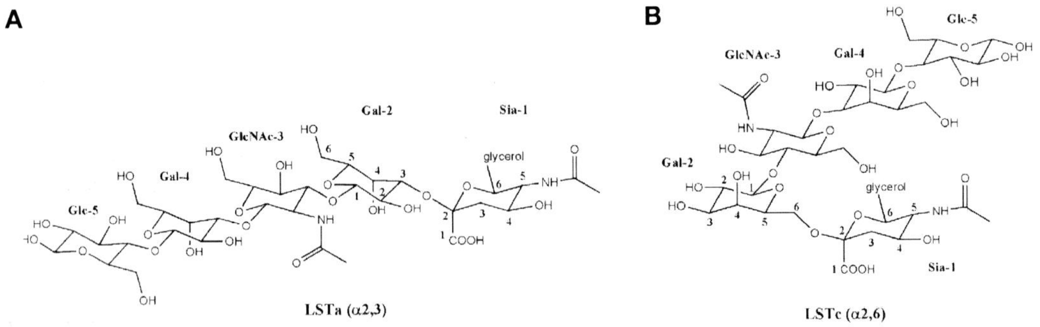

Figure 3.4 Chemical structures of (A) LSTa and (B) LSTc, analogues for avian and human receptors, respectively. In the complex structures with influenza virus HA, LSTa tends to adopt an extended conformation while LSTc is in a folded conformation.

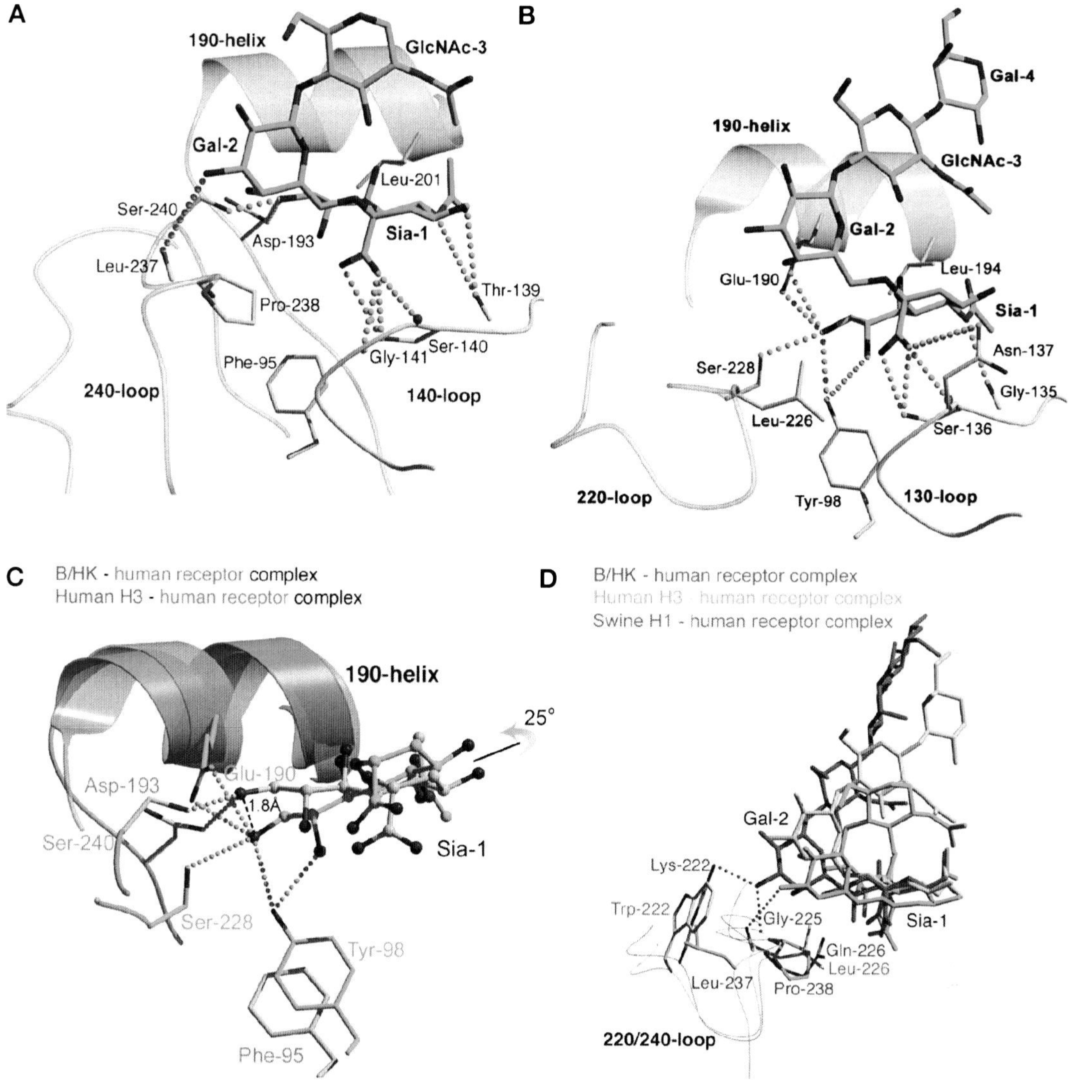

Figure 3.5 Human receptor complexes. (A) The interaction of human receptor analogue in the receptor-binding site of B/HK HA. The hydrogen-bonding interactions are shown as dashed lines in cyan. The asialo-portion of the receptor refers to the sugar rings other than the Sia-1. (B) The interaction of human receptor analogue in the receptor-binding site of X:31 HA. The hydrogen-bonding interactions are shown as dashed lines in cyan. It is clear that the receptor-binding site of X:31 HA has an overall stronger interaction network with the bound human receptor analogue than B/HK HA does. (C) Comparison of interactions of B/HK HA and X:31 HA with the Sia-1 moiety. The approximately 25° tilting of the Sia-1 ring around the C5–O6 axis is indicated by an arrow. The up-shift of the glycerol moiety of the Sia-1 in B/HK HA is also labelled. (D) The superposition of B/HK (green), X:31 (yellow) and swine H1 (purple) HA in complex with human receptor analogues. It is clear that the interactions of the Gal-2 ring with the residue at HA1 222 affect the conformations of bound human analogues. All labelled residues are on HA1 and thus no subscripts are indicated. Adapted from Wang *et al.* (2007).

$Ser240_1$. In measurement, the glycerol moiety of the Sia-1 in B/HK HA is approximately 1.8 Å higher than that in X:31 HA, with an approximate 25° tilting of the Sia-1 ring around the C5-O6 axis (Fig. 3.5C).

Both Gal-2 and GlcNAc-3 of LSTc are clearly visible in the B/HK HA–LSTc complex (Fig. 3.5A). However, only a fragmented density is observed for the terminal Gal-4 and Glc-5, which are presumably disordered. The asialo portion of

LSTc does not make significant contributions to binding with B/HK HA, except for a relatively weak hydrogen bond between the Gal-2 and the main-chain carbonyl of Leu237$_1$ (at a distance of 3.9 Å) (Fig. 3.5A). The overall conformation of bound LSTc in B/HK HA is the most similar to that of LSTc in the complex with X:31 HA. Consequently, if it follows the path of LSTc in X:31 HA–LSTc complex (Fig. 3.5B), LSTc in B/HK HA would exit the receptor-binding site from the right-hand side of the 190-helix (Fig. 3.5A and B).

Interactions with avian receptor analogue

In order to study the interactions of avian receptors with influenza B/HK HA, the avian receptor analogue, α-(2,3)-linked sialo-pentasaccharide LSTa, lactoseries tetrasaccharide a (Fig. 3.4A), was used for crystallographic study (Eisen *et al.*, 1997). The full-length LSTa is clearly visible (Fig. 3.6A) in the complex structure. Since it sits approximately 0.4–0.8 Å higher in the binding site than LSTc, the Sia-1 moiety of LSTa loses two hydrogen bonds with B/HK HA that are made by the main-chain carbonyl of Thr139$_1$ and the main-chain amide of Gly141$_1$ (Fig. 3.6A). Besides the Sia-1 moiety, the Gal-2, GlcNAc-3, and Glc-5 all interact with the receptor-binding site: the Gal-2 forms a hydrogen bond with the main-chain carbonyl of Leu237$_1$; the GlcNAc-3 contributes two hydrogen bonds with the side chain of Gln197$_1$, and the Glc-5 interacts with Asp194$_1$ and Thr196$_1$. Different from LSTc, however, LSTa exits the receptor-binding site from the N-terminal end of the 190-helix (Fig. 3.6A).

The bound LSTa in B/HK HA is in a cis-conformation, most similar to that of LSTa in X:31 HA and swine H9 HA, and in sharp contrast to the trans-conformation of LSTa in avian H3 (Ha *et al.*, 2003), avian H5 (Ha *et al.*, 2001), and human 1934 H1 HA (Gamblin *et al.*, 2004). In the cis-conformation, the hydrophobic side chain of residue Pro238$_1$ and the hydrogen bond between the Gal-2 and the main-chain carboxyl of Leu237$_1$ play important roles, just as Leu226$_1$ in X:31 HA and swine H9 HA (Ha *et al.*, 2001; Weis *et al.*, 1988). In the B/HK HA–LSTa complex, the GlcNAc-3 rotates around the Gal-2-GlcNAc-3 bond by approximately 60° in relation to the GlaNAc-3 of LSTa in X:31 HA. It is this rotation that directs the rest of LSTa towards the 190-helix

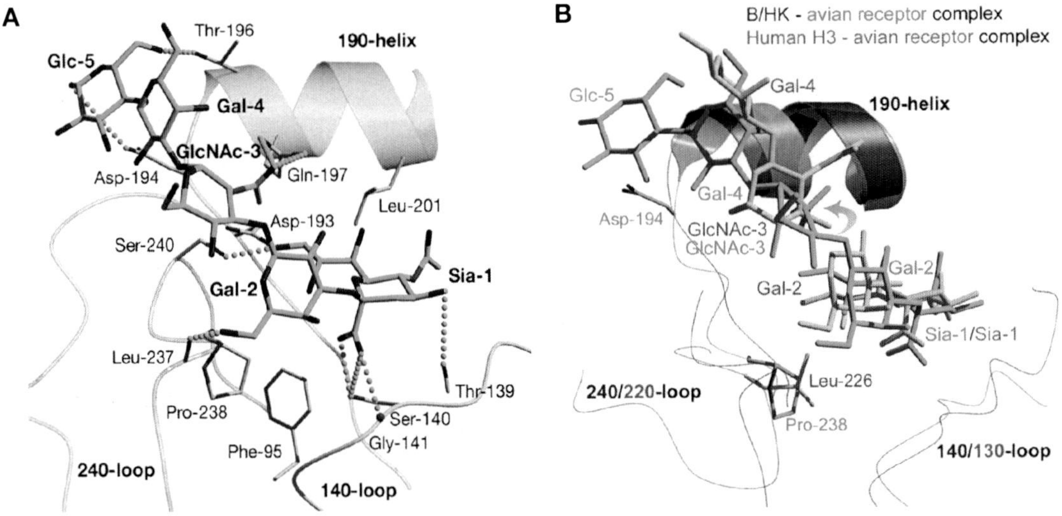

Figure 3.6 Avian receptor complexes. (A) The interaction of avian receptor analogue in the receptor-binding site of B/HK HA. The hydrogen-bonding interactions are shown as dashed lines in cyan. (B) Comparison of interactions of B/HK HA and X:31 HA with avian receptor analogues. The approximately 60° rotation of the GlcNAc-3 ring of the avian receptor analogue in the receptor-binding site of B/HK HA is indicated by an arrow. All labelled residues are on HA1 and thus no subscripts are indicated. Adapted from Wang *et al.* (2007).

and allows for an interaction between the Glc-5 and Asp194$_1$ (Fig. 3.6B). As a result, LSTa fits more snugly in the receptor-binding site of B/HK HA. In sharp contrast, LSTa in X:31 HA moves away from the 190-helix as early as the GlcNAc-3 (Fig. 3.6B). The significant contributions of the asialo portion of avian receptor to the interaction with influenza B virus HA are consistent with the substantially higher affinity of influenza B virus for larger receptors than for Neu5Ac (Matrosovich *et al.*, 1993). In marked contrast, human H1 and H3 subtypes have comparable affinities for larger receptors to those for free Neu5Ac (Matrosovich *et al.*, 1993), in agreement with the less extensive interactions made by the asialio portion of LSTa with influenza A virus HA (Eisen *et al.*, 1997; Gamblin *et al.*, 2004; Weis *et al.*, 1988).

Naturally, influenza B virus has been almost exclusively detected in humans. When adapted to growth in chicken eggs, changes in the properties of influenza B virus were noticed (Burnet and Bull, 1943). Egg adaptation frequently causes the loss of a glycosylation site at HA1 194–196 (Robertson *et al.*, 1985), and alters both the antigenicity (Robertson *et al.*, 1987; Saito *et al.*, 2004; Schild *et al.*, 1983) and receptor-binding properties of influenza B virus HA (Gambaryan *et al.*, 1999; Robertson *et al.*, 1987). The egg-adapted influenza B virus variants, with the absence of the glycans at HA1 194, tend to grow better in chicken eggs than mammalian cell-propagated variants of the same origin (Saito *et al.*, 2004). Without a complex structure of avian receptor with influenza B virus HA that is glycosylated at HA1 194 for comparison, the advantage of such egg-adapted variants is unclear. However, the B/HK HA–LSTa complex structure provides some clues why the loss of the glycosylation site at HA1 194–196, as in the case of influenza B/HK HA, is advantageous for egg-adapted influenza B virus (Saito *et al.*, 2004). The glycans attached to Asn194$_1$ in influenza B virus HA might interfere with the binding of α(2,3)-linked sialic acids that are abundant in chicken egg allantone (Gambaryan *et al.*, 1999), thereby hindering the propagation of influenza B virus. Thus, the growth environment of chicken eggs strongly selects for influenza B virus variants that do not have the glycans at residue HA1 194, with concomitant enhanced binding affinity for α(2,3)-linked sialic acids (Gambaryan *et al.*, 1999). Interestingly, the gain or loss of a glycosylation site at residue HA1 194 seems to be an effective tool used by naturally occurring influenza B virus to modify its antigenicity (Nakagawa *et al.*, 2000; Nakagawa *et al.*, 2004), as observed in egg-adapted influenza A virus (Gambaryan *et al.*, 1999; Skehel and Wiley, 2000).

Different binding modes of human receptors on influenza A and B virus HA glycoproteins

Superposition on the Sia-1 moiety of human receptor analogue LSTc reveals two different conformations of the bound analogue in different HA glycoproteins (Fig. 3.5D): a slightly more extended conformation observed in the 1930 H1 and 1934 H1 subtypes, and a more folded conformation as seen in the avian H3, human H3 (X:31), and swine H9 subtypes, and B/HK HA. In the more extended conformation, the Gal-2 moiety of LSTc sits more deeply in the receptor-binding site and forms hydrogen bonds with the main-chain carbonyl of residue 225$_1$ and with the side chain of residue Lys222$_1$. In marked contrast, the Gal-2 moiety in the more folded conformation is positioned by 1–2 Å higher in the binding site (Gamblin *et al.*, 2004). Comparison of the residues that interact with LSTc revealed the structural basis for the observed conformational differences of LSTc (Table 3.1). In H3 HA, residue HA1 226 has been implicated in determining the preferential binding with α(2,6)-linked or α(2,3)-linked receptor analogues: Leu226$_1$ prefers α(2,6)-linked human receptors, whereas Gln226$_1$ prefers α(2,3)-linked avian receptors (Rogers *et al.*, 1985; Rogers and Paulson, 1983; Skehel and Wiley, 2000). However, despite the difference at HA1 226 between them, the α(2,6)-linked receptor LSTc binds to avian H3 and human H3 HA in a folded conformation, hinting an important role of Trp222$_1$ that is common between them. Consistent with this finding, swine 1930 H1 and human 1934 H1 HA, both with Gln226$_1$ and Lys222$_1$, bind to human receptor analogues in a more extended conformation, different from that of the avian H3 and human H3 HA. Thus, HA1 222 is probably a major determinant for the conformations of bound α(2,6)-linked receptor analogues. A hydrophilic residue,

Table 3.1 Comparison of HA1 residues interacting with human receptor analogues in different HA proteins

Strain	Residue at 226	Residue at 222	Conformation of LSTc
Swine 1930 H1	Gln	Lys	Extended
Human 1934 H1	Gln	Lys	Extended
Avian H3	Gln	Trp	Folded
Human H3	Leu	Trp	Folded
Swine H9	Leu	Leu	Folded
Human B/HK	Pro (238)	Leu (237)	Folded

such as lysine at HA1 222, would offer favourable hydrogen bonds with the Gal-2 so as to pull it into a lower position. On the other hand, a hydrophobic residue at position HA1 222, tryptophan in avian H3 and human H3 HA, or leucine in swine H9 HA and human B/HK HA, tends to push the Gal-2 up into a higher position, possibly to avoid unfavourable interactions between hydrophilic groups on the Gal-2 and the hydrophobic groups on residue HA1 222 (Fig. 3.5D). For B/HK HA, a weak hydrogen bond with the main-chain carbonyl of $Leu237_1$ (Fig. 3.5D) brings the Gal-2 into a slightly lower position than its equivalence in the human H3 and swine H9 HA.

Lower receptor binding affinity of influenza B virus and host range

One important unique feature of influenza B virus HA in receptor binding (Wang *et al.*, 2007) is the presence of $Phe95_1$ in all influenza B virus HA sequences in sharp contrast to a tyrosine (Tyr) at equivalent position in all influenza A virus HA sequences (Fig. 3.3A). Previously, a $Tyr98_1 \rightarrow Phe$ mutation of X:31 HA abolished erythrocyte binding, and the mutant virus could not infect mutant MDCK cells with reduced levels of cell-surface sialic acid, underscoring the importance of the hydrogen bonds that $Tyr98_1$ makes with both $His183_1$ and the receptor inside the receptor-binding site (Martin *et al.*, 1998; Skehel and Wiley, 2000). The presence of $Phe95_1$ in influenza B virus HA results in fewer interactions between the Sia-1 moiety and the receptor-binding site (Fig. 3.3A). This provides an explanation for the experimental observation that influenza B viruses in general have lower binding affinity for Neu5Ac than the H1 and H3 subtypes of influenza A virus (Matrosovich *et al.*, 1993). Since binding specificity to cell surface receptors is an important determinant of host range restriction (Connor *et al.*, 1994), I wonder whether the lower receptor binding affinity of influenza B virus HA is a contributing factor for the limited host range of influenza B virus, mostly in humans and seals. Further study is needed to clarify this important aspect of influenza B virology.

Membrane fusion mechanism

A general scheme for low-pH sensitivity of influenza A and B virus HA and influenza C virus HEF

A series of crystallographic studies have revealed that extensive structural rearrangement takes place along HA upon exposure to low pH (Bullough *et al.*, 1994; Chen *et al.*, 1998; Wilson *et al.*, 1981). In this process, the overall stability of the HA glycoprotein appears to play a central role, as mutants of HA with elevated fusion pH (easier to undergo conformational change) all harboured destabilizing mutations at chain or subunit interfaces (Daniels *et al.*, 1985). In consistent with this finding, a close correlation was noted between the fusion pH and the stability of HA ectodomain maintained by electrostatic interactions between HA1 and HA2 (Huang *et al.*, 2002; Huang *et al.*, 2003; Korte *et al.*, 2007). The removal of a highly conserved tetrad salt-bridge network formed by $Glu89_1$, $Arg109_1$, $Arg269_1$ and $Glu67_2$ at the HA1/HA2 interface destabilized X:31 HA and resulted in membrane fusion at higher pH (Rachakonda *et al.*, 2007). Our new structure of B/HK HA has allowed a systematic examination of possible interactions in membrane fusion of influenza A and B virus HA and influenza C virus HEF. Such an examination revealed a universal scheme for the low-pH sensitivity of these membrane-fusion inducing proteins, in that there are ionizable clusters at all regions where structural rearrangement is expected to occur upon exposure to low pH, particularly at HA1/HA2 and inter-subunit interface (Fig. 3.7A, regions A–D). Although individual residues are generally not conserved among influenza virus HA/HEF, the ionizable clusters that are composed of both basic and acidic residues or of fully buried unpaired basic residues are highly conserved. These ionizable clusters are probably the structural basis for the pH-dependence and sensitivity to ionic strength of influenza virus HA/HEF (Rachakonda *et al.*, 2007). Those ionizable clusters on B/HK HA are discussed in detail here.

- Region A: This region encompasses the intersubunit R–R subdomain interface at the membrane-distal end of the trimeric HA molecule (Fig. 3.7A). In this region of B/HK HA, we found two layers of completely buried residues, Lys209$_1$ and His220$_1$, one from each subunit, to interact at the three-fold axis with a pairwise distance of 2.9 Å and 3.7 Å, respectively. We did not find any electron density for counterions near these residues. The distance between Lys209$_1$ and His220$_1$

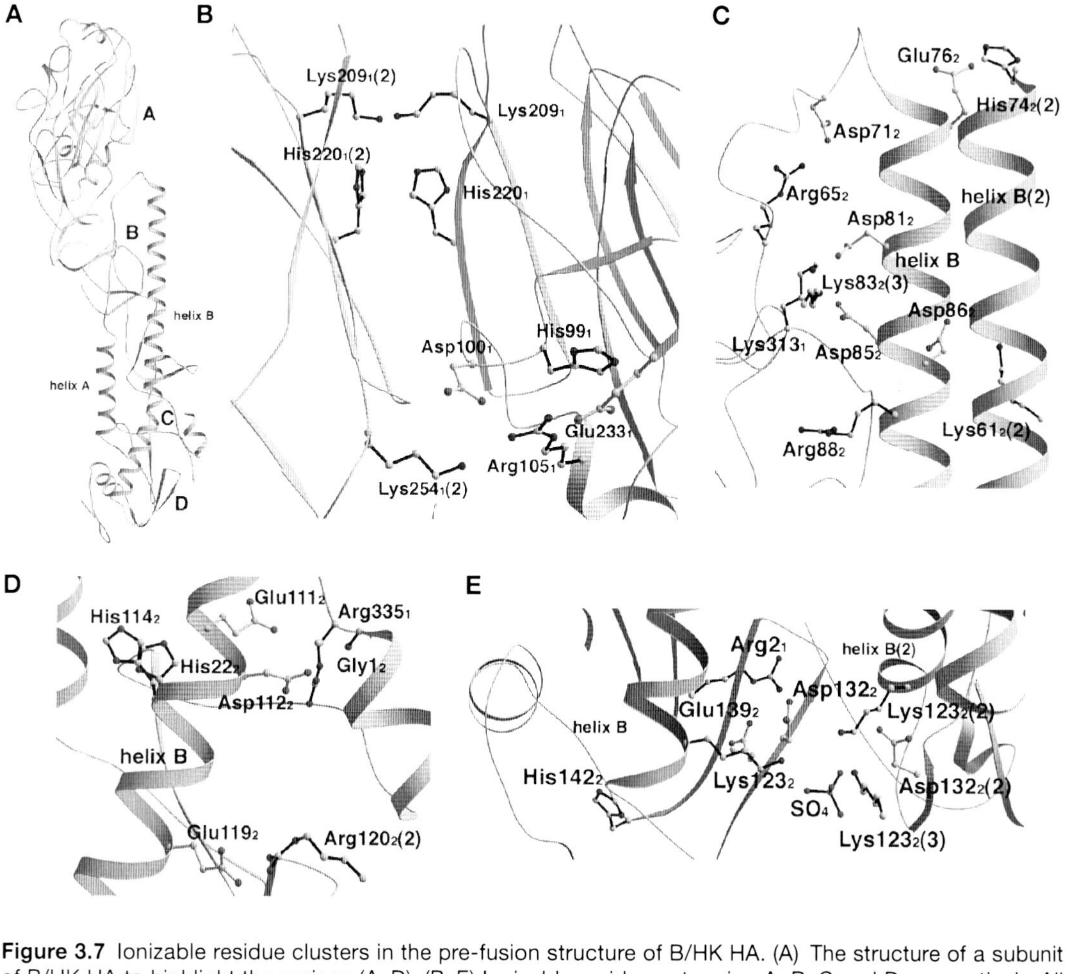

Figure 3.7 Ionizable residue clusters in the pre-fusion structure of B/HK HA. (A) The structure of a subunit of B/HK HA to highlight the regions (A–D). (B–E) Ionizable residues at region A, B, C and D, respectively. All ionizable residues are explicitly shown in ball-and-stick. Basic residues (Lys/Arg/His) are in black bonds and acidic residues (Glu/Asp) are in white bonds. The subscripts, X1 and X2, of each residue indicate that the residue X is on HA1 and HA2, respectively, while the residues from a neighbouring subunit are labelled with a '(2)' after the residues, for instance Lys83$_2$(2). Adapted from Wang *et al.* (2008).

in the same subunit is 3.4 Å. Underneath His220$_1$ is a cluster of five ionizable residues, Arg105$_1$, His99$_1$, Asp100$_1$, Glu233$_1$, with Lys254$_1$ on a neighbouring subunit (Fig. 3.7B). This region is only found in influenza B virus HA, but not in influenza A virus HA or influenza C virus HEF glycoproteins, possibly because of the much more extensive R–R subdomain interface in the former.

- Region B: This region covers the interface of the inter-helix loop of HA2 with the R-subdomain of HA1 and helix B of HA2 of the same subunit (Fig. 3.7A). Four clusters of ionizable residues are found in B/HK HA: Arg65$_2$ and Asp71$_2$; Asp81$_2$, Asp85$_2$, Lys313$_1$, and Lys83$_2$ on a neighbouring subunit; Glu76$_2$ with His74$_2$ on a neighbouring subunit; Asp86$_2$ with Lys61$_2$ on a neighbouring subunit. Moreover, Arg88$_2$ is buried without pairing with any other ions (Fig. 3.7C). Of particular significance, previous mutagenesis study has demonstrated that a cluster similarly located in X:31 HA, Glu89$_1$, Arg109$_1$, Arg269$_1$ and Glu67$_2$, contributes substantially to the overall stability of the neutral-pH protein (Rachakonda *et al.*, 2007).
- Region C: This region covers the fusion peptide and its surrounding regions (Fig. 3.7A). In influenza A virus HA, His17$_1$, Asp109$_2$ and Asp112$_2$ have been implicated in low-pH induced membrane fusion (Chen *et al.*, 1999; Ha *et al.*, 2002; Kampmann *et al.*, 2006; Skehel and Wiley, 2000; Wiley and Skehel, 1987). In B/HK HA, in the absence of residue His17$_1$ and Asp109$_2$, two histidine residues, His22$_2$ and His114$_2$, are completely buried at a location very close to His17$_1$ in X:31 HA without pairing with any other ions (Fig. 3.7D). Moreover, the extreme N-terminus of HA2 and the residue Asp112$_2$ are found to be part of an ionizable residue cluster that also includes Glu111$_2$ and Arg335$_1$. In addition, there is another ionizable cluster composed of Glu119$_2$ with Arg120$_2$ on a neighbouring subunit (Fig. 3.7D).
- Region D: This region encompasses the N- and C-terminal segments of both HA1 and HA2 (Fig. 3.7A). In B/HK HA, we found two ionizable residue clusters of Arg2$_1$ and Glu139$_2$, and Asp132$_2$ and Lys123$_2$ (Fig. 3.7E). In the latter cluster, Asp132$_2$ and Lys123$_2$ interact at a pairwise distance of 3.2 Å, while the three Lys123$_2$ residues, one from each subunit, coordinate a sulphate ion at the three-fold axis. The presence of the sulphate ion in this structure could be the result of crystallization, as similar clusters found in the structures of influenza A virus HA do not contain a counterion, even though some of them are at much higher crystallographic resolutions (for instance, H5 and H9 HAs are of 1.9 and 1.8 Å, respectively (Ha *et al.*, 2002)). In those cases, three pairs of HA2 123 and nearby acidic residues interact near the three-fold axis. Thus, the interactions between three pairs of basic residues at HA2 123 and nearby acidic residues might be a general phenomenon for this region. In addition, His142$_2$ is buried in the molecule without pairing with any other ion.

Within the conserved ionizable clusters, the drop in pH may, in theory, protonate the acidic residues to be neutral, and the basic residues to be positively charged. The excessive positive charges in the HA protein would substantially destabilize the structure, thus triggering a large-scale conformational change to facilitate the host invasion process by influenza A, B and C viruses.

Summary and remaining challenges

Although a close relative to influenza A virus, influenza B virus differs in several significant ways: (a) antigenic shift has not been observed in influenza B virus, probably due to the lack of reservoirs outside of human population; (b) distinct antigenic variants of influenza B virus co-circulate while antigenically similar viruses are isolated many years apart (Bao-Lan *et al.*, 1983; Chakraverty, 1972; Rota *et al.*, 1990, 1992; Yamashita *et al.*, 1988); (c) influenza B virus HA have an overall 3–5 times slower amino acid mutation rate than influenza A virus HA, with an average of 0.3% per year (1.5 residues per year) (Air *et al.*, 1990; Bootman and Robertson, 1988; Chakraverty, 1972; Cox and Bender, 1995; Hay *et al.*, 2001; Kanegae *et al.*, 1990; Krystal *et al.*, 1982;

Krystal *et al.*, 1983; Schild *et al.*, 1973; Verhoeyen *et al.*, 1983; Yamashita *et al.*, 1988), and the exact mechanism for such a slower mutation rate remains unknown; (d) there presents a much lower positive selective pressure for changes in the sequence of influenza B virus HA (Air *et al.*, 1990; Nunes *et al.*, 2008; Pechirra *et al.*, 2005; Shen *et al.*, 2009; Yamashita *et al.*, 1988) than influenza A virus HA (Bush *et al.*, 1999b; Fitch *et al.*, 1997; Yang *et al.*, 2000); (e) the antigenic structure of influenza B virus HA is distinct from those of influenza A virus HA. Many subtypes of influenza A virus HA contain multiple distinct antigenic sites on the membrane-distal domain. In contrast, influenza B virus HA has one, and possibly two, large, continuous antigenic sites formed by four major epitopes that are proximally located. A single substitution in the antigenic site(s) of influenza B virus HA was found to be sufficient for a variant to gain advantage in rapidly circulating among humans (Nakagawa *et al.*, 2001a; Nakagawa *et al.*, 2002; Rota *et al.*, 1990). The more compact antigenic structure of influenza B virus HA may be partially responsible for the observed slower mutation rate and lower positive selective pressure; (f) influenza B virus infects almost exclusively humans; (g) influenza B virus has an overall lower affinity for sialic acid receptors than influenza A virus H1 and H3 subtypes (Matrosovich *et al.*, 1993). However, it remains to be determined whether there exists a correlation between the limited host range (Connor *et al.*, 1994) and the lower receptor-binding affinity of influenza B virus.

Acknowledgements

The author gratefully thanks support of a Beginning-Grant-in-Aid award from American Heart Association (0865186F) and of a grant from National Institutes of Health (R01-AI067839).

References

Abed, Y., Coulthart, M.B., Li, Y., and Boivin, G. (2003). Evolution of surface and nonstructural-1 genes of influenza B viruses isolated in the Province of Quebec, Canada, during the 1998–2001 period. Virus Genes *27*, 125–135.

Air, G.M., and Compans, R.W. (1983). Influenza B and Influenza C Viruses. In Genetics of Influenza Viruses, P. Palese, and D.W. Kingsbury, eds. (Vienna, Springer-Verlag), pp. 21–69.

Air, G.M., Gibbs, A.J., Laver, W.G., and Webster, R.G. (1990). Evolutionary changes in influenza B are not primarily governed by antibody selection. Proc. Natl. Acad. Sci. U.S.A. *87*, 3884–3888.

Bao-Lan, L., Webster, R.G., Brown, L.E., and Nerome, K. (1983). Heterogeneity of influenza B viruses. Bull. World. Health. Organ. *61*, 681–687.

Barbey-Martin, C., Gigant, B., Bizebard, T., Calder, L.J., Wharton, S.A., Skehel, J.J., and Knossow, M. (2002). An antibody that prevents the hemagglutinin low pH fusogenic transition. Virology *294*, 70–74.

Berton, M.T., Naeve, C.W., and Webster, R.G. (1984). Antigenic structure of the influenza B virus hemagglutinin: nucleotide sequence analysis of antigenic variants selected with monoclonal antibodies. J. Virol. *52*, 919–927.

Berton, M.T., and Webster, R.G. (1985). The antigenic structure of the influenza B virus hemagglutinin: operational and topological mapping with monoclonal antibodies. Virology *143*, 583–594.

Bizebard, T., Barbey-Martin, C., Fleury, D., Gigant, B., Barrere, B., Skehel, J.J., and Knossow, M. (2001). Structural studies on viral escape from antibody neutralization. Curr. Top. Microbiol. Immunol. *260*, 55–64.

Bizebard, T., Gigant, B., Rigolet, P., Rasmussen, B., Diat, O., Bosecke, P., Wharton, S.A., Skehel, J.J., and Knossow, M. (1995). Structure of influenza virus haemagglutinin complexed with a neutralizing antibody. Nature *376*, 92–94.

Bootman, J.S., and Robertson, J.S. (1988). Sequence analysis of the hemagglutinin of B/Ann Arbor/1/*86*, an epidemiologically significant variant of influenza B virus. Virology *166*, 271–274.

Bullough, P.A., Hughson, F.M., Skehel, J.J., and Wiley, D.C. (1994). Structure of influenza haemagglutinin at the pH of membrane fusion. Nature *371*, 37–43.

Buonagurio, D.A., Nakada, S., Desselberger, U., Krystal, M., and Palese, P. (1985). Noncumulative sequence changes in the hemagglutinin genes of influenza C virus isolates. Virology *146*, 221–232.

Buonagurio, D.A., Nakada, S., Fitch, W.M., and Palese, P. (1986). Epidemiology of influenza C virus in man: multiple evolutionary lineages and low rate of change. Virology *153*, 12–21.

Burnet, F.M., and Bull, D.R. (1943). Changes in influenza virus associated with adaptation to passage in chick embryos. Aus.t J. Exp. Biol. Med. Sci. *21*, 55–69.

Bush, R.M., Bender, C.A., Subbarao, K., Cox, N.J., and Fitch, W.M. (1999a). Predicting the evolution of human influenza A. Science *286*, 1921–1925.

Bush, R.M., Fitch, W.M., Bender, C.A., and Cox, N.J. (1999b). Positive selection on the H3 hemagglutinin gene of human influenza virus A. Mol. Biol. Evol. *16*, 1457–1465.

Caton, A.J., Brownlee, G.G., Yewdell, J.W., and Gerhard, W. (1982). The antigenic structure of the influenza virus A/PR/8/34 hemagglutinin (H1 subtype). Cell *31*, 417–427.Chakraverty, P. (1972). Antigenic relationship between influenza B viruses. Bull. World. Health. Organ. *45*, 755–766.

Chen, J., Lee, K.H., Steinhauer, D.A., Stevens, D.J., Skehel, J.J., and Wiley, D.C. (1998). Structure of the hemagglutinin precursor cleavage site, a determinant of influenza pathogenicity and the origin of the labile conformation. Cell *95*, 409–417.

Chen, J., Skehel, J.J., and Wiley, D.C. (1999). N- and C-terminal residues combine in the fusion-pH influenza hemagglutinin HA(2) subunit to form an N cap that terminates the triple- stranded coiled coil. Proc. Natl. Acad. Sci. U.S.A. *96*, 8967–8972.

Chen, J.M., Guo, Y.J., Wu, K.Y., Guo, J.F., Wang, M., Dong, J., Zhang, Y., Li, Z., and Shu, Y.L. (2007). Exploration of the emergence of the Victoria lineage of influenza B virus. Arch. Virol. *152*, 415–422.

Chen, R., and Holmes, E.C. (2008). The Evolutionary Dynamics of Human Influenza B Virus. J Mol Evol (More details?)

Chi, X.S., Bolar, T.V., Zhao, P., Rappaport, R., and Cheng, S.M. (2003). Cocirculation and evolution of two lineages of influenza B viruses in europe and Israel in the 2001–2002 season. J. Clin. Microbiol. *41*, 5770–5773.

Connor, R.J., Kawaoka, Y., Webster, R.G., and Paulson, J.C. (1994). Receptor specificity in human, avian, and equine H2 and H3 influenza virus isolates. Virology *205*, 17–23.

Cox, N.J., and Bender, C.A. (1995). The molecular epidemiology of influenza viruses. Semin. Virol. *6*, 359–370.

Cox, N.J., and Subbarao, K. (2000). Global epidemiology of influenza: past and present. Annu. Rev. Med. *51*, 407–421.

Daniels, R.S., Douglas, A.R., Skehel, J.J., and Wiley, D.C. (1983). Analyses of the antigenicity of influenza haemagglutinin at the pH optimum for virus-mediated membrane fusion. J. Gen. Virol. 64 (Pt 8), 1657–1662.

Daniels, R.S., Downie, J.C., Hay, A.J., Knossow, M., Skehel, J.J., Wang, M.L., and Wiley, D.C. (1985). Fusion mutants of the influenza virus hemagglutinin glycoprotein. Cell *40*, 431–439.

Daum, L.T., Canas, L.C., Klimov, A.I., Shaw, M.W., Gibbons, R.V., Shrestha, S.K., Myint, K.S., Acharya, R.P., Rimal, N., Reese, F., *et al.* (2006). Molecular analysis of isolates from influenza B outbreaks in the U.S. and Nepal, 2005. Arch. Virol. *151*, 1863–1874.

Eisen, M.B., Sabesan, S., Skehel, J.J., and Wiley, D.C. (1997). Binding of the influenza A virus to cell-surface receptors: structures of five hemagglutinin–sialyloligosaccharide complexes determined by X- ray crystallography. Virology 232, 19–31.

Fitch, W.M., Bush, R.M., Bender, C.A., and Cox, N.J. (1997). Long term trends in the evolution of H(3) HA1 human influenza type A. Proc. Natl. Acad. Sci. U.S.A. *94*, 7712–7718.

Fleury, D., Barrere, B., Bizebard, T., Daniels, R.S., Skehel, J.J., and Knossow, M. (1999). A complex of influenza hemagglutinin with a neutralizing antibody that binds outside the virus receptor binding site. Nat. Struct. Biol. *6*, 530–534.

Fouchier, R.A., Munster, V., Wallensten, A., Bestebroer, T.M., Herfst, S., Smith, D., Rimmelzwaan, G.F., Olsen, B., and Osterhaus, A.D. (2005). Characterization of a novel influenza A virus hemagglutinin subtype (H16) obtained from black-headed gulls. J. Virol. *79*, 2814–2822.

Gambaryan, A.S., Robertson, J.S., and Matrosovich, M.N. (1999). Effects of egg-adaptation on the receptor-binding properties of human influenza A and B viruses. Virology *258*, 232–239.

Gamblin, S.J., Haire, L.F., Russell, R.J., Stevens, D.J., Xiao, B., Ha, Y., Vasisht, N., Steinhauer, D.A., Daniels, R.S., Elliot, A., *et al.* (2004). The structure and receptor binding properties of the 1918 influenza hemagglutinin. Science *303*, 1838–1842.

Gerhard, W., Yewdell, J., Frankel, M.E., and Webster, R. (1981). Antigenic structure of influenza virus haemagglutinin defined by hybridoma antibodies. Nature *290*, 713–717.

Gigant, B., Barbey-Martin, C., Bizebard, T., Fleury, D., Daniels, R., Skehel, J.J., and Knossow, M. (2000). A neutralizing antibody Fab-influenza haemagglutinin complex with an unprecedented 2:1 stoichiometry: characterization and crystallization. Acta Crystallogr. D Biol. Crystallogr. *56*, 1067–1069.

Ha, Y., Stevens, D.J., Skehel, J.J., and Wiley, D.C. (2001). X-ray structures of H5 avian and H9 swine influenza virus hemagglutinins bound to avian and human receptor analogs. Proc. Natl. Acad. Sci. U.S.A. *98*, 11181–11186.

Ha, Y., Stevens, D.J., Skehel, J.J., and Wiley, D.C. (2002). H5 avian and H9 swine influenza virus haemagglutinin structures: possible origin of influenza subtypes. EMBO J. *21*, 865–875.

Ha, Y., Stevens, D.J., Skehel, J.J., and Wiley, D.C. (2003). X-ray structure of the hemagglutinin of a potential H3 avian progenitor of the 1968 Hong Kong pandemic influenza virus. Virology *309*, 209–218.

Hay, A.J., Gregory, V., Douglas, A.R., and Lin, Y.P. (2001). The evolution of human influenza viruses. Phil. Trans. R. Soc. Lond. B Biol. Sci. *356*, 1861–1870.

Hiromoto, Y., Saito, T., Lindstrom, S.E., Li, Y., Nerome, R., Sugita, S., Shinjoh, M., and Nerome, K. (2000). Phylogenetic analysis of the three polymerase genes (PB*1*, PB2 and PA) of influenza B virus. J. Gen. Virol. *81*, 929–937.

Hovanec, D.L., and Air, G.M. (1984). Antigenic structure of the hemagglutinin of influenza virus B/Hong Kong/8/73 as determined from gene sequence analysis of variants selected with monoclonal antibodies. Virology *139*, 384–392.

Huang, Q., Opitz, R., Knapp, E.W., and Herrmann, A. (2002). Protonation and stability of the globular domain of influenza virus hemagglutinin. Biophys. J. *82*, 1050–1058.

Huang, Q., Sivaramakrishna, R.P., Ludwig, K., Korte, T., Bottcher, C., and Herrmann, A. (2003). Early steps of the conformational change of influenza virus hemagglutinin to a fusion active state: stability and energetics of the hemagglutinin. Biochim. Biophys. Acta *1614*, 3–13.

Ikonen, N., Pyhala, R., Axelin, T., Kleemola, M., and Korpela, H. (2005). Reappearance of influenza B/Victoria/2/87-lineage viruses: epidemic activity, genetic diversity and vaccination efficacy in the Finnish Defence Forces. Epidemiol. Infect. *133*, 263–271.

Jian, J.W., Lai, C.T., Kuo, C.Y., Kuo, S.H., Hsu, L.C., Chen, P.J., Wu, H.S., and Liu, M.T. (2008). Genetic analysis and evaluation of the reassortment of influenza B viruses isolated in Taiwan during the 2004–2005 and 2006–2007 epidemics. Virus. Res. *131*, 243–249.

Kampmann, T., Mueller, D.S., Mark, A.E., Young, P.R., and Kobe, B. (2006). The Role of histidine residues in low-pH-mediated viral membrane fusion. Structure *14*, 1481–1487.

Kanegae, Y., Sugita, S., Endo, A., Ishida, M., Senya, S., Osako, K., Nerome, K., and Oya, A. (1990). Evolutionary pattern of the hemagglutinin gene of influenza B viruses isolated in Japan: cocirculating lineages in the same epidemic season. J. Virol. *64*, 2860–2865.

Katagiri, S., Ohizumi, A., and Homma, M. (1983). An outbreak of type C influenza in a children's home. J. Infect. Dis. *148*, 51–56.

Kaverin, N.V., Rudneva, I.A., Govorkova, E.A., Timofeeva, T.A., Shilov, A.A., Kochergin-Nikitsky, K.S., Krylov, P.S., and Webster, R.G. (2007). Epitope mapping of the hemagglutinin molecule of a highly pathogenic H5N1 influenza virus by using monoclonal antibodies. J. Virol. *81*, 12911–12917.

Kaverin, N.V., Rudneva, I.A., Ilyushina, N.A., Lipatov, A.S., Krauss, S., and Webster, R.G. (2004). Structural differences among hemagglutinins of influenza A virus subtypes are reflected in their antigenic architecture: analysis of H9 escape mutants. J. Virol. *78*, 240–249.

Kaverin, N.V., Rudneva, I.A., Ilyushina, N.A., Varich, N.L., Lipatov, A.S., Smirnov, Y.A., Govorkova, E.A., Gitelman, A.K., Lvov, D.K., and Webster, R.G. (2002). Structure of antigenic sites on the haemagglutinin molecule of H5 avian influenza virus and phenotypic variation of escape mutants. J. Gen. Virol. *83*, 2497–2505.

Kinnunen, L., Ikonen, N., Poyry, T., and Pyhala, R. (1992). Evolution of influenza B/Victoria/2/87-like viruses: occurrence of a genetically conserved virus under conditions of low epidemic activity. J. Gen. Virol. 73 (Pt 3), 733–736.

Knossow, M., Gaudier, M., Douglas, A., Barrere, B., Bizebard, T., Barbey, C., Gigant, B., and Skehel, J.J. (2002). Mechanism of neutralization of influenza virus infectivity by antibodies. Virology *302*, 294–298.

Korte, T., Ludwig, K., Huang, Q., Rachakonda, P.S., and Herrmann, A. (2007). Conformational change of influenza virus hemagglutinin is sensitive to ionic concentration. Eur. Biophys. J. *36*, 327–335.

Krystal, M., Elliott, R.M., Benz, E.W., Jr., Young, J.F., and Palese, P. (1982). Evolution of influenza A and B viruses: conservation of structural features in the hemagglutinin genes. Proc. Natl. Acad. Sci. U.S.A. *79*, 4800–4804.

Krystal, M., Young, J.F., Palese, P., Wilson, I.A., Skehel, J.J., and Wiley, D.C. (1983). Sequential mutations in hemagglutinins of influenza B virus isolates: definition of antigenic domains. Proc. Natl. Acad. Sci. U.S.A. *80*, 4527–4531.

Li, W.C., Shih, S.R., Huang, Y.C., Chen, G.W., Chang, S.C., Hsiao, M.J., Tsao, K.C., and Lin, T.Y. (2008). Clinical and genetic characterization of severe influenza B-associated diseases during an outbreak in Taiwan. J. Clin. Virol. *42*, 45–51.

Lindstrom, S.E., Hiromoto, Y., Nerome, R., Omoe, K., Sugita, S., Yamazaki, Y., Takahashi, T., and Nerome, K. (1998). Phylogenetic analysis of the entire genome of influenza A (H3N2) viruses from Japan: evidence for genetic reassortment of the six internal genes. J. Virol. *72*, 8021–8031.

Lindstrom, S.E., Hiromoto, Y., Nishimura, H., Saito, T., Nerome, R., and Nerome, K. (1999). Comparative analysis of evolutionary mechanisms of the hemagglutinin and three internal protein genes of influenza B virus: multiple cocirculating lineages and frequent reassortment of the NP, M, and NS genes. J. Virol. *73*, 4413–4426.

Lubeck, M.D., Schulman, J.L., and Palese, P. (1980). Antigenic variants of influenza viruses: marked differences in the frequencies of variants selected with different monoclonal antibodies. Virology *102*, 458–462.

Lugovtsev, V.Y., Vodeiko, G.M., and Levandowski, R.A. (2005). Mutational pattern of influenza B viruses adapted to high growth replication in embryonated eggs. Virus. Res. *109*, 149–157.

Lugovtsev, V.Y., Vodeiko, G.M., Strupczewski, C.M., Ye, Z., and Levandowski, R.A. (2007). Generation of the influenza B viruses with improved growth phenotype by substitution of specific amino acids of hemagglutinin. Virology *365*, 315–323.

Luo, C., Morishita, T., Satou, K., Tateno, Y., Nakajima, K., and Nobusawa, E. (1999). Evolutionary pattern of influenza B viruses based on the HA and NS genes during 1940 to 1999: origin of the NS genes after 1997. Arch. Virol. *144*, 1881–1891.

Martin, J., Wharton, S.A., Lin, Y.P., Takemoto, D.K., Skehel, J.J., Wiley, D.C., and Steinhauer, D.A. (1998). Studies of the binding properties of influenza hemagglutinin receptor-site mutants. Virology *241*, 101–111.

Matrosovich, M.N., Gambaryan, A.S., Tuzikov, A.B., Byramova, N.E., Mochalova, L.V., Golbraikh, A.A., Shenderovich, M.D., Finne, J., and Bovin, N.V. (1993). Probing of the receptor-binding sites of the H1 and H3 influenza A and influenza B virus hemagglutinins by synthetic and natural sialosides. Virology *196*, 111–121.

Matsuzaki, Y., Katsushima, N., Nagai, Y., Shoji, M., Itagaki, T., Sakamoto, M., Kitaoka, S., Mizuta, K., and Nishimura, H. (2006). Clinical features of influenza C virus infection in children. J. Infect. Dis. *193*, 1229–1235.

McCullers, J.A., Wang, G.C., He, S., and Webster, R.G. (1999). Reassortment and insertion-deletion are strategies for the evolution of influenza B viruses in nature. J. Virol. 73, 7343–7348.

Mikheeva, A., and Ghendon, Y.Z. (1982). Intrinsic interference between influenza A and B viruses. Arch. Virol. 73, 287–294.

Murphy, B.R., and Webster, R.G. (2001). Orthomyxoviruses. In Fields – Virology, D.M. Knipe, and P.M. Howley, eds. (Boston, Lippincott Williams & Wilkins).

Muyanga, J., Matsuzaki, Y., Sugawara, K., Kimura, K., Mizuta, K., Ndumba, I., Muraki, Y., Tsuchiya, E.,

Hongo, S., Kasolo, F.C., *et al.* (2001). Antigenic and genetic analyses of influenza B viruses isolated in Lusaka, Zambia in 1999. Arch. Virol. *146*, 1667–1679.

Nakagawa, N., Kubota, R., Maeda, A., Nakagawa, T., and Okuno, Y. (2000). Heterogeneity of influenza B virus strains in one epidemic season differentiated by monoclonal antibodies and nucleotide sequences. J. Clin. Microbiol. *38*, 3467–3469.

Nakagawa, N., Kubota, R., Maeda, A., and Okuno, Y. (2004). Influenza B virus victoria group with a new glycosylation site was epidemic in Japan in the 2002–2003 season. J. Clin. Microbiol. *42*, 3295–3297.

Nakagawa, N., Kubota, R., Morikawa, S., Nakagawa, T., Baba, K., and Okuno, Y. (2001a). Characterization of new epidemic strains of influenza B virus by using neutralizing monoclonal antibodies. J. Med. Virol. *65*, 745–750.

Nakagawa, N., Kubota, R., Nakagawa, T., and Okuno, Y. (2001b). Antigenic variants with amino acid deletions clarify a neutralizing epitope specific for influenza B virus Victoria group strains. J. Gen. Virol. *82*, 2169–2172.

Nakagawa, N., Kubota, R., Nakagawa, T., and Okuno, Y. (2003). Neutralizing epitopes specific for influenza B virus Yamagata group strains are in the 'loop'. J. Gen. Virol. *84*, 769–773.

Nakagawa, N., Kubota, R., and Okuno, Y. (2005). Variation of the conserved neutralizing epitope in influenza B virus victoria group isolates in Japan. J. Clin. Microbiol. *43*, 4212–4214.

Nakagawa, N., Nukuzuma, S., Haratome, S., Go, S., Nakagawa, T., and Hayashi, K. (2002). Emergence of an influenza B virus with antigenic change. J. Clin. Microbiol. *40*, 3068–3070.

Nakagawa, N., Suzuoki, J., Kubota, R., Kobatake, S., and Okuno, Y. (2006). Discovery of the neutralizing epitope common to influenza B virus victoria group isolates in Japan. J. Clin. Microbiol. *44*, 1564–1566.

Nerome, R., Hiromoto, Y., Sugita, S., Tanabe, N., Ishida, M., Matsumoto, M., Lindstrom, S.E., Takahashi, T., and Nerome, K. (1998). Evolutionary characteristics of influenza B virus since its first isolation in 1940: dynamic circulation of deletion and insertion mechanism. Arch. Virol. *143*, 1569–1583.

Nunes, B., Pechirra, P., Coelho, A., Ribeiro, C., Arraiolos, A., and Rebelo-de-Andrade, H. (2008). Heterogeneous selective pressure acting on influenza B Victoria- and Yamagata-like hemagglutinins. J. Mol. Evol. *67*, 427–435.

Okamatsu, M., Sakoda, Y., Kishida, N., Isoda, N., and Kida, H. (2008). Antigenic structure of the hemagglutinin of H9N2 influenza viruses. Arch. Virol. *153*, 2189–2195.

Osterhaus, A.D., Rimmelzwaan, G.F., Martina, B.E., Bestebroer, T.M., and Fouchier, R.A. (2000). Influenza B virus in seals. Science *288*, 1051–1053.

Oxford, J.S., Klimov, A.I., Corcoran, T., Ghendon, Y.Z., and Schild, G.C. (1984). Biochemical and serological studies of influenza B viruses: comparisons of historical and recent isolates. Virus. Res. *1*, 241–258.

Oxford, J.S., Newman, R., Corcoran, T., Bootman, J., Major, D., Yates, P., Robertson, J., and Schild, G.C. (1991). Direct isolation in eggs of influenza A (H1N1) and B viruses with haemagglutinins of different antigenic and amino acid composition. J. Gen. Virol. 72 (Pt 1), 185–189.

Oxford, J.S., Schild, G.C., Corcoran, T., Newman, R., Major, D., Robertson, J., Bootman, J., Higgins, P., al-Nakib, W., and Tyrrell, D.A. (1990). A host-cell-selected variant of influenza B virus with a single nucleotide substitution in HA affecting a potential glycosylation site was attenuated in virulence for volunteers. Arch. Virol. *110*, 37–46.

Palese, P., and Young, J.F. (1982). Variation of influenza A, B, and C viruses. Science *215*, 1468–1474.

Pechirra, P., Nunes, B., Coelho, A., Ribeiro, C., Goncalves, P., Pedro, S., Castro, L.C., and Rebelo-de-Andrade, H. (2005). Molecular characterization of the HA gene of influenza type B viruses. J. Med. Virol. *77*, 541–549.

Peng, G., Hongo, S., Kimura, H., Muraki, Y., Sugawara, K., Kitame, F., Numazaki, Y., Suzuki, H., and Nakamura, K. (1996). Frequent occurrence of genetic reassortment between influenza C virus strains in nature. J. Gen. Virol. 77 (Pt 7), 1489–1492.

Peng, G., Hongo, S., Muraki, Y., Sugawara, K., Nishimura, H., Kitame, F., and Nakamura, K. (1994). Genetic reassortment of influenza C viruses in man. J. Gen. Virol. 75 (Pt 12), 3619–3622.

Puzelli, S., Frezza, F., Fabiani, C., Ansaldi, F., Campitelli, L., Lin, Y.P., Gregory, V., Bennett, M., D'Agaro, P., Campello, C., *et al.* (2004). Changes in the hemagglutinins and neuraminidases of human influenza B viruses isolated in Italy during the 2001–02, 2002–03, and 2003–04 seasons. J. Med. Virol. *74*, 629–640.

Rachakonda, P.S., Veit, M., Korte, T., Ludwig, K., Bottcher, C., Huang, Q., Schmidt, M.F., and Herrmann, A. (2007). The relevance of salt bridges for the stability of the influenza virus hemagglutinin. FASEB J. *21*, 995–1002.

Raymond, F.L., Caton, A.J., Cox, N.J., Kendal, A.P., and Brownlee, G.G. (1986). The antigenicity and evolution of influenza H1 haemagglutinin, from 1950–1957 and 1977–1983: two pathways from one gene. Virology *148*, 275–287.

Rivera, K., Thomas, H., Zhang, H., Bossart-Whitaker, P., Wei, X., and Air, G.M. (1995). Probing the structure of influenza B hemagglutinin using site-directed mutagenesis. Virology *206*, 787–795.

Robertson, J.S., Bootman, J.S., Newman, R., Oxford, J.S., Daniels, R.S., Webster, R.G., and Schild, G.C. (1987). Structural changes in the haemagglutinin which accompany egg adaptation of an influenza A(H1N1) virus. Virology *160*, 31–37.

Robertson, J.S., Bootman, J.S., Nicolson, C., Major, D., Robertson, E.W., and Wood, J.M. (1990). The hemagglutinin of influenza B virus present in clinical material is a single species identical to that of mammalian cell-grown virus. Virology *179*, 35–40.

Robertson, J.S., Naeve, C.W., Webster, R.G., Bootman, J.S., Newman, R., and Schild, G.C. (1985). Alterations in the hemagglutinin associated with adaptation of influenza B virus to growth in eggs. Virology *143*, 166–174.

Rogers, G.N., Daniels, R.S., Skehel, J.J., Wiley, D.C., Wang, X.F., Higa, H.H., and Paulson, J.C. (1985).

Host-mediated selection of influenza virus receptor variants. Sialic acid-alpha 2,6Gal-specific clones of A/duck/Ukraine/1/63 revert to sialic acid-alpha 2,3Gal-specific wild type in ovo. J. Biol. Chem. *260*, 7362–7367.

Rogers, G.N., Herrler, G., Paulson, J.C., and Klenk, H.D. (1986). Influenza C virus uses 9-O-acetyl-N-acetylneuraminic acid as a high affinity receptor determinant for attachment to cells. J. Biol. Chem. *261*, 5947–5951.

Rogers, G.N., and Paulson, J.C. (1983). Receptor determinants of human and animal influenza virus isolates: differences in receptor specificity of the H3 hemagglutinin based on species of origin. Virology *127*, 361–373.

Rosenthal, P.B., Zhang, X., Formanowski, F., Fitz, W., Wong, C.H., Meier-Ewert, H., Skehel, J.J., and Wiley, D.C. (1998). Structure of the haemagglutinin-esterase-fusion glycoprotein of influenza C virus. Nature *396*, 92–96.

Rota, P.A., Hemphill, M.L., Whistler, T., Regnery, H.L., and Kendal, A.P. (1992). Antigenic and genetic characterization of the haemagglutinins of recent cocirculating strains of influenza B virus. J. Gen. Virol. *73*, 2737–2742.

Rota, P.A., Wallis, T.R., Harmon, M.W., Rota, J.S., Kendal, A.P., and Nerome, K. (1990). Cocirculation of two distinct evolutionary lineages of influenza type B virus since 1983. Virology *175*, 59–68.

Russell, R.J., Gamblin, S.J., Haire, L.F., Stevens, D.J., Xiao, B., Ha, Y., and Skehel, J.J. (2004). H1 and H7 influenza haemagglutinin structures extend a structural classification of haemagglutinin subtypes. Virology *325*, 287–296.

Saito, T., Nakaya, Y., Suzuki, T., Ito, R., Saito, H., Takao, S., Sahara, K., Odagiri, T., Murata, T., Usui, T., *et al.* (2004). Antigenic alteration of influenza B virus associated with loss of a glycosylation site due to host-cell adaptation. J. Med. Virol. *74*, 336–343.

Schild, G.C., Oxford, J.S., de Jong, J.C., and Webster, R.G. (1983). Evidence for host-cell selection of influenza virus antigenic variants. Nature *303*, 706–709.

Schild, G.C., Pereira, M.S., Chakraverty, P., Coleman, M.T., Dowdle, W.R., and Chang, W.K. (1973). Antigenic variants of influenza B virus. Br. Med. J. *4*, 127–131.

Schulze, I.T. (1997). Effects of glycosylation on the properties and functions of influenza virus hemagglutinin. J. Infect. Dis. 176 *Suppl. 1*, S24–28.

Schweiger, B., Zadow, I., and Heckler, R. (2002). Antigenic drift and variability of influenza viruses. Med. Microbiol. Immunol. (Berl.) *191*, 133–138.

Shaw, M.W., Xu, X., Li, Y., Normand, S., Ueki, R.T., Kunimoto, G.Y., Hall, H., Klimov, A., Cox, N.J., and Subbarao, K. (2002). Reappearance and global spread of variants of influenza B/Victoria/2/87 lineage viruses in the 2000–2001 and 2001–2002 seasons. Virology *303*, 1–8.

Shen, J., Kirk, B.D., Ma, J., and Wang, Q. (2009). Diversifying selective pressure on influenza B virus hemagglutinin. J. Med. Virol. *81*, 114–124.

Skehel, J.J., Stevens, D.J., Daniels, R.S., Douglas, A.R., Knossow, M., Wilson, I.A., and Wiley, D.C. (1984). A carbohydrate side chain on hemagglutinins of Hong Kong influenza viruses inhibits recognition by a monoclonal antibody. Proc. Natl. Acad. Sci. U.S.A. *81*, 1779–1783.

Skehel, J.J., and Wiley, D.C. (2000). Receptor binding and membrane fusion in virus entry: the influenza hemagglutinin. Annu. Rev. Biochem. *69*, 531–569.

Stevens, J., Blixt, O., Tumpey, T.M., Taubenberger, J.K., Paulson, J.C., and Wilson, I.A. (2006). Structure and receptor specificity of the hemagglutinin from an H5N1 influenza virus. Science *312*, 404–410.

Stevens, J., Corper, A.L., Basler, C.F., Taubenberger, J.K., Palese, P., and Wilson, I.A. (2004). Structure of the uncleaved human H1 hemagglutinin from the extinct 1918 influenza virus. Science *303*, 1866–1870.

Stuart-Harris, C.H., Schild, G.C., and Oxford, J.S. (1985). Influenza: The viruses and the Disease, 2nd edn (Baltimore, MD, Edward Arnold).

Taubenberger, J.K., and Morens, D.M. (2008). The pathology of influenza virus infections. Annu. Rev. Pathol. *3*, 499–522.

Tsai, H.P., Wang, H.C., Kiang, D., Huang, S.W., Kuo, P.H., Liu, C.C., Su, I.J., and Wang, J.R. (2006). Increasing appearance of reassortant influenza B virus in Taiwan from 2002 to 2005. J. Clin. Microbiol. *44*, 2705–2713.

Verhoeyen, M., Van Rompuy, L., Jou, W.M., Huylebroeck, D., and Fiers, W. (1983). Complete nucleotide sequence of the influenza B/Singapore/222/79 virus hemagglutinin gene and comparison with the B/Lee/40 hemagglutinin. Nucleic Acids Res. *11*, 4703–4712.

Vlasak, R., Krystal, M., Nacht, M., and Palese, P. (1987). The influenza C virus glycoprotein (HE) exhibits receptor-binding (hemagglutinin) and receptor-destroying (esterase) activities. Virology *160*, 419–425.

Wang, Q., Cheng, F., Lu, M., Tian, X., and Ma, J. (2008). Crystal Structure of Unliganded Influenza B Virus Hemagglutinin. J. Virol. *82*, 3011–3020.

Wang, Q., Tian, X., Chen, X., and Ma, J. (2007). Structural basis for receptor specificity of influenza B virus hemagglutinin. Proc. Natl. Acad. Sci. U.S.A. *104*, 16874–16879.

Webster, R.G. (1998). Influenza: an emerging disease. Emerg. Infect. Dis. *4*, 436–441.

Webster, R.G., Bean, W.J., Gorman, O.T., Chambers, T.M., and Kawaoka, Y. (1992). Evolution and ecology of influenza A viruses. Microbiol. Rev. *56*, 152–179.

Webster, R.G., and Berton, M.T. (1981). Analysis of antigenic drift in the haemagglutinin molecule of influenza B virus with monoclonal antibodies. J. Gen. Virol. *54*, 243–251.

Webster, R.G., Laver, W.G., Air, G.M., and Schild, G.C. (1982). Molecular mechanisms of variation in influenza viruses. Nature *296*, 115–121.

Weis, W., Brown, J.H., Cusack, S., Paulson, J.C., Skehel, J.J., and Wiley, D.C. (1988). Structure of the influenza virus haemagglutinin complexed with its receptor, sialic acid. Nature 333, 426–431.

WHO-MEMORANDUM (1980). A revision of the system of nomenclature for influenza viruses. Bull. World Health Org. *58*, 585–591.

Wiley, D.C., and Skehel, J.J. (1987). The structure and function of the hemagglutinin membrane glycoprotein of influenza virus. Annu. Rev. Biochem. *56*, 365–394.

Wiley, D.C., Wilson, I.A., and Skehel, J.J. (1981). Structural identification of the antibody-binding sites of Hong Kong influenza haemagglutinin and their involvement in antigenic variation. Nature *289*, 373–378.

Wilson, I.A., Skehel, J.J., and Wiley, D.C. (1981). Structure of the haemagglutinin membrane glycoprotein of influenza virus at 3 Å resolution. Nature *289*, 366–373.

Xu, G., Horiike, G., Suzuki, T., Miyamoto, D., Kumihashi, H., and Suzuki, Y. (1996). A novel strain, B/Gifu/2/73, differs from other influenza B viruses in the receptor binding specificities toward sialo-sugar chain linkage. Biochem. Biophys. Res. Commun. *224*, 815–818.

Xu, G., Suzuki, T., Tahara, H., Kiso, M., Hasegawa, A., and Suzuki, Y. (1994). Specificity of sialyl-sugar chain mediated recognition by the hemagglutinin of human influenza B virus isolates. J Biochem (Tokyo) *115*, 202–207.

Xu, X., Lindstrom, S.E., Shaw, M.W., Smith, C.B., Hall, H.E., Mungall, B.A., Subbarao, K., Cox, N.J., and Klimov, A. (2004). Reassortment and evolution of current human influenza A and B viruses. Virus. Res. *103*, 55–60.

Yamada, S., Suzuki, Y., Suzuki, T., Le, M.Q., Nidom, C.A., Sakai-Tagawa, Y., Muramoto, Y., Ito, M., Kiso, M., Horimoto, T., *et al.* (2006). Haemagglutinin mutations responsible for the binding of H5N1 influenza A viruses to human-type receptors. Nature *444*, 378–382.

Yamashita, M., Krystal, M., Fitch, W.M., and Palese, P. (1988). Influenza B virus evolution: co-circulating lineages and comparison of evolutionary pattern with those of influenza A and C viruses. Virology *163*, 112–122.

Yang, Z. (2007). PAML 4: phylogenetic analysis by maximum likelihood. Mol. Biol. Evol. *24*, 1586–1591.

Yang, Z., Nielsen, R., Goldman, N., and Pedersen, A.M. (2000). Codon-substitution models for heterogeneous selection pressure at amino acid sites. Genetics *155*, 431–449.

Zhang, X., Rosenthal, P.B., Formanowski, F., Fitz, W., Wong, C.H., Meier-Ewert, H., Skehel, J.J., and Wiley, D.C. (1999). X-ray crystallographic determination of the structure of the influenza C virus haemagglutinin-esterase-fusion glycoprotein. Acta Cryst. D *55*, 945–961.

Influenza A Virus Nucleoprotein

4

Yizhi Jane Tao and Qiaozhen Ye

Abstract

The (−)RNA genome of the influenza A virus, eight segments in total, is encapsidated in the form of ribonucleoprotein (RNP) complexes. The nucleoprotein (NP), the major protein component of RNPs, binds along the entire length of each genomic RNA segment at a 24-nt interval, forming the double-helical RNP structures found in mature viruses. The viral polymerase, consisting of PA, PB1, and PB2 subunits, binds to the two RNA termini of the RNP. As one of the most abundant proteins made in infected cells, influenza virus NP has essential roles in many important viral processes, including intracellular trafficking of the viral genome, viral RNA replication, viral genome packaging, and virus assembly. The recently determined crystal structures of two NP trimers show an overall fold and an external RNA binding mode that are different from rhabdovirus NP, as confirmed by a new cryo-EM reconstruction of a mini-RNP. Site-directed mutagenesis and RNA binding assays have confirmed that a positively charged groove plays an important role in NP:RNA binding. Other exciting results from recent studies of NP oligomerization, NP RNA binding, NP intracellular trafficking, NP phosphorylation, and NP-supported viral RNA replication are also discussed. These new findings have led to a more detailed model for RNP structure and assembly.

Introduction

The segmented genome of influenza viruses is organized in the form of ribonucleoprotein (RNP) complexes like in all negative-stranded RNA viruses (Fig. 4.1). Each influenza virus RNP contains one viral RNA (vRNA) molecule that binds to multiple copies of the viral nucleoprotein (NP) at the spacing of one NP per 24-nt of RNA. The 5′ and 3′ ends of the viral RNA, which are partially complementary in sequence and highly conserved among every gene segment of all influenza A viruses (Compans *et al.*, 1972; Lamb and Krug, 2001), provide the binding site for a viral polymerase molecule, a heterotrimer consisting of three proteins PA, PB1, and PB2. Electron microscopy studies have revealed that the influenza virus RNPs are double-helical, rod-like structures with an outer diameter of ~12 nm (Klumpp *et al.*, 1997; Noda *et al.*, 2006). The length of the RNP varies, presumably depending on the size of the associated RNA. The rod-shaped structure of RNP is maintained even after the removal of bound RNA (Ruigrok and Baudin, 1995), suggesting that NP, the major component of RNP, plays an important role in organizing the RNP's double-helical structure through protein–protein interactions. Isolated RNPs are active transcription units and can synthesize full-length mRNAs if cap primers and nucleotide substrates are provided *in vitro* (Plotch *et al.*, 1979; Siegert *et al.*, 1973).

The RNPs of an influenza virus play important roles in virus replication by mediating the intracellular trafficking of the viral genome, viral genome packaging, and virus assembly. After cell entry, the RNPs are dissociated from the viral matrix protein because of the low endosomal pH and are actively imported into the cell nucleus (Lamb and Krug, 2001; Portela and Digard, 2002). After

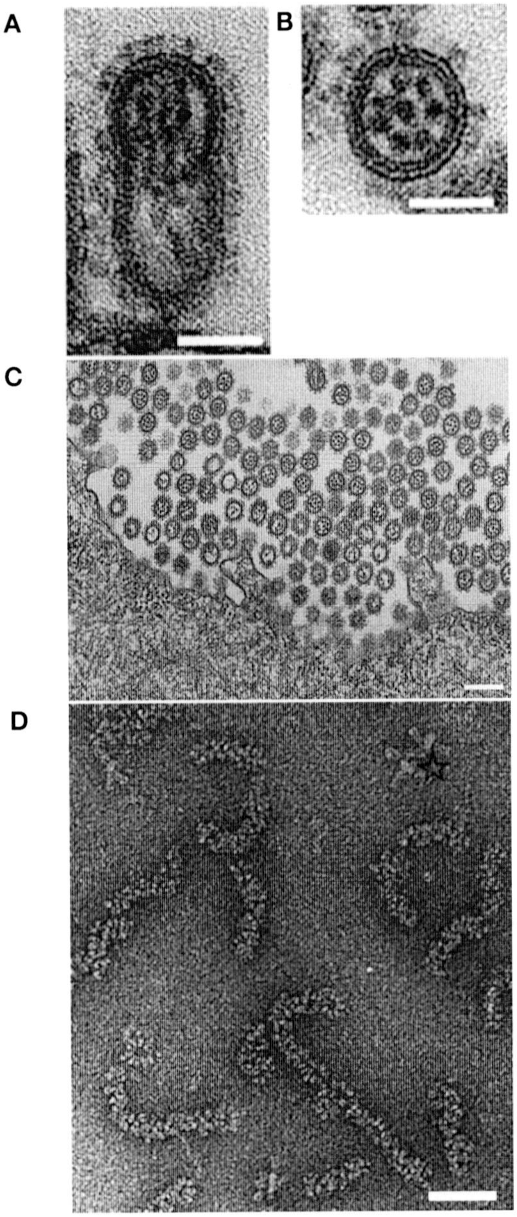

Figure 4.1 Influenza ribonucleoprotein complex (RNP). (A) Electron micrograph of thin-sectioned A/WSN/33 (H1N1) virions budding from Madin-Darby canine kidney (MDCK) cells (Noda *et al.*, 2006). Rod-like structures, 12 nm in width, are associated with the viral envelope at the distal end of the budding virion. (B) and (C) Electron micrograph of transversely sectioned virus particles. Eight electron-dense dots, representing eight RNP rods, were observed in each virion. (D) Electron micrograph of RNP purified from A/PR/8/34 virus samples (Klumpp *et al.*, 1997). Scale bars, 50 nm in (A), (C) and (D); 200 nm in (B). (A–C) are taken from Noda *et al* (2006) with permission from Nature Publishing Group, and (D) is taken from Klumpp *et al* (1997) by permission of the European Molecular Biology Organization.

translation of viral mRNAs, viral proteins necessary for the assembly of new RNPs (NP and the three polymerase subunits PA, PB1, and PB2), along with the viral matrix protein M1 and the viral nuclear export protein NS2, are imported into the nucleus. Both replication products, cRNA (the full-length copy of vRNA) and vRNA, are encapsidated by NP, but mRNAs are not. It has been proposed that the viral polymerase binding to the conserved v/c RNA terminal sequences may act as a nucleation signal for subsequent encapsidation of the v/c RNA by NP. Late in infection, vRNPs are exported from the nucleus into the cytoplasm via the cellular Crm1-mediated nuclear export pathway (Elton *et al.*, 2001; Ma *et al.*, 2001; Watanabe *et al.*, 2001), facilitated by NP and two other viral proteins, M1 and NS2 (Akarsu *et al.*, 2003). The viral genome is incorporated into the assembling viruses in the form of RNPs. A recent electron microscopy study shows that a newly assembled influenza A virus contains exactly eight RNP structures that are packed along the long axis of the virus particle (Noda *et al.*, 2006). This observation argues strongly that viral RNPs are selectively packaged during assembly, most likely using packaging signals residing in the sequences of different vRNA segments. This result also raises the intriguing question of how RNP is organized and how RNA sequence information is presented on the surface of RNPs, such that it allows specific communications among the different RNP segments. Answers to these questions and others regarding the biological activities of viral RNPs require a clear understanding of how influenza virus NP self-oligomerizes and binds RNA.

Crystal structure of NP

Influenza A virus NP, an integral component of viral RNPs, is a polypeptide of 498 amino acids. Phylogenetic analysis of influenza A virus strains has revealed that the NP gene is relatively well conserved, as the maximal pair-wise amino acid difference reached only 10.8% (Gorman *et al.*, 1990). The sequence of influenza A virus NP also shows considerable conservation when compared to the NP of other influenza viruses (~35% identical to influenza B and ~20% identical to influenza C viruses), suggesting that the structure and function of NP is likely to be broadly conserved among all influenza viruses.

The crystal structures of influenza A virus NP from strains A/WSN/33 (H1N1) and A/HK/483 (H5N1) have been determined to 3.2 Å and 3.3 Å resolution, respectively, both in the RNA-free state (Ng *et al.*, 2008; Ye *et al.*, 2006) (Fig. 4.2). Although NP is capable of forming oligomeric ring structures of various sizes, both H1N1 and H5N1 NP were crystallized as trimers. For H1N1, monomeric NP samples from gel filtration chromatography were able to grow crystals similar to those of oligomeric NP, suggesting that NP–NP interaction is dynamic and that NP crystallization is probably accompanied by NP oligomer–monomer conversion.

The structure of H1N1 NP shows that it is mostly α-helical and that each monomer has a curved, crescent-like, shape (Ye *et al.*, 2006) (Fig. 4.2A and B). There are six cysteine residues in each NP monomer, but none are close enough for disulphide bond formation. The overall structure of NP can be divided into two domains, a head and a body, that are joined together by the polypeptide chain at three regions. The head and body domains are formed by non-contiguous polypeptide regions. The head domain is formed by residues 150–272 and 438–452, and the body domain consists of the three polypeptide segments 21–149, 273–396, and 453–489. Consequently, the overall fold and topology of the influenza virus NP are different from that of the rhabdovirus nucleocapsid proteins in which the two domains are formed by co-linear polypeptide sequences (Albertini *et al.*, 2006; Green *et al.*, 2006).

Interaction between adjacent molecules in an NP trimer involves two interfaces (Fig. 4.2B). The first interface, likely the most import one, is mediated by a tail loop, formed by residues 402–428, at the back of the crescent-shaped molecule (Fig. 4.3). With a rather extended conformation, the tail loop from one subunit is inserted into the body domain of a neighbouring molecule in an anticlockwise direction when viewed along the three-fold axis from the side of the head domain. Intramolecular interactions for the tail loop are not possible because the distance is too far for the tail loop to reach the binding pocket located on the opposite side of the same molecule. The tail loop contains three β-strands with the last two forming an intra-molecular β-hairpin. The tail loop from one molecule makes extensive interactions with a neighbouring subunit through several types of molecular forces, including intermolecular β-sheets, hydrophobic interactions, and salt bridges. The interactions mediated by the tail loop are almost equivalent among the three subunits in an NP trimer. In addition, because the tail loop is loosely attached to the rest of the NP through highly flexible linkers, the tail loop can make similar intersubunit interactions in both closed rings and helical RNP-like structures, as has been proposed for rhabodovirus nucleocapsid proteins. The second interface between neighbouring NP molecules is made by the C-terminal region of each monomer (residues 440 to end). However, contacts made by this interface are far fewer in number compared to the tail loop and are non-equivalent among the three subunits. Variations are also expected in NP oligomeric rings of different sizes. Therefore, the second interface is unlikely to play a major role in RNP organization.

H1N1 and H5N1 NP share 94% sequence identity. Aligning the 398 common Cα atoms from the two crystal structures resulted in a root-mean-square (RMS) deviation of 1.0 Å (Ng *et al.*, 2008) (Fig. 4.2C). Several differences were noted between the two structures including: (1) the three subunits of the H5N1 NP trimer are related by a crystallographic 3-fold axis, so the intermolecular contacts within each trimer are identical; (2) the H1N1 NP tail loop interaction is preserved in H5N1 NP but the tail loop adopts two different orientations, with the H1N1 tail loop points towards the body domain whereas the H5N1 tail loop points away from the body domain; (3) residues 397–401 and 429–437, which connect the extended tail loop (residues 402–428) to the main body, are disordered in H1N1 NP but are ordered in H5N1 NP; and (4) a flexible, basic loop (residues 72–92) in the RNA binding groove (see below) that is important for RNA binding is partially ordered in H5N1 but is completely disordered in H1N1 NP.

To determine whether the tail loop is essential for NP oligomerization, three H1N1 NP mutants were made, including two single residue mutants, R416A and E339A, and a deletion

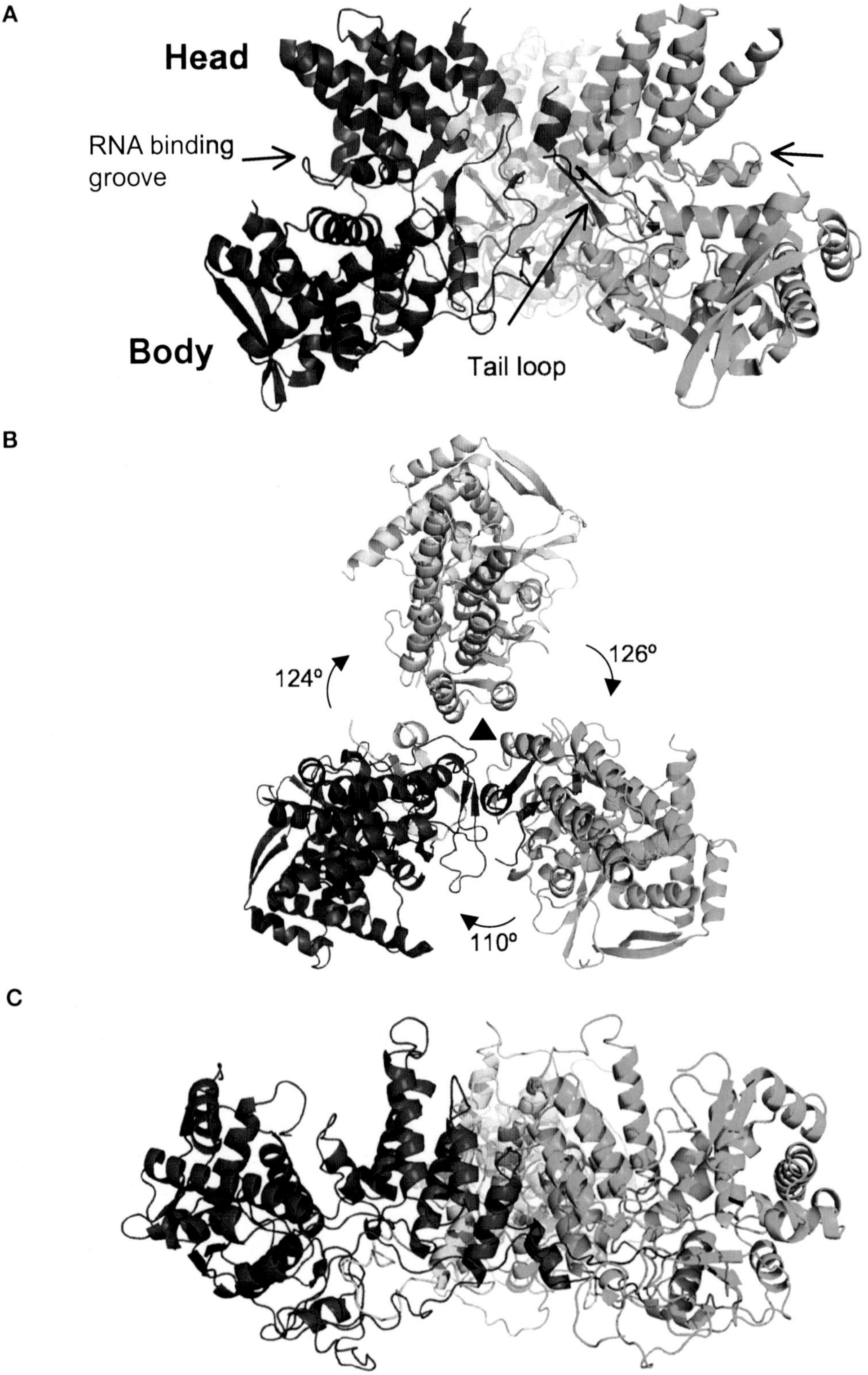

Figure 4.2 Crystal structure of the influenza A virus. (A) A/WSN/33 H1N1 NP, side view. Three subunits A, B and C from a NP trimer are shown in different colours. Different structural elements are assigned on the red subunit. (B) A/WSN/33 H1N1 NP, top view. (C) A/HK/97 H5N1 NP, side view.

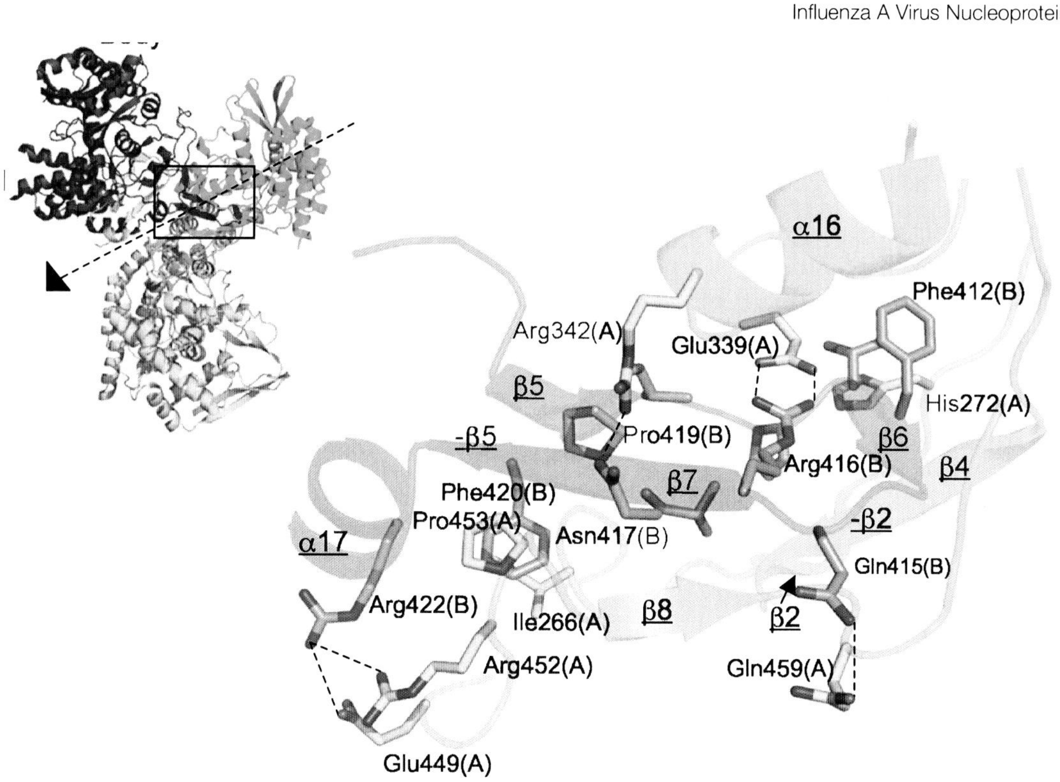

Figure 4.3 Intersubunit interactions mediated by the tail loop. Subunit A and B are shown in green and red, respectively. Amino acid side chains are drawn only for these residues making important interactions. The last capitalized letter in amino acid labels is used to show the molecular origin of that residue. The tail loop region is marked on a NP trimer displayed to the left. At this orientation, the NP trimer is tilted with the head domain pointing into the paper.

mutant Δ402–428 that removes the tail loop (Ye *et al.*, 2006). Residues R416 and E339 form an inter-subunit salt bridge. When the oligomerization behaviour of R416A, E339A, and Δ402–428 was evaluated by gel filtration chromatography, all three mutants did not self-associate and formed predominantly monomers. Electron microscopy of the R416A mutant protein confirmed that trimers were no longer formed, in contrast to wild-type NP where trimers and larger oligomers were frequently observed. Proper self-association is important for the biological activities of NP, because the R416A mutant was shown to be defective in viral RNA synthesis (Elton *et al.*, 1999b; Mena *et al.*, 1999). It has also been shown that the R416A NP mutant has a significantly reduced RNA binding affinity (Elton *et al.*, 1999b), as the displacement of the tail loop from its binding pocket may cause significant structural rearrangements in NP, resulting in a disrupted RNA binding groove. The essential biological function of the tail loop suggests that chemical compounds which competitively displace the tail loop from its binding pocket would interfere with viral genome replication and therefore serve as promising leads for anti-influenza drug development.

NP RNA binding

Influenza virus NP has a RNA binding activity with little or no sequence specificity (Baudin *et al.*, 1994; Kingsbury *et al.*, 1987; Scholtissek and Becht, 1971; Yamanaka *et al.*, 1990). In contrast, NP from another segmented (–) RNA virus, bunyavirus, often displays a higher affinity for the 5′-termini of the genome segments (Osborne and Elliott, 2000; Severson *et al.*, 2001). It has been estimated that one influenza A virus NP binds to every 24 nucleotides of RNA (Compans *et al.*, 1972; Ortega *et al.*, 2000). Filter binging assays using NP purified from influenza viruses

(A/PR/8/34, H1N1) yielded a K_d in the low- to mid-nanomolar range (Baudin *et al.*, 1994). A later study using partially purified recombinant NP gave a K_d in the mid-nanomolar range (Elton *et al.*, 1999b). It is puzzling that results from both studies showed that the length of RNA substrate (ranging from 15 nt to 890 nt long) had no obvious effect on NP's RNA binding affinity, contradicting with the expectation for a protein that co-operatively binds RNA. A recent study using a 24-nt RNA and surface plasmon resonance gave a K_d of 23 nM for purified recombinant NP (A/HK/97, H5N1) (Ng *et al.*, 2008). RNA binding affinity of NP is likely to be affected by NP self-oligomerization, which has been found to be dependent on the type of virus strain, solvent condition, and temperature (see below). A full understanding of NP RNA binding activity will likely require the simultaneous characterization of NP oligomerization and RNA binding stoichiometry.

The NP crystal structure shows that a deep groove located at the exterior of NP oligomers between the head and body domain is most likely the RNA binding site (Ye *et al.*, 2006) (Fig. 4.4). The surface of the groove is made up of a large number of basic residues, including R65, R150, R152, R156, R174, R175, R195, R199, R213, R214, R221, R236, R355, K357, R361, and R391, that are likely to function in RNA binding by interacting with the phosphodiester backbone. An aromatic residue, Y148, is found at one end of the groove and may stack with a nucleotide base. Eleven out of the 16 basic residues observed at the putative RNA binding surface are conserved among influenza A, B, and C viruses (of strains A/WSN/33, B/Ann Arbor/1/66, and C/California/78, respectively), further supporting a role in RNA binding. The aromatic residue Y148 is also conserved. This result is consistent with an earlier conclusion that NP contains a N-terminal region that displays RNA-binding activity, but high-affinity binding requires the concerted action of other regions distributed throughout the protein polypeptide (Albo *et al.*, 1995; Elton *et al.*, 1999b; Kobayashi *et al.*, 1994; Medcalf *et al.*, 1999). Site-directed mutagenesis targeted at basic residues in the potential RNA binding groove of H5N1 NP identified a number of basic amino acid residues including R74, R75, R174, R175, R221, R150, R152, R156, and R162, all of which are important for RNA binding (Ng *et al.*, 2008). With RNA bound to the groove at the outer periphery of oligomeric NP, the distance between two neighbouring binding sites in an NP trimer is approximately 70 Å, consistent with the stoichiometry of one NP per 24 nucleotides of RNA as previously determined.

The NP structure also indicates that RNA would be largely exposed on the external surface of NP molecules in the RNP (Figs 4.4 and 4.5), different from the rhabdovirus nucleocapsid proteins which have their RNA binding groove facing the interior (Albertini *et al.*, 2006; Green *et al.*, 2006). NP from bornavirus, another non-segmented (−)RNA virus, also has its RNA binding groove facing the interior of the oligomeric NP ring even in the absence of bound RNA (Rudolph *et al.*, 2003). When an excessive amount of RNA was added to purified influenza A virus NP, electron microscopy showed that recombinant NP oligomeric rings retain their overall molecular structure upon the formation of NP–RNA complexes (Ye *et al.*, 2006), indicating that RNA-bound NP is likely to oligomerize through the tail loop like in RNA-free NP. Interestingly, incubating NP samples with RNA, even those short ones (i.e. 24 nt in length) with only one NP binding site, promotes NP oligomerization, resulting in the formation of larger NP oligomeric ring structures (Ng *et al.*, 2008). The molecular basis of the enhanced oligomerization effect is not well understood, but it may be related to potential NP conformational changes and/or simultaneous binding of one RNA to multiple NP molecules. During vRNA encapsidation, the polymerization of NP into large RNP-like structures is likely to be a direct consequence of cooperative binding of NP to long RNA molecules. Because the oligomerization tail loop is flexibly attached to the NP body domain, we anticipate that many of the inter-subunit interactions observed in the NP crystal structure would be preserved in the continuous NP:RNA helical fibre in RNPs, and that similar NP–NP interactions would also be maintained when the bound RNA temporarily dissociates during viral RNA replication/transcription.

Chemical modification experiments have suggested that NP binds to the phosphate backbone and leaves the Watson-Crick positions of the

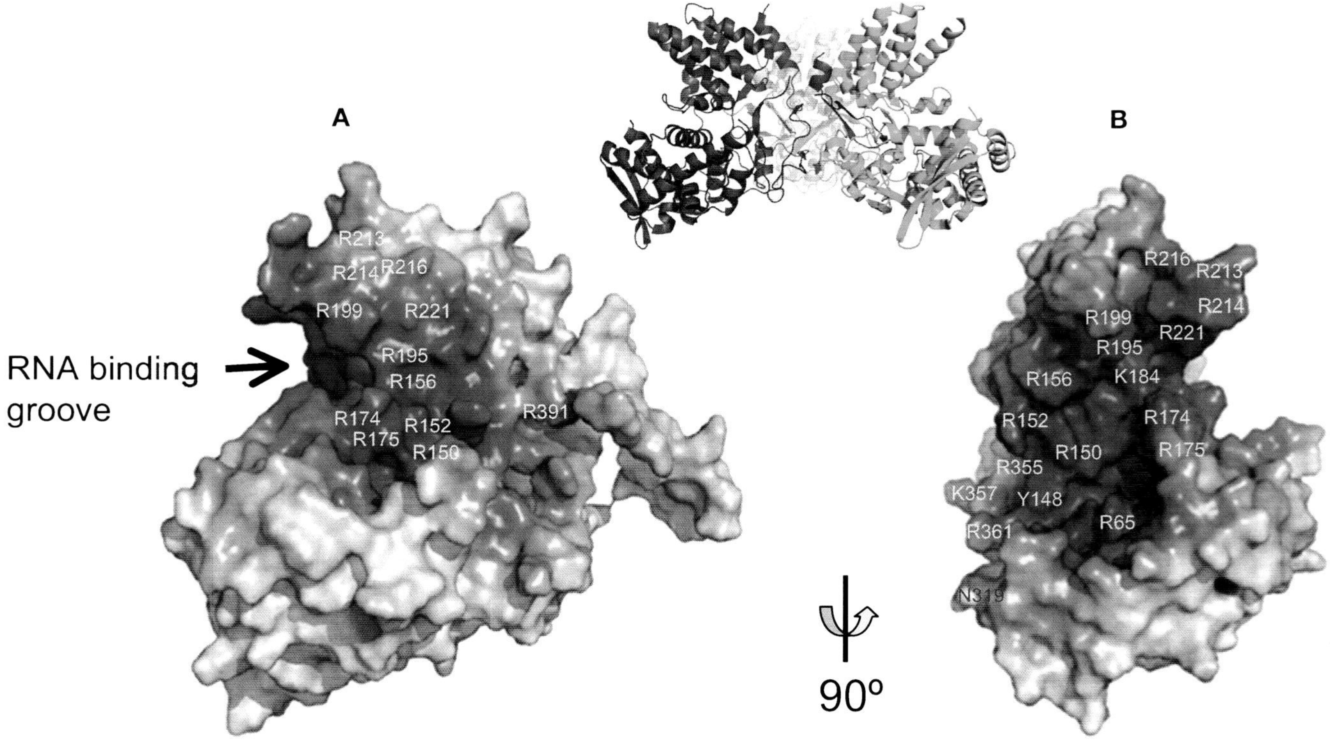

Figure 4.4 NP RNA-binding groove. (A and B) Electrostatic potential distribution for NP identifies a potential RNA-binding site. (A) and (B) are in similar orientations to the red and green subunits in Fig. 4.2A, respectively. Positive and negative electrostatic potentials are shown in blue and red, respectively.

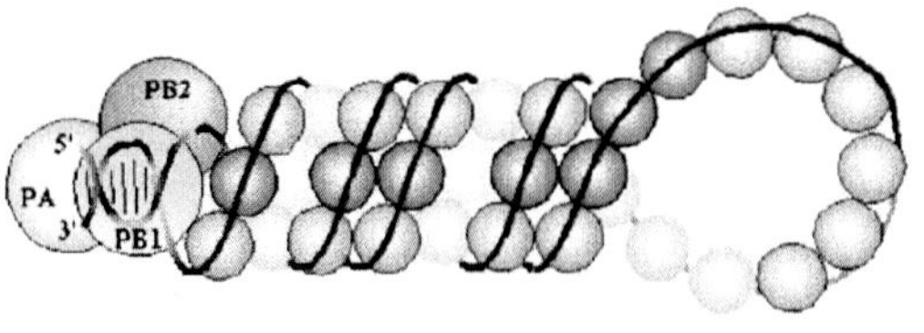

Figure 4.5 Influenza A virus RNP model. Blue spheres represent NP monomers, and the black line shows associated vRNA molecule. Influenza RNP folds into a double-helical hairpin structure. A short duplex formed between the 5′ and the 3′ ends provides the binding site for the heterotrimeric RNA-dependent RNA polymerase. Taken from Portela et al (2002) with permission from the Society for General Microbiology.

bases open to the solvent (Baudin *et al.*, 1994). We expect that vRNA from influenza A virus RNP to be highly accessible to solvent, different from vRNA in rhabdovirus RNP that is largely sequestered within protein cavities or clefts. Many different properties were previously noted for influenza RNPs and the RNPs of non-segmented RNA viruses like VSV and rabies virus. For example, the viral RNA in influenza virus RNPs is digested by RNase treatment, whereas the viral RNA in the RNPs of parainfluenza viruses and rhabdoviruses is completely resistant to RNase digestion (Baudin *et al.*, 1994; Duesberg, 1969; Pons *et al.*, 1969). In addition, polyvinylsulphate (PVS), a negatively charged polymer, is able to completely displace RNA from influenza virus RNP, whereas it has no effect on RNPs from VSV (Goldstein and Pons, 1970). The length of bound RNA is also different for NP from influenza viruses and non-segmented (−)RNA viruses, with 24 nt for influenza A virus, 9 nt for rhabdovirus, and 7 nt for paramyxoviruses.

The influenza NP crystal structure displays a rather different morphology compared to the NP structure from a 27-Å electron microscopic (EM) model of a mini-RNP, which was reconstituted *in vivo* from recombinant proteins and a model genomic vRNA (Martin-Benito *et al.*, 2001). The 3-D reconstruction shows that the influenza virus mini-RNP has a circular shape and contains nine NP monomers, two of which are connected with the polymerase complex. Each NP monomer appears to have an elongated, banana-like shape, different from the one seen in crystal structures. To explain the apparent differences, it was proposed that upon RNA binding the structure of influenza A virus NP may dramatically change its conformation to such that its structure closely resembles that of rhabdovirus NP (Luo *et al.*, 2007). A cryo-EM reconstruction recently calculated for the mini-RNP at 12 Å resolution showed that the structure of vRNA-bound NP closely resembles the crystal structure, which can be fitted into the 3-D EM reconstruction without major adjustment of the model (Coloma *et al.*). Due to the limited resolution, the new reconstruction of the mini-RNP is still unable to pinpoint the exact location of bound vRNA. Therefore, an atomic or near-atomic structure of a NP–RNA complex, either from cryo-EM or X-ray crystallography, is needed to resolve the detailed molecular interaction network involved in NP RNA binding as well as any conformational changes that may have occurred in NP upon RNA binding.

NP self-oligomerization

Oligomeric/polymeric NP was widely found in both virus-infected cells and *in vitro*. Early studies showed that influenza virus RNPs retain their high-order structure after the removal of RNA (Ruigrok and Baudin, 1995), indicating that the NP–NP interactions are both important and sufficient for maintaining RNP structures once they are assembled (Kingsbury and Webster, 1969; Pons *et al.*, 1969). NP is likely to form the core scaffold of the RNP, although the vRNA and the viral polymerase may play important roles in guiding the ordered assembly of RNP. Indeed, it has been reported that recombinant NP purified from *E. coli* was able to form RNP-like structures in presence of long RNA molecules. NP–NP interactions have been estimated to have a K_d of ~200 nM in a binding assay using crude samples (Elton *et al.*, 1999a). Removing the C-terminal 23 residues of NP resulted in stronger NP oligomerization, but the biological significance of this effect is not clear. Some mutations altered NP oligomerization (e.g. R416A) also affected the ability of NP to support influenza virus gene expression in an *in vivo* assay, suggesting proper NP–NP interaction is necessary for its biological function (Elton *et al.*, 1999a; Ye *et al.*, 2006).The oligomerization behaviour of NP was found to be dependent on the solvent pH and salt concentration (Prokudina *et al.*, 2005; Prokudina *et al.*, 2001; Ruigrok and Baudin, 1995). In addition, the efficiency of intracellular NP oligomerization also depends on the host origin of an influenza A virus strain. In cells infected with avian influenza A virus strains, the viral NP was almost completely oligomerized and only traces of monomeric NP were detected by polyacrylamide gel electrophoresis (PAGE) in unboiled samples. However, in the cells infected with human influenza A virus strains, a significant amount of non-oligomerized monomeric NP was detected in unboiled samples (Prokudina *et al.*, 2001). Some difference is observed for the antigenic properties of monomeric and polymeric NPs, suggesting that they may adopt somewhat different structures (Prokudina *et al.*, 2008). The presence or absence of nucleic acids also affects NP self-oligomerization. Our laboratory has observed that recombinant NP overexpressed in both *E. coli* and insect cells form insoluble aggregates presumably by non-specific binding to host nucleic acids, and that the insoluble NP can be converted to soluble oligomers (i.e. trimers, tetramers, etc.) when treated with large amounts of RNase and DNase.

While the tail loop interaction is able to mediate NP polymerization in a linear fashion, the formation of a double-helical RNP structure as previously proposed would require additional interactions between the two antiparallel RNA strands, a situation similar to basepairing in the stabilization of a dsDNA helix. The molecular nature of such an inter-strand interaction is unknown for influenza virus RNP, and NP dimers are rarely observed in NP samples purified from viruses and recombinant sources. Such inter-strand interactions could potentially be mediated by RNA, NP, or both. Additional viral proteins may also function to tether the two antiparallel RNA strands together by interacting with RNA and/or RNA-bound NP.

NP phosphorylation

An analysis of NP from 29 different strains of influenza A viruses showed that all NPs are phosphorylated *in vivo* (Kistner *et al.*, 1989). Moreover, NP exhibits a dynamic phosphorylation pattern during viral replication (Kistner *et al.*, 1989). An earlier study showed that the treatment of cells with a protein kinase C inhibitor increased the amounts of nuclear NP, whereas treatment of cells with a phosphorylation stimulator increased the amounts of cytoplasmic NP, suggesting a role of phosphorylation in nucleo-cytoplasmic trafficking of NP (Bui *et al.*, 2002; Neumann *et al.*, 1997). The importance of phosphorylation in the nuclear import of the simian virus 40 large tumour antigen (T40) has also been reported (Rihs *et al.*, 1991). NP has a complex network of serine and threonine residues, whose phosphorylation statuses change during the virus replication cycle (Kistner *et al.*, 1989). A serine at position 3 has been shown to be a major phosphorylation site of A/Victoria/75 NP (Arrese and Portela, 1996). Its phosphorylation reduces the nuclear accumulation of NP, and thus might contribute to preventing the re-entry of the RNPs into the nucleus (Bullido *et al.*, 2000). S3 is highly conserved in influenza A virus strains except for A/WSN/33, which has a threonine at the equivalent position.

About 12 years ago, Ruigrok and Baudin, two pioneers of RNP structural studies, raised an intriguing issue: 'It will be interesting to know whether NP in the infected cells is monomeric and, if so, whether other proteins are involved in preventing NP polymerization' (Ruigrok and Baudin, 1995). This question is especially relevant because NP has a non-specific RNA binding activity and recombinant NP was found to bind to host nucleic acids. Rhabdoviruses encode a phosphoprotein (P) that prevents non-specific aggregation of NP by forming a NP–P complex as the precursor for encapsidating newly synthesized RNA (Howard and Wertz, 1989). A protein analogous to the rhabdovirus P protein, however, is missing from influenza viruses. It has also been reported that rhabdovirus NP phosphorylation reduces non-specific RNA binding (Toriumi and Kawai, 2004; Yang *et al.*, 1999). Treatment of the influenza virus NP with alkaline phosphatase enhanced its RNA binding activity (Yamanaka *et al.*, 1990). It is possible that a yet unidentified phosphorylation at the C-terminal region of the NP could affect the functionality of the protein with respect to its role in virus-specific RNA synthesis (Arrese and Portela, 1996). Therefore, influenza virus NP phosphorylation may be involved in regulating the RNA binding and self-oligomerization activity of NP.

The role of NP in viral RNA replication

Besides its structural role in organizing RNP, NP is also required for RNA replication (Lamb and Krug, 2001). Whereas viral transcription produces fully capped and polyadenylated mRNAs, viral RNA replication, which is also catalysed by viral polymerase, generates vRNAs and cRNAs that are full-length copies of the viral genome. To switch from the synthesis of mRNA to that of vRNA and cRNA, it is necessary to change from capped RNA-primed initiation to unprimed initiation and to prevent premature termination. Infected cell extracts depleted of non-nucleocapsid NP molecules are not able to synthesize either vRNA or cRNA (Shapiro and Krug, 1988). Non-nucleocapsid NP molecules promote anti-termination, probably by binding to nascent RNA and preventing the template from backslipping and reiteratively copying the poly(U) sequence (Beaton and Krug, 1986; Shapiro and Krug, 1988). One role of NP during viral RNA replication is to block the termination at the U tract, thereby enabling the polymerase to synthesize cRNA, a complete copy of the vRNA template. NP is also required for the switch from capped RNA-primed initiation to unprimed initiation, and three models have been proposed to describe the role of NP in this switch by considering NP's ability to bind RNA or to interact with the viral polymerase (Krug *et al.*, 1975; St Angelo *et al.*, 1987; Vreede *et al.*, 2004). Two of these models attribute the switch to the RNA-binding activity of NP. In the first model, denoted as the 'stabilization model,' the role of NP is to stabilize nascent cRNA and vRNA transcripts that would otherwise be degraded in the absence of NP (Vreede *et al.*, 2004). In the second model, denoted as the 'template modification model,' NP binds to the template RNA and alters its structure to favour unprimed RNA synthesis (Portela and Digard, 2002). The third model, denoted as the 'polymerase modification model,' proposes a direct protein–protein interaction between NP and one or more subunits of the viral polymerase, resulting in a modification of the polymerase in favour of unprimed initiation (Portela and Digard, 2002).

By establishing an *in vitro* replication assay using the purified proteins, it has recently been shown that RNA binding activity is not needed for its stimulatory effects on viral RNA replication, suggesting that NP is likely to regulate the activity of viral RNA polymerase by direct polymerase binding (Newcomb *et al.*, 2008). The same study also showed that NP lacking RNA-binding activity binds directly to the viral polymerase (Newcomb *et al.*, 2008). NP co-immunoprecipitated with all three polymerase proteins, and the level of viral RNA synthesis correlated with the efficiency of NP protein binding to the trimeric viral polymerase (Labadie *et al.*, 2007). It has also been proposed that NP–polymerase interaction may be a key determinant of influenza A virus host range, and some viral and cellular factors may play a role in regulating the interaction (Naffakh *et al.*, 2008). It has been observed that restricted growth of the avian influenza A virus in human cells correlates with poor association of the NP with the polymerase compared to what is observed for the human isolate (Naffakh *et al.*, 2008). The NP-polymerase association is restored by a Glu-to-Lys substitution at residue 627 of PB2, which is invariantly an E in avian influenza A viruses and a K in human influenza A viruses. According to a recently determined crystal structure, residue 627 is located at the centre of a highly basic surface patch (Tarendeau *et al.*, 2008). It would be interesting to determine whether the presence of a K rather than an E at PB2 627 enhances NP-dependent unprimed viral RNA replication in an *in vitro* replication assay using purified proteins.

Transfection assays have shown that mutating four NP amino acids results in reduced viral RNA replication (Mena *et al.*, 1999). These amino acids are located in the body domain of NP, suggesting that the site(s) that functionally interact with the trimeric viral polymerase are in this domain (Ye *et al.*, 2006). It is not known which polymerase protein(s) in trimeric polymerase complexes interact with NP, thereby resulting in the activation of unprimed cRNA and vRNA synthesis. Previous experiments that assayed the binding of NP to individual polymerase proteins (not in functional polymerase complexes) found that NP co-immonoprecipitated with PB1 and PB2 (Biswas *et al.*, 1998). Three NP regions, amino acids 1–160, 256–340 and 340–498, have also been implicated in binding to the PB1 and PB2 subunits of the viral polymerase (Biswas *et al.*, 1998). The majority of these amino acids (1–149, 273–396, 453–498) are localized in the body domain of NP. It is possible that conserved amino acid regions in the body domain that are on the surface of NP mediate NP–polymerase interactions that are crucial for viral RNA replication. The NP structures will enable investigators to make rational mutations to determine the roles of specific NP amino acids in functional interactions with viral RNA polymerase.

In addition to the 'polymerase modification model', the 'stabilization model' has also received a lot of attention lately (Vreede *et al.*, 2004). Because virion-derived vRNPs are able to synthesize both cRNA and mRNA in the absence of additional viral or cellular factors, it was proposed that viral polymerase and NP expressed during an *in vivo* viral infection are required to protect and replicate nascent cRNA but do not affect the infecting viral polymerase. cRNA, which is synthesized in infected cells, would be degraded by cellular pathways unless free viral polymerase and NP are present to bind the nascent transcript, thereby stabilizing it (Vreede and Brownlee, 2007). Perhaps these different regulation models for NP-supported viral RNA replication are not necessarily mutually exclusive and instead may suggest multiple roles that are assumed by NP in viral RNA synthesis. Future mechanistic studies on NP and the polymerase will shed new light on the various functional requirements for the synthesis of full-length c/vRNAs.

NP nucleo-trafficking

There is a unidirectional transport of the RNPs to the nucleus at the start of the infection process and an opposite unidirectional export of RNPs at the end of the infection (Boulo *et al.*, 2007). One of NP's functions is to mediate nuclear trafficking of itself and perhaps also RNPs. Two nuclear localization signals (NLS) have been identified in NP, a non-conventional NLS near the N-terminus and a bipartite NLS located near the middle of the polypeptide (Neumann *et al.*, 1997; Wang *et al.*, 1997; Weber *et al.*, 1998). In addition, systematic deletion analysis of NP suggests the presence of another potential NLS located between amino acids 320 and 400 (Bullido *et al.*, 2000). The interaction of NP with host cell importin-α results in the nuclear import of NP and infecting RNPs (O'Neill *et al.*, 1995; Wang *et al.*, 1997). It has been reported that an adaptive mutation, N319K, in NP resulted in enhanced binding of these proteins to importin-α1 in mammalian cells, and that enhanced binding was paralleled by transient nuclear accumulation and cytoplasmic depletion of importin-α1 as well as increased transport of NP into the nucleus (Gabriel, 2008). As infection proceeds to the late stage, RNP export becomes dominant (Martin and Helenius, 1991; Whittaker *et al.*, 1996). In newly made RNPS, NLS signals of NP are probably masked by interaction with M1, which in turn interacts with NS2 and then the host exportin CRM1, leading to nuclear export of viral RNPs that are necessary for the budding of new virus particles. Free NP does not bind to matrix protein directly. A nuclear export signal (NES) in NP has also been proposed to mediate NP or RNP export through direct interaction with CRM1 (Elton *et al.*, 2001). NP contains a possible cytoplasmic accumulation signal (CAS, residues 327–345), which interacts with F-actin and causes cytoplasmic retention of NP late in infection. Purified NP binds F-actin *in vitro* with a K_d of 1 mM and a stoichiometry of one NP per actin subunit (Digard *et al.*, 1999). Indeed, cytoplasmic NP has been shown to associate with the cytoskeleton late in infection (Avalos *et al.*, 1997; Husain and Gupta, 1997). In addition, there is evidence that phosphorylation plays a role in regulating RNP transport (Neumann *et al.*, 1997).

The NP structure indicates that the non-conventional NLS (nNLS: 3SxGTKRSYxxM) (Cros *et al.*, 2005; Neumann *et al.*, 1997; Wang *et al.*, 1997), previously identified at the N-terminus of NP, can function as its nuclear localization signal. The nNLS sequence is located at the outer periphery of the NP trimer, where it is highly accessible to solvent. In addition, this region is disordered in electron density maps, and this flexibility should allow the nNLS to fit easily into the substrate recognition pocket on importin-α. The classical bipartite NLS (cNLS: 198KRGINDRNFWRGENGRKTR) located near the middle of the polypeptide is unsuitable for functioning as a NLS, because the NP crystal structure indicates that the 198KR and the 213RKTR motifs are ~15Å apart, a distance far shorter than the minimal separating distance of 28Å needed for efficient binding of a bipartite NLS. The NP crystal structure also shows that the CAS region is buried internally in the NP structure and in this conformational state would not be able to bind F-actin.

In purified RNP, the nNLS is indeed much more accessible compared with the bipartite cNLS based on results from dot blotting and immunogold labelling of vRNPs with sequence-specific antibodies (Wu *et al.*, 2007b). A regular periodicity of repeat was observed with gold particles labelling for nNLS, indicative of a highly regular helical conformation present in the vRNPs. However, it appears that the complete inhibition of vRNP nuclear import requires the addition of peptides that mimic nNLS as well as cNLS from NP, although nNLS on NP appears to be the main contributor to the nuclear import of vRNPs (Wu *et al.*, 2007a). Another study performed on overexpressed NP shows that mutation of the unconventional NLS completely impaired the nuclear import of NP (Cros *et al.*, 2005). A peptide bearing the unconventional NLS could inhibit the nuclear import of the NP as well as the ribonucleoprotein complexes (Cros *et al.*, 2005).

Other roles of NP

NP has also been shown to interact with a number of host factors. This is not surprising as NP is strongly expressed during all stages of the virus infection cycle. In addition to its interaction with F-actin, importin-α, and CRM1, it has been reported that NP interacts with UAP56, a cellular splicing factor in the DEAD-box family of RNA-dependent ATPases (Momose *et al.*, 2001). Deletion analysis of NP indicated that a UAP56-binding site is located in the N-terminal 20 amino acids of NP (Momose *et al.*, 2001). UAP56 obtained from an uninfected cell nuclear extract, as well as recombinant UAP56, stimulated influenza virus RNA synthesis catalysed by micrococcal nuclease-treated RNPs isolated from purified virus. It was proposed that UAP56 acts as a chaperone for non-RNA bound NP to facilitate its binding to RNA (Momose *et al.*, 2001). It is expected that more NP-interacting host factors will be identified, as host adaptation often results in mutations in NP, although the detailed mechanisms are unclear in most cases. In addition, influenza A virus NP, along with M1, is the major target for human virus-specific cytotoxic T-lymphocyte (CTL) responses (Berkhoff *et al.*, 2007; Boon *et al.*, 2006). It has been shown that vaccination with NP leads to development of protective immunity against diverse influenza virus through its ability to stimulate cellular immunity (Barefoot *et al.*, 2009; Ohba *et al.*, 2007; Roose *et al.*, 2009).

Conclusion

Nucleoproteins are generally found in (−)-strand RNA viruses. These (−)RNA viruses include the vesicular stomatitis virus (VSV) and the rabies virus in the Rhabdoviridae family, the borna disease virus in the Bornaviridae family, and the measles virus in the Paramyxoviridae family. Recent studies, however, showed that influenza A virus NP exhibits fundamental differences in both overall structure and important properties such as RNA binding and self-oligomerization when compared to many other non-segmented (−)RNA viruses. These differences are perhaps not surprising considering their different genome structures, different mechanisms for gene transcription, and viral RNA replication. In addition, the packaging of a multi-segment viral genome, such as those of influenza viruses, imposes a different set of challenges compared to viruses with a non-segmented viral genome. Therefore it is comprehensible that even though influenza and other non-segmented viruses may originate from a common ancestor, different properties, both structural and functional, were selected for during evolution. Further studies on are likely to reveal additional similarities and differences in the structure, function, and assembly of RNPs among the different negative-strand RNA viruses.

New results from the biochemical and structural characterization of influenza virus NP and RNP strongly agree with an earlier RNP structural model in which vRNA is extensively coated with NP to form a double-helical hairpin structure, with the two free RNA termini bound to the viral polymerase at the hairpin (Fig. 4.5). The NP crystal structures not only substantiate this model but also reveal mechanisms by which NP may oligomerize. The NP crystal structure also provides direct evidence that vRNA is likely to be largely exposed on the external surface of RNP, thus allowing easy access to genetic sequence information. While functioning as a template for RNA synthesis, vRNA may briefly dissociate from the internal NP scaffold, which is likely to remain intact for the vRNA to re-associate after the viral polymerase has slid through. The external RNA binding mode also facilitates the recognition of specific virus packaging signals in vRNA molecules during virus assembly. While these findings have shed some light on the structure and properties of RNP, many important questions remain unanswered. For example, how is NP kept in a soluble form that is competent for RNA binding prior to RNP assembly, what are the molecular interactions involved in NP RNA binding, is the viral polymerase needed for the specific encapsidation of viral RNA, how is NP's RNA binding and other activities regulated by phosphorylation and other yet undiscovered mechanisms, does NP play an active role in the selective packaging of the eight vRNA segments, and what is NP's role in unprimed viral RNA replication? To address these questions, we expect that a broad spectrum of biochemical, structural, biophysical, and virological methods will be needed to study NP:NP, NP:RNA, NP:polymerase, and polymerase–RNA interactions in order to elucidate the detailed mechanisms of RNP assembly as well as NP-supported viral RNA synthesis.

Acknowledgements

We thank Aaron Collier for valuable discussions. Research in the authors' laboratory is supported by the National Institutes of Health, the Robert Welch Foundation, and the Hamill Foundation.

References

Akarsu, H., Burmeister, W.P., Petosa, C., Petit, I., Muller, C.W., Ruigrok, R.W., and Baudin, F. (2003). Crystal structure of the M1 protein-binding domain of the influenza A virus nuclear export protein (NEP/NS2). EMBO J. 22, 4646–4655.

Albertini, A.A., Wernimont, A.K., Muziol, T., Ravelli, R.B., Clapier, C.R., Schoehn, G., Weissenhorn, W., and Ruigrok, R.W. (2006). Crystal structure of the rabies virus nucleoprotein–RNA complex. Science *313*, 360–363.

Albo, C., Valencia, A., and Portela, A. (1995). Identification of an RNA binding region within the N-terminal third of the influenza A virus nucleoprotein. J. Virol. *69*, 3799–3806.

Arrese, M., and Portela, A. (1996). Serine 3 is critical for phosphorylation at the N-terminal end of the nucleoprotein of influenza virus A/Victoria/3/75. J. Virol. *70*, 3385–3391.

Avalos, R.T., Yu, Z., and Nayak, D.P. (1997). Association of influenza virus NP and M1 proteins with cellular cytoskeletal elements in influenza virus-infected cells. J. Virol. *71*, 2947–2958.

Barefoot, B.E., Sample, C.J., and Ramsburg, E.A. (2009). Recombinant vesicular stomatitis virus expressing influenza nucleoprotein induces CD8 T-cell responses that enhance antibody-mediated protection after lethal challenge with influenza virus. Clin. Vaccine. Immunol. *16*, 488–498.

Baudin, F., Bach, C., Cusack, S., and Ruigrok, R.W. (1994). Structure of influenza virus RNP. I. Influenza virus nucleoprotein melts secondary structure in panhandle RNA and exposes the bases to the solvent. EMBO J. *13*, 3158–3165.

Beaton, A.R., and Krug, R.M. (1986). Transcription antitermination during influenza viral template RNA synthesis requires the nucleocapsid protein and the absence of a 5′ capped end. Proc. Natl. Acad. Sci. U.S.A. *83*, 6282–6286.

Berkhoff, E.G. Geelhoed-Mieras, M.M., Fouchier, R.A., Osterhaus, A.D., and Rimmelzwaan, G.F. (2007). Assessment of the extent of variation in influenza A virus cytotoxic T-lymphocyte epitopes by using virus-specific CD8+ T-cell clones. J. Gen. Virol. *88*, 530–535.

Biswas, S.K., Boutz, P.L., and Nayak, D.P. (1998). Influenza virus nucleoprotein interacts with influenza virus polymerase proteins. J. Virol. *72*, 5493–5501.

Boon, A.C., de Mutsert, G., Fouchier, R.A., Osterhaus, A.D., and Rimmelzwaan, G.F. (2006). The hypervariable immunodominant NP418–426 epitope from the influenza A virus nucleoprotein is recognized by cytotoxic T lymphocytes with high functional avidity. J. Virol. *80*, 6024–6032.

Boulo, S., Akarsu, H., Ruigrok, R.W., and Baudin, F. (2007). Nuclear traffic of influenza virus proteins and ribonucleoprotein complexes. Virus. Res. *124*, 12–21.

Bui, M., Myers, J.E., and Whittaker, G.R. (2002). Nucleocytoplasmic localization of influenza virus nucleoprotein depends on cell density and phosphorylation. Virus. Res. *84*, 37–44.

Bullido, R., Gomez-Puertas, P., Albo, C., and Portela, A. (2000). Several protein regions contribute to determine the nuclear and cytoplasmic localization of the influenza A virus nucleoprotein. J. Gen. Virol. *81*, 135–142.

Coloma, R., Valpuesta, J. M., Arranz, R., Carrascosa, J. L., Ortin, J., and Martin-Benito, J. The structure of a biologically active influenza virus ribonucleoprotein complex. PLoS Pathogens In press (More details?)

Compans, R.W., Content, J., and Duesberg, P.H. (1972). Structure of the ribonucleoprotein of influenza virus. J. Virol. *10*, 795–800.

Cros, J.F., Garcia-Sastre, A., and Palese, P. (2005). An unconventional NLS is critical for the nuclear import of the influenza A virus nucleoprotein and ribonucleoprotein. Traffic *6*, 205–213.

Digard, P., Elton, D., Bishop, K., Medcalf, E., Weeds, A., and Pope, B. (1999). Modulation of nuclear localization of the influenza virus nucleoprotein through interaction with actin filaments. J. Virol. 73, 2222–2231.

Duesberg, P.H. (1969). Distinct subunits of the ribonucleoprotein of influenza virus. J. Mol. Biol. 42, 485–499.

Elton, D., Medcalf, E., Bishop, K., and Digard, P. (1999a). Oligomerization of the influenza virus nucleoprotein: identification of positive and negative sequence elements. Virology 260, 190–200.

Elton, D., Medcalf, L., Bishop, K., Harrison, D., and Digard, P. (1999b). Identification of amino acid residues of influenza virus nucleoprotein essential for RNA binding. J. Virol. 73, 7357–7367.

Elton, D., Simpson-Holley, M., Archer, K., Medcalf, L., Hallam, R., McCauley, J., and Digard, P. (2001). Interaction of the influenza virus nucleoprotein with the cellular CRM1-mediated nuclear export pathway. J. Virol. 75, 408–419.

Goldstein, E.A., and Pons, M.W. (1970). The effect of polyvinylsulfate on the ribonucleoprotein of influenza virus. Virology 41, 382–384.

Gorman, O.T., Bean, W.J., Kawaoka, Y., and Webster, R.G. (1990). Evolution of the nucleoprotein gene of influenza A virus. J. Virol. 64, 1487–1497.

Green, T.J., Zhang, X., Wertz, G.W., and Luo, M. (2006). Structure of the vesicular stomatitis virus nucleoprotein–RNA complex. Science 313, 357–360.

Howard, M., and Wertz, G. (1989). Vesicular stomatitis virus RNA replication: a role for the NS protein. J. Gen. Virol. 70, 2683–2694.

Husain, M., and Gupta, C.M. (1997). Interactions of viral matrix protein and nucleoprotein with the host cell cytoskeletal actin in influenza viral infection. Curr. Sci. 73, 40–47.

Kingsbury, D.W., Jones, I.M., and Murti, K.G. (1987). Assembly of influenza ribonucleoprotein *in vitro* using recombinant nucleoprotein. Virology 156, 396–403.

Kingsbury, D.W., and Webster, R.G. (1969). Some properties of influenza virus nucleocapsids. Journal of Virology 4, 219–225.

Kistner, O., Muller, K., and Scholtissek, C. (1989). Differential phosphorylation of the nucleoprotein of influenza A viruses. J. Gen. Virol. 70, 2421–2431.

Klumpp, K., Ruigrok, R.W., and Baudin, F. (1997). Roles of the influenza virus polymerase and nucleoprotein in forming a functional RNP structure. EMBO J. 16, 1248–1257.

Kobayashi, M., Toyoda, T., Adyshev, D.M., Azuma, Y., and Ishihama, A. (1994). Molecular dissection of influenza virus nucleoprotein: deletion mapping of the RNA binding domain. J. Virol. 68, 8433–8436.

Krug, R.M., Ueda, M., and Palese, P. (1975). Temperature-sensitive mutants of influenza WSN virus defective in virus-specific RNA synthesis. J. Virol. 16, 790–796.

Labadie, K., Dos Santos Afonso, E., Rameix-Welti, M.A., van der Werf, S., and Naffakh, N. (2007). Host-range determinants on the PB2 protein of influenza A viruses control the interaction between the viral polymerase and nucleoprotein in human cells. Virology 362, 271–282.

Lamb, R.A., and Krug, R.M. (2001). Orthomyxoviridae: the viruses and their replication. In Field's Virology, D.M. Knipe, and P.M. Howley, eds. (Lippincott-Raven), pp. 1487–1531.

Luo, M., Green, T.J., Zhang, X., Tsao, J., and Qiu, S. (2007). Structural comparisons of the nucleoprotein from three negative strand RNA virus families. Virol. J. 4, 72.

Ma, K., Roy, A.M., and Whittaker, G.R. (2001). Nuclear export of influenza virus ribonucleoproteins: identification of an export intermediate at the nuclear periphery. Virology 282, 215–220.

Martin, K., and Helenius, A. (1991). Nuclear transport of influenza virus ribonucleoproteins: the viral matrix protein (M1) promotes export and inhibits import. Cell 67, 117–130.

Martin-Benito, J., Area, E., Ortega, J., Llorca, O., Valpuesta, J.M., Carrascosa, J.L., and Ortin, J. (2001). Three-dimensional reconstruction of a recombinant influenza virus ribonucleoprotein particle. EMBO Rep 2, 313–317.

Medcalf, L., Poole, E., Elton, D., and Digard, P. (1999). Temperature-sensitive lesions in two influenza A viruses defective for replicative transcription disrupt RNA binding by the nucleoprotein. J. Virol. 73, 7349–7356.

Mena, I., Jambrina, E., Albo, C., Perales, B., Ortin, J., Arrese, M., Vallejo, D., and Portela, A. (1999). Mutational analysis of influenza A virus nucleoprotein: identification of mutations that affect RNA replication. J. Virol. 73, 1186–1194.

Momose, F., Basler, C.F., O'Neill, R.E., Iwamatsu, A., Palese, P., and Nagata, K. (2001). Cellular splicing factor RAF-2p48/NPI-5/BAT1/UAP56 interacts with the influenza virus nucleoprotein and enhances viral RNA synthesis. J. Virol. 75, 1899–1908.

Naffakh, N., Tomoiu, A., Rameix-Welti, M.A., and van der Werf, S. (2008). Host restriction of avian influenza viruses at the level of the ribonucleoproteins. Annu. Rev. Microbiol. 62, 403–424.

Neumann, G., Castrucci, M.R., and Kawaoka, Y. (1997). Nuclear import and export of influenza virus nucleoprotein. J. Virol. 71, 9690–9700.

Newcomb, L.L., Kuo, R.L., Ye, Q., Jiang, Y., Tao, Y.J., and Krug, R.M. (2008). Interaction of the influenza A virus nucleocapsid protein with the viral RNA polymerase potentiates unprimed viral RNA replication. J. Virol. (More details?)

Ng, A.K., Zhang, H., Tan, K., Li, Z., Liu, J.H., Chan, P.K., Li, S.M., Chan, W.Y., Au, S.W., Joachimiak, A., *et al.* (2008). Structure of the influenza virus A H5N1 nucleoprotein: implications for RNA binding, oligomerization, and vaccine design. FASEB J. 22, 3638–3647.

Noda, T., Sagara, H., Yen, A., Takada, A., Kida, H., Cheng, R.H., and Kawaoka, Y. (2006). Architecture of ribonucleoprotein complexes in influenza A virus particles. Nature 439, 490–492.

Ohba, K., Yoshida, S., Zahidunnabi Dewan, M., Shimura, H., Sakamaki, N., Takeshita, F., Yamamoto, N., and Okuda, K. (2007). Mutant influenza A virus nucleoprotein is preferentially localized in the cytoplasm and its immunization in mice shows higher immunogenicity and cross-reactivity. Vaccine *25*, 4291–4300.

O'Neill, R.E., Jaskunas, R., Blobel, G., Palese, P., and Moroianu, J. (1995). Nuclear import of influenza virus RNA can be mediated by viral nucleoprotein and transport factors required for protein import. J. Biol. Chem. *270*, 22701–22704.

Ortega, J., Martin-Benito, J., Zurcher, T., Valpuesta, J.M., Carrascosa, J.L., and Ortin, J. (2000). Ultrastructural and functional analyses of recombinant influenza virus ribonucleoproteins suggest dimerization of nucleoprotein during virus amplification. J. Virol. *74*, 156–163.

Osborne, J.C., and Elliott, R.M. (2000). RNA binding properties of bunyamwera virus nucleocapsid protein and selective binding to an element in the 5′ terminus of the negative-sense S segment. J. Virol. *74*, 9946–9952.

Plotch, S.J., Bouloy, M., and Krug, R.M. (1979). Transfer of 5′-terminal cap of globin mRNA to influenza viral complementary RNA during transcription *in vitro*. Proc. Natl. Acad. Sci. U.S.A. *76*, 1618–1622.

Pons, M.W., Schulze, I.T., Hirst, G.K., and Hauser, R. (1969). Isolation and characterization of the ribonucleoprotein of influenza virus. Virology *39*, 250–259.

Portela, A., and Digard, P. (2002). The influenza virus nucleoprotein: a multifunctional RNA-binding protein pivotal to virus replication. J. Gen. Virol. *83*, 723–734.

Prokudina, E.N., Semenova, N., Chumakov, V., and Stitz, L. (2008). An antigenic epitope of influenza virus nucleoprotein (NP) associated with polymeric forms of NP. Virol. J. *5*, 37.

Prokudina, E.N., Semenova, N.P., and Chumakov, V.M. (2005). Stability of intracellular influenza virus nucleocapsid protein oligomers. Arch. Virol. *150*, 833–839.

Prokudina, E.N., Semenova, N.P., Rudneva, I.A., Chumakov, V.M., and Yamnikova, S.S. (2001). Avian and human influenza a virus strains possess different intracellular nucleoprotein oligomerization efficiency. Acta. Virol. *45*, 201–207.

Rihs, H.P., Jans, D.A., Fan, H., and Peters, R. (1991). The rate of nuclear cytoplasmic protein transport is determined by the casein kinase II site flanking the nuclear localization sequence of the SV40 T-antigen. EMBO J. *10*, 633–639.

Roose, K., Fiers, W., and Saelens, X. (2009). Pandemic preparedness: Toward a universal influenza vaccine. Drug. News. Perspect. *22*, 80–92.

Rudolph, M.G., Kraus, I., Dickmanns, A., Eickmann, M., Garten, W., and Ficner, R. (2003). Crystal structure of the borna disease virus nucleoprotein. Structure *11*, 1219–1226.

Ruigrok, R.W., and Baudin, F. (1995). Structure of influenza virus ribonucleoprotein particles. II. Purified RNA-free influenza virus ribonucleoprotein forms structures that are indistinguishable from the intact influenza virus ribonucleoprotein particles. J. Gen. Virol. 76 (Pt 4), 1009–1014.

Scholtissek, C., and Becht, H. (1971). Binding of ribonucleic acids to the RNP-antigen protein of influenza viruses. J. Gen. Virol. *10*, 11–16.

Severson, W.E., Xu, X., and Jonsson, C.B. (2001). cis-Acting signals in encapsidation of Hantaan virus S-segment viral genomic RNA by its N protein. J. Virol. *75*, 2646–2652.

Shapiro, G.I., and Krug, R.M. (1988). Influenza virus RNA replication *in vitro*: synthesis of viral template RNAs and virion RNAs in the absence of an added primer. J. Virol. *62*, 2285–2290.

Siegert, W., Bauer, G., and Hofschneider, P.H. (1973). Direct evidence for messenger activity of influenza virion RNA. Proc. Natl. Acad. Sci. U.S.A. *70*, 2960–2963.

St Angelo, C., Smith, G.E., Summers, M.D., and Krug, R.M. (1987). Two of the three influenza viral polymerase proteins expressed by using baculovirus vectors form a complex in insect cells. J. Virol. *61*, 361–365.

Tarendeau, F., Crepin, T., Guilligay, D., Ruigrok, R.W., Cusack, S., and Hart, D.J. (2008). Host determinant residue lysine 627 lies on the surface of a discrete, folded domain of influenza virus polymerase PB2 subunit. PLoS Pathog. *4*, e1000136.

Toriumi, H., and Kawai, A. (2004). Association of rabies virus nominal phosphoprotein (P) with viral nucleocapsid (NC) is enhanced by phosphorylation of the viral nucleoprotein (N). Microbiol. Immunol. *48*, 399–409.

Vreede, F.T., and Brownlee, G.G. (2007). Influenza virion-derived viral ribonucleoproteins synthesize both mRNA and cRNA *in vitro*. J. Virol. *81*, 2196–2204.

Vreede, F.T., Jung, T.E., and Brownlee, G.G. (2004). Model suggesting that replication of influenza virus is regulated by stabilization of replicative intermediates. J. Virol. *78*, 9568–9572.

Wang, P., Palese, P., and O'Neill, R.E. (1997). The NPI-1/NPI-3 (karyopherin alpha) binding site on the influenza a virus nucleoprotein NP is a nonconventional nuclear localization signal. J. Virol. *71*, 1850–1856.

Watanabe, K., Takizawa, N., Katoh, M., Hoshida, K., Kobayashi, N., and Nagata, K. (2001). Inhibition of nuclear export of ribonucleoprotein complexes of influenza virus by leptomycin B. Virus. Res. *77*, 31–42.

Weber, F., Kochs, G., Gruber, S., and Haller, O. (1998). A classical bipartite nuclear localization signal on Thogoto and influenza A virus nucleoproteins. Virology *250*, 9–18.

Whittaker, G., Bui, M., and Helenius, A. (1996). Nuclear trafficking of influenza virus ribonuleoproteins in heterokaryons. J. Virol. *70*, 2743–2756.

Wu, W.W., Sun, Y.H., and Pante, N. (2007a). Nuclear import of influenza A viral ribonucleoprotein complexes is mediated by two nuclear localization sequences on viral nucleoprotein. Virol. J. *4*, 49.

Wu, W.W., Weaver, L.L., and Pante, N. (2007b). Ultrastructural analysis of the nuclear localization sequences on influenza A ribonucleoprotein complexes. J. Mol. Biol. *374*, 910–916.

Yamanaka, K., Ishihama, A., and Nagata, K. (1990). Reconstitution of influenza virus RNA–nucleoprotein complexes structurally resembling native viral ribonucleoprotein cores. J. Biol. Chem. *265*, 11151–11155.

Yang, J., Koprowski, H., Dietzschold, B., and Fu, Z.F. (1999). Phosphorylation of rabies virus nucleoprotein regulates viral RNA transcription and replication by modulating leader RNA encapsidation. J. Virol. 73, 1661–1664.

Ye, Q., Krug, R.M., and Tao, Y.J. (2006). The mechanism by which influenza A virus nucleoprotein forms oligomers and binds RNA. Nature *444*, 1078–1082.

Influenza A Virus Haemagglutinin Glycoproteins

5

David A. Steinhauer

Abstract

The influenza A virus haemagglutinin glycoprotein (HA) is the principal mediator of viral entry into host cells. It is responsible for attachment of virions to sialic acid-containing receptors on the host cell surface, and for inducing membrane fusion between viral envelopes and cellular endosomal membranes following endocytosis. HA serves a classic example of a type I membrane glycoprotein, with a cleaved N-terminal signal sequence, a membrane anchor domain near the C-terminus, and post-translational modifications resulting from the addition of N-linked oligosaccharide side chains to the ectodomain, and acylation of cysteine residues in the cytoplasmic tail region. HA spikes on the viral surface are also the major target for neutralizing antibodies, and as such, the antigenic properties of the HA are of fundamental significance for the design of influenza vaccines. The depth of knowledge relating to high-resolution atomic structures of HA in various forms have made it a prototype for the investigation of viral glycoproteins in general, and this chapter outlines some of the special features that have derived from structure-function studies on HA.

Introduction

The influenza A virus haemagglutinin glycoprotein (HA) is one of the most well characterized membrane proteins in all of biology. It is the surface antigen of influenza A virus particles that is responsible for the receptor binding and membrane fusion functions required for virus entry into host cells, and it also serves as the principal target for neutralizing antibody responses. The significance of the functions of HA for the biology of influenza A viruses, and our capacity to control them, is evident on many levels. The binding characteristics of HA for cell surface receptors at sites of infection in avian and mammalian hosts act as a contributing factor for cross-species transmission and the emergence of viruses with pandemic potential. The properties associated with proteolytic activation of HA membrane fusion activity serve as a major determinant for virus pathogenicity, and the structural rearrangements of HA that drive the membrane fusion process provide an ideal target for the design of antiviral drugs. In addition, the antigenic properties of HA and the mechanisms by which neutralizing antibodies interact with it, make HA the viral component of primary concern for vaccine considerations. Therefore, structural and functional studies on HA have been central to our understanding of the molecular mechanisms by which influenza A viruses infect cells and transmit among hosts, and have provided insights on the interplay between receptor binding, virus neutralization, and antigenic drift. An expanded comprehension of the biological properties of HA will be essential for efforts to design and implement improved vaccines, and for the development of novel viral intervention strategies to combat the continuing threat posed by these unpredictable pathogens.

Influenza A viruses

The viruses that cause human influenza belong to three genera of the family *Orthomyxoviridae*: influenza A, influenza B, and influenza C. During most years, influenza A viruses cause a majority of the clinical cases of human influenza. Unlike influenza B and influenza C viruses, which are predominantly restricted to human hosts, avian species serve as the natural reservoir for influenza A viruses. The lipid envelope of influenza A viruses contain two species of glycoprotein spikes that serve as the major antigens, the HA, and the viral neuraminidase (NA), which is the receptor destroying enzyme that functions in the release and dissemination of virus particles at the end of the virus life cycle. Influenza A viruses are commonly typed based on the antigenic properties of their HA and NA antigens. For example, HA glycoproteins that cross-react to polyclonal antisera are classified within a given influenza A virus subtype, and viruses are typed by antigenic relatedness to standard reference strains. In wild waterfowl, 16 antigenically distinct subtypes of HA and nine subtypes of NA are known to exist. At present, direct sequence data are available only for influenza A viruses that have infected humans since 1918, and since then, only viruses of the H1N1, H2N2, and H3N2 antigenic subtypes have become established in humans and circulated widely. For several years prior to 1918, serological studies suggest that H3 subtypes constituted the major source of influenza A disease in humans (Dowdle, 1999). However, in 1918 H1N1 viruses that had emerged initially from avian sources began to spread rapidly in humans and caused a devastating pandemic that resulted in tens of millions of deaths over a two-year period (Ahmed *et al.*, 2007; Taubenberger *et al.*, 2005). These H1N1 viruses continued to circulate in humans until 1957 when they were displaced by H2N2 subtype viruses, which caused another pandemic, albeit not as severe as that of 1918 and 1919. The H2N2 viruses were, in turn, displaced in 1968 by H3N2 viruses. In 1977 H1N1 subtypes reappeared in humans without displacing the H3N2 strains, and both H3N2 and H1N1 viruses continue to circulate to this day. Between pandemics, the prevailing viruses of a given subtype often accumulate mutations from one year to the next in a process referred to as antigenic drift. This allows influenza A viruses of the same subtype to re-infect individuals, and for this reason, seasonal influenza causes problems on a nearly annual basis. Even these seasonal epidemics of influenza A are quite serious both medically and economically, and can cause on the order of hundreds of thousands of deaths worldwide in an average year.

The influenza A genome is composed of eight distinct segments of negative sense RNA, and both the HA and the NA are encoded by individual gene segments. Human pandemics, such as those exemplified by the 1918, 1957, and 1968 outbreaks, can result from the direct introduction of avian viruses into humans, or from the acquisition by human viruses of avian gene segments encoding HA (and sometimes NA) following genetic reassortment. Whereas the 1918 pandemic appears to have been caused by an entirely avian H1N1 subtype virus that developed the capacity to transmit in humans (Taubenberger *et al.*, 2005), the 1957 and 1968 pandemics were caused by reassortant viruses. In 1957, the circulating human viruses acquired H2 subtype HA, N2 subtype NA, and PB1 gene segments from avian sources, and in 1968 an avian H3 HA gene and yet another avian PB1 gene replaced those of the circulating human H2N2 viruses (Kawaoka *et al.*, 1989; Scholtissek *et al.*, 1978). For all three pandemics of the past century, it is possible that the pig may have played a role as an intermediate host or 'mixing vessel' for viral strains prior to emergence in humans (Scholtissek, 1994; Scholtissek *et al.*, 1985). The emergence in humans of influenza A viruses with antigenically novel surface antigens is referred to as antigenic shift. The mechanisms by which such avian viruses or avian/mammalian virus reassortants develop the capacity to emerge and efficiently transmit among humans is not well understood, but it is likely that properties of the viral HA, particularly those related to receptor binding and antigenicity, feature significantly in the emergence of pandemic viruses.

Overview of HA structures

Influenza viruses are pleomorphic in size and morphology. Many laboratory strains are fairly spherical in shape with a diameter of approximately 100 nm, particularly if cultured in embryonated chicken eggs, but clinical isolates tend to be more filamentous (Choppin *et al.*, 1960; Chu *et al.*, 1949; Mosley and Wyckoff, 1946). Regardless of morphology, the surface of influenza A virus particles is densely packed with HA and NA spikes. It has been estimated that spherical laboratory strains of influenza contain approximately 500 HA and 100 NA spikes per virus particle (Ruigrok *et al.*, 1984; Schulze, 1972; Tiffany and Blough, 1970), although these numbers and the relative ratios of HA to NA can vary among different virus strains, subtypes, and possibly even conditions of replication. The HA spikes are trimeric molecules that are roughly conical in shape, while the NA molecules are tetrameric oligomers that contain a thin stem structure and a distal knob region giving it a mushroom-like appearance. Fig. 5.1 shows an electron micrograph (EM) of a spherical influenza A virion demonstrating the distribution of glycoprotein spikes on the surface of the viral envelope, and an example of individual HA and NA spikes are indicated.

The HA of A/Aichi/2/68 virus, a prototype of the H3N2 subtype viruses, has been the most extensively studied and this 'Aichi HA' will serve as the 'reference HA' for most examples cited in this chapter. The primary structure of the Aichi HA is illustrated in Fig. 5.2. HA is a classical type I membrane glycoprotein with an N-terminal signal sequence that is cleaved in the ER, a hydrophobic transmembrane anchor domain near its C-terminus, and a short cytoplasmic tail sequence. HA folds and forms non-covalently

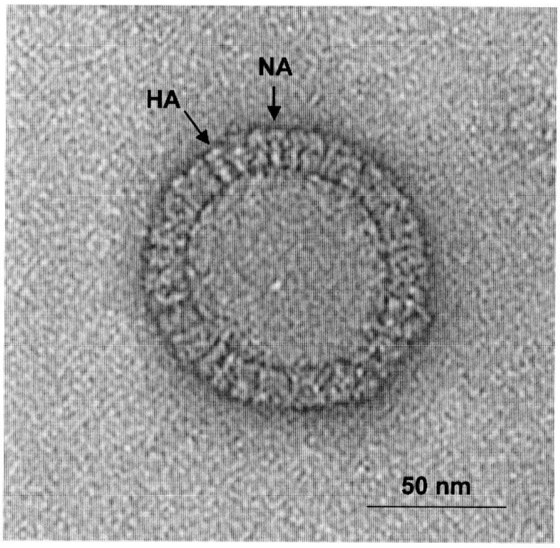

Figure 5.1 Negative stain electron micrograph of a spherical influenza A virus particle. The densely packed glycoprotein spikes can be observed protruding from the viral membrane, and an example of a haemagglutinin (HA) and neuraminidase (NA) are indicated.

associated homotrimers in the ER. It is post-translationally modified by the addition of acyl chains at three cysteine residues in the cytoplasmic tail and its transmembrane interface (Kordyukova *et al.*, 2008; Schmidt, 1982), and is N-glycosylated at seven asparagine residues in the ectodomain (Ward and Dopheide, 1981; Wilson *et al.*, 1981). The mature trimeric ectodomain extends by over 130 angstroms from the viral membrane and varies in radius from between 15 to 40 angstroms along its length. The stem of the molecule is fibrous in nature and features a long central triple-stranded coiled coil formed by helices from each monomer. The membrane distal head domains are globular in structure, and contain the receptor binding sites and major antigenic regions (Wiley *et al.*, 1981; Wilson *et al.*, 1981).

The HA takes on three separate conformations during the virus life cycle and these will be comprehensively examined here (Bizebard *et al.*, 1995; Bullough *et al.*, 1994; Chen *et al.*, 1998a,1999; Wilson *et al.*, 1981). These conformations include the precursor polypeptide structure, HA0, the neutral pH structure that HA assumes following proteolytic cleavage of HA0, and the low pH structure that the HA folds into as a result of the conformational changes that occur during membrane fusion. These three structures are illustrated from a lateral perspective in Fig. 5.3. When viewed down the three-fold axis of symmetry, the trimeric nature of the HA structure is revealed more distinctly, as shown in Fig. 5.4 for the precursor HA0 and cleaved HA molecules.

Among the notable features of the HA0 precursor are the three surface loop structures (one from each monomer) in the membrane-proximal half of the stem region of the molecule (Chen *et al.*, 1998a). Proteolytic cleavage within these loop domains is required for virus infectivity as it allows the HA to assume a conformation that can subsequently be triggered to mediate membrane fusion (Appleyard and Maber, 1974; Klenk *et al.*, 1975; Lazarowitz and Choppin, 1975). The left panels of Figs. 5.3 and 5.4 show the alternative views of HA0 and indicate the location of the cleavage site from each perspective. The proteolytic activation of each HA0 polypeptide generates the HA1 (328 residues) and HA2 (221 residues) subunits, which are covalently linked by a single disulphide bond between the conserved

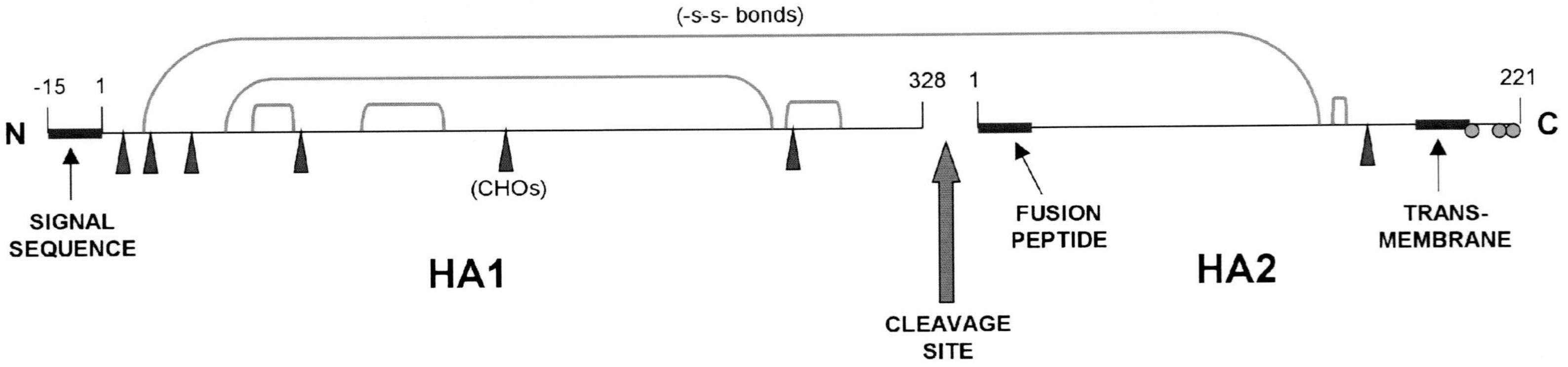

Figure 5.2 Primary structure representing the Aichi HA following cleavage into the HA1 and HA2 subunits. The signal sequence at the N-terminus is not actually a component of the cleaved HA, as it is removed soon after HA synthesis in the ER. The mature HA contains 328 residues in the HA1 subunit and 221 residues in the HA2 subunit. The arginine at position 329 of HA0 that is recognized by activating proteases is removed following cleavage into HA1 and HA2 by carboxypeptidase activity (Garten and Klenk, 1983). The red lines indicate hydrophobic domains, which include the fusion peptide (HA2 residues 1–23) and the transmembrane domain (HA2 residues 185–210). The locations of carbohydrate attachments sites (CHOs) are indicated by arrows pointing up from under the linear representation of the polypeptide chains (HA1 positions 8, 22, 38, 81, 165, and 285, and HA2 position 154). The blue brackets above the polypeptide represent the six disulphide linkages (HA1 residues 52–277, 64–76, 97–139, 281–305; HA2 residues 144–148; and HA1 14-HA2 137). The green circles near the C-terminal end of the HA2 polypeptide chain represent the acylation sites at cysteine residues at HA2 positions 210, 217, and 220.

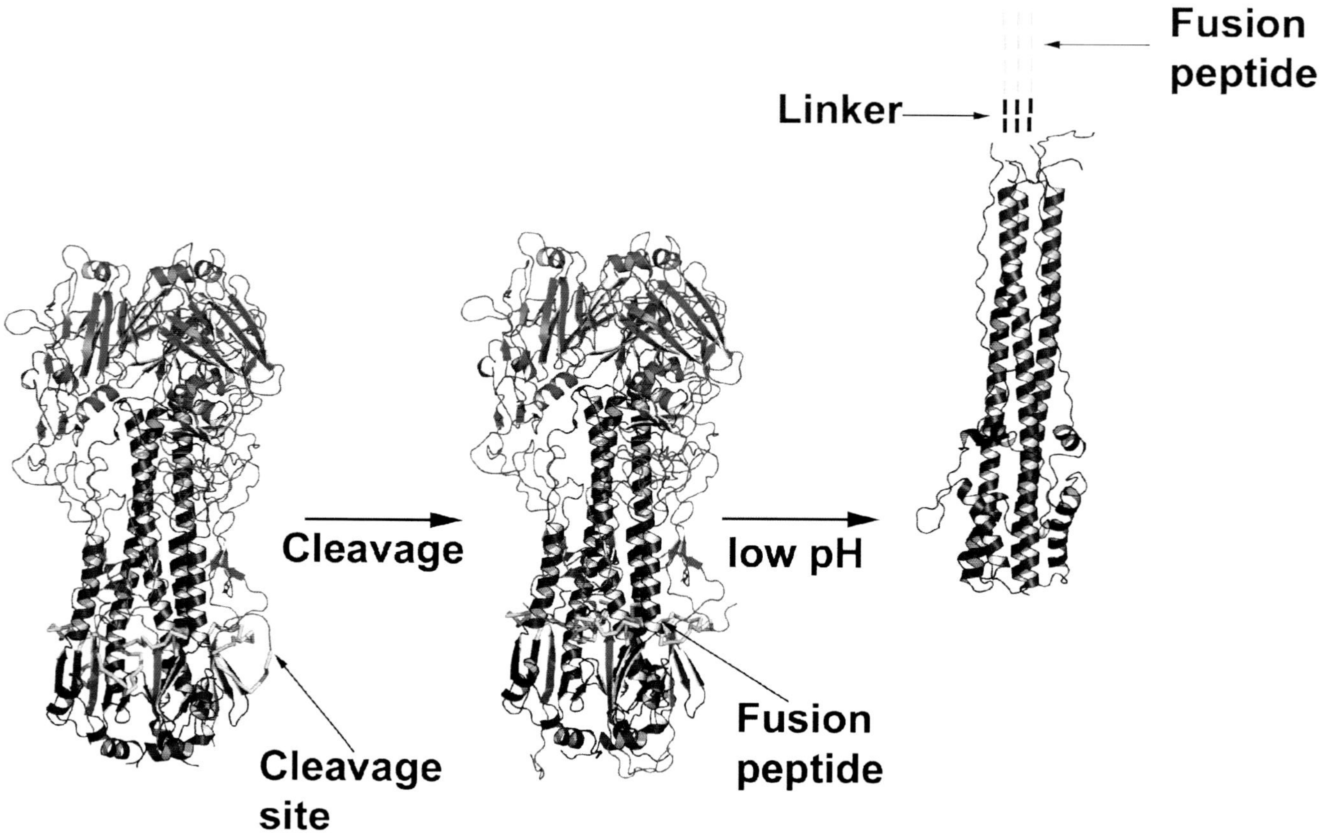

Figure 5.3 Ribbon diagrams of the three conformations of the HA trimer. The HA0 precursor structure is shown on the left. Residues that will constitute the HA1 subunit following cleavage are depicted in blue, and the HA2 subunits in red. In all three structures, the residues that ultimately form the fusion peptide at the N-terminus of HA2 are shown in yellow. In the left panel, the cleavage loop of one of the HA0 monomers can be seen extending from the trimer surface and the cleavage site is indicated. The centre panel shows that most of the HA structure remains unchanged following cleavage, the major structural consequence being the relocation of N-terminal HA2 'fusion peptide' residues from the bottom of the cleavage loop to the interior of the trimer (yellow). The right panel shows the structure of the HA2 trimer following the acid-induced conformational changes required for membrane fusion, which are described in detail in the text and in Fig. 5.7. In this thermostable rod-like structure, the fusion peptide and the viral transmembrane domain are located at the same end. Dashed lines indicate that the structure is unknown for both the fusion peptide and the 10-residue peptide that links it to HA2 residues of known structure.

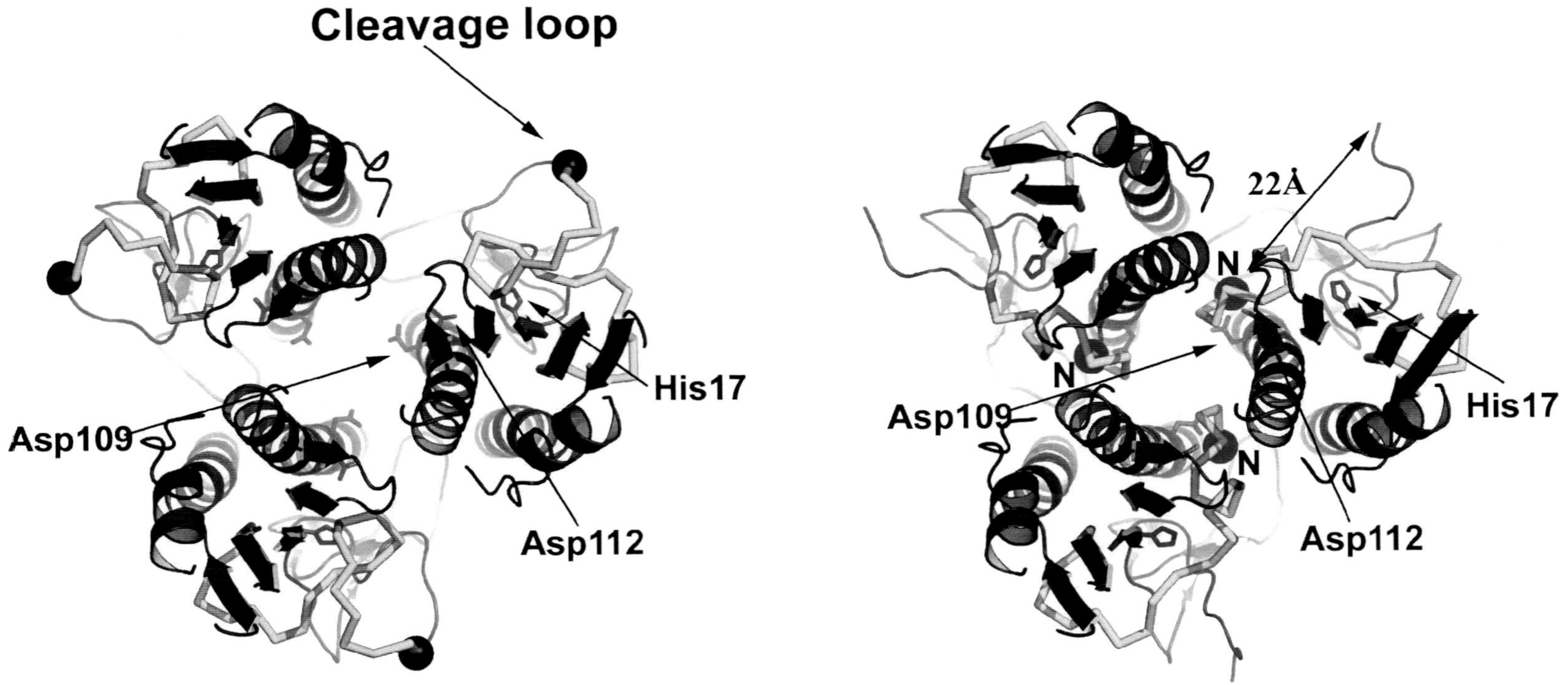

Figure 5.4 Structure of HA0 and cleaved neutral pH HA viewed down the three-fold axis of symmetry. Colours are as described for Figure 3 and the site of protease cleavage in the loop is indicated. This view highlights more clearly the relocation of fusion peptide residues at the N-terminus of HA2 (yellow) from the surface loop of uncleaved HA0 to the interior of the trimer, whereupon the N-terminus of HA2 and the C-terminus of HA1 are separated by 22Å. The location in HA0 of the glycine residues that will constitute the HA2 N-termini following cleavage are shown as a filled black circles in the left panel, and as filled grey circles labelled N in the cleaved HA structure on the right. Also depicted are the locations of ionizable residues HA1 His 17 and HA2 Asp 109 and $Asp11_2$, which are buried by fusion peptide residues following cleavage.

cysteine residues at HA1 position 14 and HA2 position 137. The structure of the HA0 precursor and that of cleaved HA are very similar, with only six residues at the C-terminus of the newly formed HA1 subunit and 12 residues at the N-terminus of HA2 relocating as a result (Chen *et al.*, 1998a). However, the structural changes carry particular significance for activating membrane fusion potential. Specifically, the highly conserved N-terminus of HA2, which occupies the membrane-proximal half of the cleavage loop in the HA0 structure, is repositioned to the interior of the trimer to fill a cavity that is present in HA0. The contacts formed between the HA2 N-terminal domain and ionizable residues in the cavity are thought to prime the HA for fusion, as protonation of residues in this region may play a role in triggering the irreversible conformational changes that lead to the fusion process. During virus entry, this occurs due to the acidification of endosomes, where HA structural rearrangements result in the formation of a thermostable rod-like structure that draws the viral and endosomal membranes together as shown in the right panel of Fig. 5.3. The functional significance of each of these three structures for receptor binding, membrane fusion, pathogenicity, and antigenicity are discussed in detail below.

Receptor binding

In general, avian influenza A viruses do not replicate well in humans, and vice versa (Beare and Webster, 1991; Hinshaw *et al.*, 1981; Hinshaw *et al.*, 1983). However, it is known that avian viruses of various subtypes can transmit to humans (Bender *et al.*, 1999; Claas *et al.*, 1998; Fouchier *et al.*, 2004; Lin *et al.*, 2000; Lipatov *et al.*, 2004; Peiris *et al.*, 1999, 2004; Shortridge *et al.*, 1998; Suarez *et al.*, 1998; Subbarao *et al.*, 1998), and should such an event lead to the establishment of a new subtype lineage in the human population the consequences could be catastrophic. Among the factors that can contribute to the capacity to replicate and transmit efficiently in a new host are the binding properties of the HA, and the types of potential receptor molecules present on the surfaces of host cells. Influenza A viruses attach to host cells due to interactions between the HA and glycans on the host cell surface that contain terminal sialic acids (Gottschalk and Lind, 1949; Klenk *et al.*, 1955). Sialic acids are N- or O-substituted derivatives of neuraminic acid, a monosaccharide containing a nine-carbon backbone. These are present at the termini of oligosaccharides attached to cell surface glycoproteins and glycolipids that are widely distributed in host tissues. The type of linkage by which sialic acid is attached to the penultimate sugar of influenza virus receptors is regarded as one of the fundamental determinants of species specificity, with avian viruses normally demonstrating a preference for receptors containing α2,3-linked sialic acid and human viruses generally favouring receptors with α2,6 linkages (Connor *et al.*, 1994; Rogers *et al.*, 1983a, 1985).

A number of studies have addressed the types of sialic acid linkage on glycans of various cells and tissues, primarily using techniques based on lectin immunohistochemistry. These reports show that to an extent, the distribution and density of receptors of a specific linkage type at sites of replication correlates with species specificity of influenza viruses. For example, a preponderance of α2,3-linked sialosides exist in the epithelial cells of the intestinal tract of birds, where avian influenza virus replication generally takes place (Ito *et al.*, 1998, 2000). The upper respiratory tract of pigs contains both α2,3- and α2,6-linked receptors to allow for efficient replication of both avian and human influenza strains (Ito *et al.*, 1998). In humans, most reports suggest that cells of the upper respiratory tract contain predominantly α2,6 linkages, whereas both α2,3- and α2,6-linked receptors can be detected in the lower respiratory tract, depending on cell type (Ibricevic *et al.*, 2006; Kogure *et al.*, 2006; Matrosovich *et al.*, 2004; Nicholls *et al.*, 2007; Shinya *et al.*, 2006; van Riel *et al.*, 2006). This may explain why although human viruses generally have α2,6 linkage specificity, avian viruses have been known to replicate and cause disease in humans on occasion. It is reasonable to speculate that such α2,3-specific viruses will transmit poorly to human contacts if they do not acquire mutations that allow for efficient replication in the upper respiratory tract to aid in transmission via respiratory droplets. Studies using the ferret model with receptor specificity mutants of 1918 H1N1 viruses support this hypothesis (Tumpey *et al.*, 2007). However, the situation regarding the distribution of functional receptors and the tissue specificity for infection by particular viruses is not clear. For example, recent studies suggest that human nasopharyngeal, adenoid, and tonsillar tissues may be permissive for infection by avian H5N1 viruses with apparent α2,3-specificity (Nicholls *et al.*, 2007). It has also been proposed that productive binding by influenza might involve the use of high-affinity second receptors following initial attachment to sialic acid-containing glycans (Rapoport *et al.*, 2006; Stray *et al.*, 2000). It appears that the complexities surrounding the identification and natural distribution of functional influenza receptors may be underappreciated (Nicholls *et al.*, 2008), and that while useful, determinations based solely on lectin binding may not provide a definitive representation of this.

When the X-ray crystal structure of the bromelain-released soluble ectodomain of HA (BHA) was first determined (Wilson *et al.*, 1981), a shallow cavity of conserved amino acids at the membrane distal tip of each monomer was postulated as the receptor binding site. The observation that mutations in this region lead to changes in receptor binding properties supported this hypothesis (Rogers *et al.*, 1983a), and numerous aspects of HA–ligand interactions have since been revealed based on the X-ray crystal structures of HA complexes with receptor analogues (Eisen *et al.*, 1997; Gamblin *et al.*, 2004; Ha *et al.*, 2001; Russell *et al.*, 2004; Sauter *et al.*, 1992; Watowich *et al.*, 1994; Weis *et al.*, 1988). The data now encompass representative HAs of several of the 16 antigenic subtypes (H1, H3, H5, H7, and H9) bound to a variety of receptor analogues of different length and linkage specificity. In all HA–receptor complexes the location of sialic acid in the receptor binding site is virtually superimposable, with the positioning of sialic acid relative to conserved residues with which it makes contacts varying by only about 1Å in spatial location and about 10° in tilt angle. For several of the complexes as many as five sugars of the receptor molecule can be observed in the electron density map. Notably, the structural configuration of the internal asialo receptor constituents varies depending on the receptor chemistry and the subtype and/or strain of the virus.

The location of the HA receptor binding sites on the trimer and a magnified view of the site for one monomer are shown for Aichi HA in Fig. 5.5. Viewed from a lateral perspective as depicted in the right panel of the figure, the sialic acid binding domain is a shallow depression bordered by a short α-helix, referred to as the 190 helix, at the membrane-distal edge. Although not depicted in this figure for reasons of clarity, the side chains of HA1 residues E190 and L194 in the 190 helix are oriented to protrude down into the pocket where they make contacts with sialic acid. The 130 loop frames the front of the site, and residues G135, S136, and N137 form hydrogen bonds with sialic acid. The left side of the pocket is composed of a chain of amino acids referred to as the 220 loop. This loop contains residues L226 and S228, which have been shown to play a role in defining receptor specificity for various strains and subtypes of influenza A. A hydrogen bonded network of non-contiguous amino acids, Y98, W153, H183, and Y195 constitute the base of the site, and all except Y195 also contact sialic acid. Several of the residues mentioned here and highlighted for Aichi HA in Fig. 5.5 are highly conserved, or display restricted variability among the different HA subtypes, suggesting that the contacts that they form with sialic acid and/or one another are relevant for receptor binding of influenza A viruses in general.

In the HA–receptor complexes that have been determined to date, the saccharides with α2,3 sialic acid linkage adopt extended configurations that exit the binding pocket from the left side, whereas those containing α2,6 linkages are more diverse in structure. The bound α2,6-linkaged receptor analogues assume folded conformations in which the third sugar resides above the sialic acid ring structure and the adjacent sugars exit the pocket either from the top of the binding site or from its right side depending on the HA it is bound to (Russell *et al.*, 2006). Fig. 5.6 illustrates examples of this using the structures of HA–receptor complexes using variants of Aichi HA with different specificity. In each panel the nature of the sialic acid glycosidic linkage to galactose (α2,3 or α2,6) and their locations are indicated. The 'WT' HA with human-like receptor specificity has leucine at position 226, and binds preferentially to the α2,6-linked pentasaccharide in a folded conformation. On the other hand, the HA with glutamine at position 226 has avian-like specificity and favours α2,3-linked receptors in

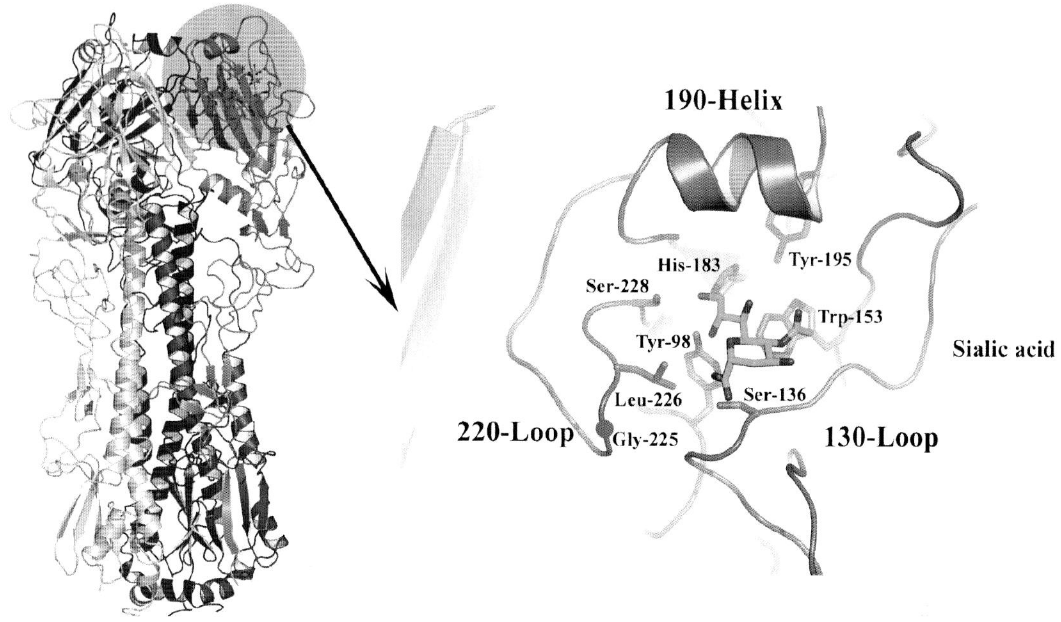

Figure 5.5 Left: Ribbon diagram of the HA trimer. One of the three monomers is coloured to show the HA1 subunit in blue and the HA2 subunit in red. The region encompassing the receptor-binding domain for the coloured monomer is shaded and the location of bound sialic acid is indicated in green. Right: Enlarged view of the receptor-binding region. The binding site is a shallow depression bordered by a short α-helix (the 190 helix) at the membrane-distal edge, the 130 loop at the front of the site, and the 220 loop at the left side. Conserved residues Y98, W153, H183, and Y195 form the base of the site, and the positions of some of the other residues discussed in the text are indicated. Sialic acid is shown in green.

extended conformations. As the structural and biological data continue to accumulate it becomes increasingly clear that many of the more variable amino acids that surround the binding pocket have the potential to impact receptor binding properties.

Since the earliest descriptions of changes in erythrocyte agglutination properties upon passage of human influenza viruses in eggs (Burnett and Bull, 1943), scores of reports have detailed various types of adaptive changes associated with HA receptor binding properties. The classic example of this involves the variants of H3 subtype human viruses that differ at position 226 as described above and illustrated in Fig. 5.6. The WT HA that binds specifically to receptors with α2,6-linked sialic acid was shown to alter its preference to α2,3-linked receptors when grown in the presence of horse serum, which contains an abundance of α2-macroglobulin, a glycoprotein rich in α2,6-linked sialic acids (Rogers and Paulson, 1983; Rogers *et al.*, 1983b). These altered binding properties were shown to result from a change of leucine to glutamine at HA1 position 226, which resides at the left side of the receptor binding site in the 220 loop. Subsequent studies on receptor binding specificity and sequence analysis of avian and human virus isolates showed that concurrent changes involving both positions 226 and 228 are also often associated with binding specificity (Connor *et al.*, 1994; Ito *et al.*, 2000; Naeve *et al.*, 1984; Vines *et al.*, 1998). Although there are numerous examples showing that residues 226 and 228 can contribute to receptor specificity in some HA subtypes, the accumulated data demonstrate that residues at a variety of different structural locations in the receptor site region can be linked to adaptive changes related to binding properties. These include observations on changes associated with adaptation of human clinical isolates for growth in eggs (Gambaryan *et al.*, 1997, 1999; Katz *et al.*, 1987; Robertson, 1993; Robertson *et al.*, 1987), variability among viruses isolated from different species (Connor *et al.*, 1994;

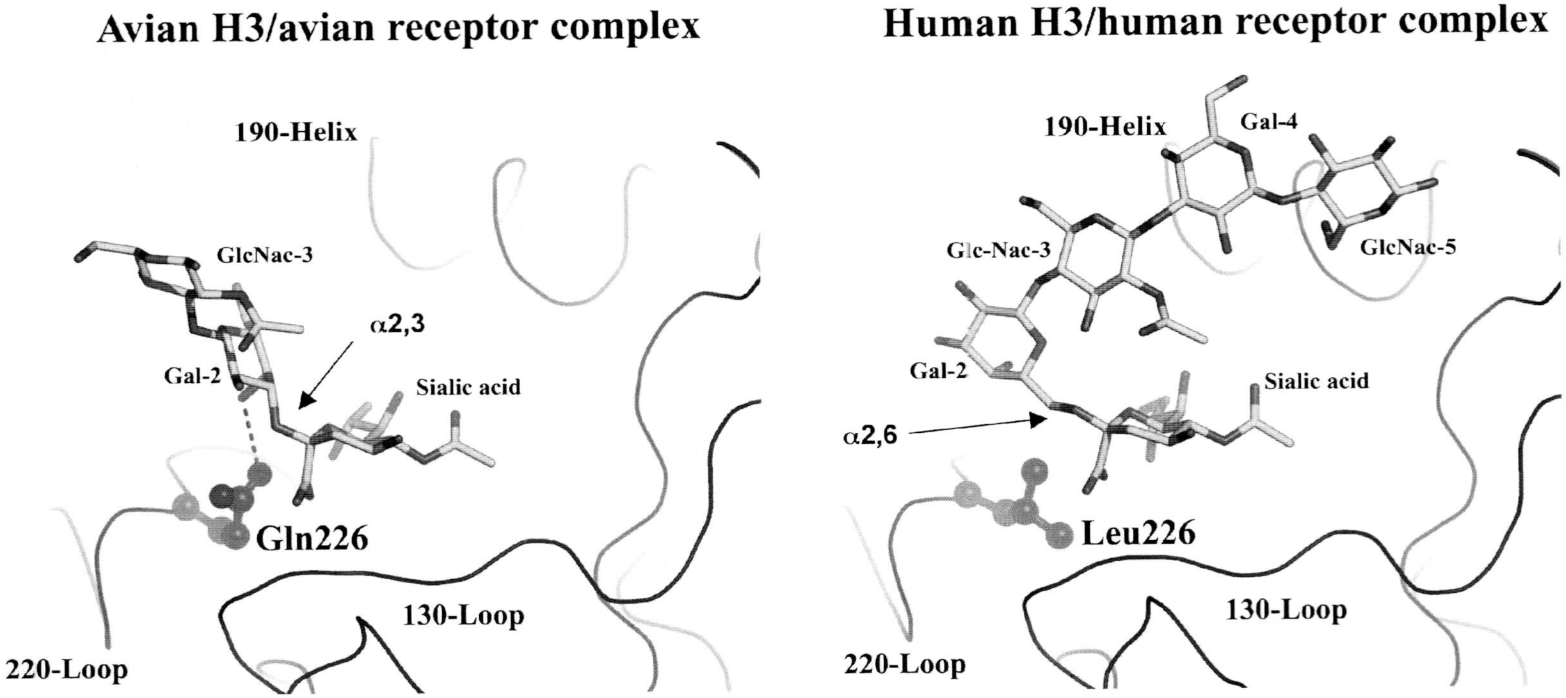

Figure 5.6 Diagrams of Aichi HA variants with different receptor specificity bound to their preferred receptor analogues. The left panel depicts an avian-like H3 subtype HA with glutamine at HA1 residue 226 bound to an α2,3-linked receptor analogue. The HA main chain traces of the 190 helix, 220 loop, and 130 loop are shown in red and the first three sugars of the receptor are shown in yellow as an extended conformation. The human-like HA with leucine at position 226 is shown on the right, bound to a α2,6-linked pentasaccharide receptor analogue in folded conformation. In this structure the fourth and fifth sugars of the pentasaccharide are visible, and are observed to exit the pocket to the 'upper right' side towards the 190 helix. In each structure the location (arrow) and type (α2,3 or α2,6) of glycosidic linkage between sialic acid and the penultimate galactose (Gal-2) of the receptor analogue are indicated.

Matrosovich *et al.*, 1998; Matrosovich *et al.*, 1999; Matrosovich *et al.*, 1997; Rogers and D'Souza, 1989; Rogers and Paulson, 1983), and changes associated with HAs in humans following the introduction into humans of pandemic viruses, as occurred in 1957 and 1968 (Matrosovich *et al.*, 2000; Ryan-Poirier *et al.*, 1998).

Studies of H1N1 viruses, including strains reconstructed from the 1918 influenza pandemic viruses, show that changes in receptor specificity can be correlated with amino acid substitutions at positions 225 and 190 (Matrosovich *et al.*, 2000; Tumpey *et al.*, 2007). These residues reside at the left side of the HA binding site in close proximity to 226 and 228. Binding studies that include analyses using glycan array technology suggest that a number of different residues may be involved with receptor specificity for pathogenic H5N1 isolates that have recently been circulating widely in avian species and causing sporadic disease in humans (Ilyushina *et al.*, 2008; Stevens *et al.*, 2008; Stevens *et al.*, 2006; Yamada *et al.*, 2006; Yang *et al.*, 2007). These studies have implicated 'familiar' residues such as 226 and 228, as well as residues 186, 193, 196, 248, and a glycosylation mutant at position 158. Residue 186 resides at the left side of the site, whereas the others are at or near the top and right hand sides where α2,6-linked receptors are thought to exit the pocket. The receptor binding data on recent H5N1 isolates is not entirely consistent, but this may reflect in part, the differences in the strains examined. What appears to be emerging from these studies is that binding specificity is a more complex phenomenon than might have been derived previously from studies on a limited number of HAs. Several residues in and around the binding region have the potential to contact receptors, which are in fact flexible and dynamic in their natural state. The binding site can be affected by oligosaccharides originating from the same polypeptide chain or from adjacent monomers. Therefore, residues at monomer–monomer interfaces that influence their spacing or orientation of head domains relative to one another also have the potential to affect binding properties (Meisner *et al.*, 2008).

It should be noted that the binding affinities of individual interactions between HA and monovalent sialosides is relatively weak compared to most receptor–ligand complexes, exhibiting dissociation constants (K_d) of approximately 2–3 mM as measured by proton-NMR (Hanson *et al.*, 1992; Sauter *et al.*, 1989). Utilizing the well-characterized receptor variants with alternative residues at position 226, these studies also demonstrated that the affinities of the HAs for monosialosides containing their preferred versus their non-preferred linkage type (α2,3 or α2,6) display only about 2-fold differences. This suggests that the biological consequences manifested from relatively minor differences in binding affinity result from the amplifying effects of multiple HA–receptor interactions. The interplay between receptor binding affinity and specificity, and the distribution of functional receptors at sites of natural infection in various potential host species, remains an area of intense research and is likely to be a significant determinant with regard to the emergence of novel pandemic viruses.

Structure of HA0 and priming of membrane fusion potential

Membrane fusion mediated by influenza requires two independent sets of conformational changes in the HA molecule. The first involves cleavage of the HA0 precursor structure into HA1 and HA2, which allows for the formation of the fusion competent neutral pH structure. The second occurs following receptor binding and internalization of viruses, due to the acidification of the endosomal environment. Although cleavage of HA0 and priming of fusion potential involves relatively minor structural changes, the generation and relocation of the N-terminal domain of the HA2 subunit are critical for membrane fusion. The membrane proximal (bottom) half of the HA0 cleavage loop becomes the N-terminus of the HA2 subunit (Figs. 5.3 and 5.4). This N-terminal polypeptide is relatively hydrophobic, and during the fusion process this 'fusion peptide' domain is relocated to interact with the endosomal membrane, which is a critical step in the fusion process. The fusion peptide domain is highly conserved among both influenza A viruses and influenza B viruses. Presumably this conservation is due partly to its role in functionally interacting with target membranes in a fashion that leads to membrane destabilization and fusion, and this is discussed in detail below. However, in addition to the requirement for functional association with target membranes, residues of the fusion peptide are also critical for activating the fusion potential of neutral pH HA.

In the HA0 structure a cavity lined with ionizable residues lies adjacent to the cleavage loop. Following proteolytic activation, the highly conserved N-terminal domain of HA2 inserts into the trimer interior and buries these ionizable residues, which are also either completely or partially conserved (Fig. 5.4). The newly formed contacts resulting from this structural relocation are thought to be important for priming the neutral pH HA such that it can subsequently be activated by acidification to mediate fusion (Chen *et al.*, 1998a; Russell *et al.*, 2004; Thoennes *et al.*, 2008). This may also explain the conserved nature of both fusion peptide and cavity-lining amino acids. Among the contacts formed when the fusion peptide is inserted into the cavity are a series of hydrogen bonds between main chain atoms of HA2 residues 2 through 6 and the highly conserved aspartic acid residues located at HA2 positions 109 and 112, which reside in a segment of the long helix of the central coiled coil that undergoes a helix-to-loop transition during fusion. Another ionizable residue, HA1 histidine 17, forms hydrogen bonds with fusion peptide residues 6 and 10 via a water molecule in H3 group HA subtypes. During fusion, protonation of ionizable residues such as these may function to initiate conformational changes by expelling the fusion peptide and allowing other molecular rearrangements to take place. It is notable that the HA0 precursor, which lacks the contacts between fusion peptide residues and aforementioned ionizable residues, does not respond to acidification in the same manner as cleaved neutral pH HA. The close association between fusion peptide residues and conserved ionizable residues that are buried following HA0 cleavage suggests that the HA2 N-terminal domain may be conserved, in part, to allow for the triggering of the neutral pH HA structure at a specific pH. Nearly all single residue substitution mutations that have been examined within the first 10 amino acids of the fusion peptide lead to an elevated pH at which conformational changes are initiated (Cross *et al.*, 2001; Daniels *et al.*, 1985; Gething *et al.*, 1986; Qiao *et al.*, 1999; Steinhauer *et al.*, 1995), and many of these mutants are functional for the subsequent aspects of membrane fusion involving interactions with target membranes. Therefore, strong selective pressure may exist on conserved fusion peptide residues not only for fusion capacity, but also for maintaining a neutral pH cleaved HA structure that is energetically favourable for triggering membrane fusion at an optimal pH. Such a structure would ideally balance requirements for stabilizing HA against premature triggering of the irreversible conformational changes with those for maintaining the capacity to induce fusion within the endosomal pathway. This balance may have an influence on the stability of viruses in the environment, and might account, in part, for the differences in fusion pH and kinetics that can be observed among different HA subtypes (Korte *et al.*, 1999; Puri *et al.*, 1990).

Acid-induced conformational changes required for fusion

The viral fusion proteins (VFPs) of enveloped viruses share a number of common functional properties (Harrison, 2008), but are often grouped into three classes, I, II, and III, based on structural and mechanistic considerations (Kielian and Rey, 2006; Weissenhorn *et al.*, 2007). Influenza HA is categorized as a class I VFP, which includes among others, the fusion proteins of orthomyxoviruses, paramyxoviruses, and retroviruses. All VFPs mediate membrane fusion as a result of conformational changes triggered by external stimuli. For influenza A viruses, which enter host cells via the endosomal pathway, this stimulus involves a decrease in the pH that results from the action of cellular proton pumps. As endosomes become acidified, a critical pH is attained at which membrane fusion is initiated. This normally occurs between pH 5.0 and 6.0 depending on the viral strain, but mutants have been identified which fuse membranes at higher or lower pH than this (for examples see Daniels *et al.*, 1985; Steinhauer *et al.*, 1991a; Thoennes *et al.*, 2008). The locations of such mutations show that they occur at interfaces of structural domains throughout the trimer. These include HA1–HA1 monomer interfaces of the membrane distal head domains, HA2–HA2 interfaces within the stem domain, HA1–HA2 interfaces within monomers, and regions in and around the fusion peptide.

The interpretation derived from the positions of these mutations in the HA structure, was that the conformational changes that accompany membrane fusion involve the relocation of intact structural domains relative to one another. These inferences were confirmed with the elucidation of high-resolution structures of low-pH HA by X-ray crystallography (Bizebard *et al.*, 1995; Bullough *et al.*, 1994; Chen *et al.*, 1999). The structural studies reveal that the HA conformational changes responsible for mediating membrane fusion can essentially be segregated into four somewhat distinct sets of molecular rearrangements, which are shown in Fig. 5.7 and detailed below:

1 The membrane distal monomeric head domains become de-trimerized, but remain tethered to the trimeric core structure of low pH HA by HA1 residues 28–43, which are disordered in the fusion pH structure (Bizebard *et al.*, 1995). This suggests that this 'linking polypeptide' is flexible, possibly to allow head domains to distance themselves from the HA2 fusion subunits during the fusion process. The structure of the head domains actually remains intact at low pH, preserving receptor binding function and the integrity of most of the antigenic sites.
2 The HA2 N-terminal fusion peptide domain is extruded from its buried position in the trimer interior, whereupon it is presumably relocated to interact with target membranes due to the other structural transitions.
3 The N-terminal segment (top) of the long helix of neutral pH HA becomes extended due to the transition into a helical structure of the 'extended chain' that links the long and short alpha helices of neutral pH structure. As a result, the triple-stranded coiled coil of the neutral pH HA trimer extends N-terminally to include residues of the linking polypeptides as well as the short helices. In essence, this functions to recruit the fusion peptide to the N-terminal end of this extended coiled coil.
4 Amino acids at HA2 positions 106–112 in the central coiled coil of neutral pH HA undergo a helix-to-loop transition, and all of the structural entities C-terminal to HA2 residue 112 are inverted by 180 degrees. This 'new loop' allows for the jack knife-like structural rearrangements that position the membrane anchor domain at the same end of the low pH rod-like structure as the fusion peptide (Chen *et al.*, 1999), and the proximity of the two membrane associating regions that result from this are thought to be crucial for the fusion process.

The order in which these structural rearrangements occur, the extent to which individual domains unfold and refold, and the detailed structure of any intermediates in the pathway, all remain to be determined. However, these conformational changes allow us to envisage a mechanism by which a bridge is formed between the viral and cellular membranes and how they might be drawn into proximity with one another. The thermostable rod-like structure adopted by HA2, in which both of the membrane-associating domains reside at one end, reflects a common theme that has been observed for many cellular and viral fusion proteins (Skehel and Wiley, 1998). For class I fusion proteins these structures all contain a central trimeric coiled coil core structure formed by interacting helices from each of the monomers. The fusion peptide domains reside at the N-terminal end of each helix of the central core, either as a direct extension of the coiled coil, or linked by a small peptide sequence. At the C-terminal end of the central core, structural elements lead to an inversion of the polypeptide chain, which then traces antiparallel and packs against the coiled coil resulting in the rod-like structure. The transmembrane domains of the fusion proteins are located C-terminal to the antiparallel polypeptide chain, placing them at the same end of the structure as the fusion peptide (Figs. 5.3 and 5.7). Presumably the close approximation of the two membrane-associating domains of the protein in this energetically stable conformational state is an important feature for the fusion process.

For several of the class I VFP rod structures, exemplified by representatives of the retrovirus and paramyxovirus fusion proteins, the antiparallel C-terminal polypeptide chains that pack against the coiled coil are mostly helical and the overall structures are commonly referred to as six-helix bundles (Baker *et al.*, 1999; Weissenhorn

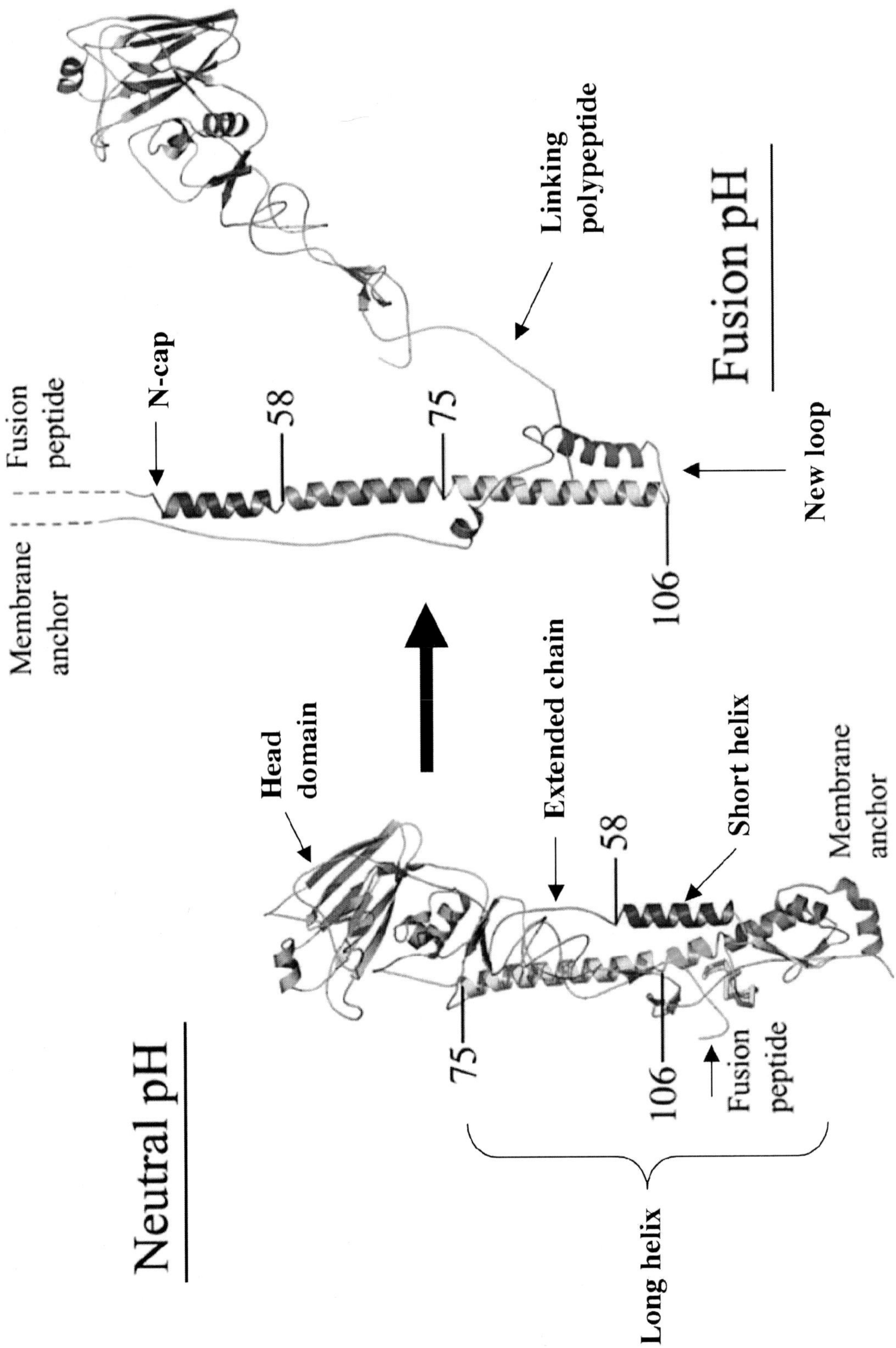
Neutral pH
Fusion pH
Head domain
Extended chain
Short helix
Membrane anchor
Fusion peptide
Long helix
58
75
106
Membrane anchor
Fusion peptide
N-cap
58
75
106
New loop
Linking polypeptide

et al., 1997; Zhao *et al.*, 2000). For influenza HA the antiparallel polypeptides exist as extended chains that pack into the grooves between the core helices. In the case of HA the membrane-proximal ends of the rod structure appears to be held together in part due to an unusual N-cap domain that terminates the core helices and includes highly conserved HA2 residues Ala-35, Ala-36, and Asp-37 (Chen *et al.*, 1999). The location of these residues is indicated in Fig. 5.7. It is thought that the N-cap structure, and the interactions it makes with the anti-parallel C-terminal polypeptide chain provide increased stability to the low pH HA structure, which may be important for driving the fusion process. This is supported by the observation that the E. coli-expressed EHA2 construct that contains all HA 2 residues between the fusion peptide and the transmembrane domain (residues 23–185), has a higher temperature of thermal denaturation than the thermolysin-derived proteolytic TBHA2 fragment, which contains only residues 38–175 (Chen *et al.*, 1995, 1998b, 1999). Mutagenesis studies on N-cap residues and residues of the antiparallel chain in this region suggest that at least some of these interactions have relevance for fusion activity (Borrego-Diaz *et al.*, 2003; Park *et al.*, 2003). In addition, the length of the peptide chains that link the fusion rod structures to the membrane associating transmembrane and fusion peptide domains may also be important for fusion function (Li *et al.*, 2008).

A number of related models have been proposed for the mechanisms by which membrane destabilization and merger occur, and Fig. 5.8 illustrates some of the basic stages that are thought to take place during the fusion process. However, many of the specifics involved remain the topic of study, and as such, a unifying detailed model for fusion is not currently available. Despite this, there is general agreement on the broad concept by which fusion proteins draw the membrane interacting domains into proximity with one another, and that fusion proceeds through a hemifusion intermediate, in which the outer leaflets of the membranes merge and lipid mixing occurs prior to full fusion and content mixing (Kemble *et al.*, 1994). The details of how hemifusion intermediates proceed to form fusion pores that dilate during the process of full membrane merger are still being defined. It seems likely that the composition and structure of the fusion peptides, and possibly the transmembrane anchor domains (Melikyan *et al.*, 1999), play a critical role in the disruption of membranes in a manner that proceeds to fusion. For reviews that offer detailed models on various aspects membrane fusion see the following references (Chernomordik and Kozlov, 2008; Colman and Lawrence, 2003; Eckert and Kim, 2001; Harrison, 2008; Kielian and Rey, 2006; Lamb and Jardetzky, 2007; Moscona, 2005; Ray and Doms, 2006; Sieczkarski and Whittaker, 2005; Skehel and Wiley, 2000; Smith and Helenius, 2004; Weissenhorn *et al.*, 2007; White *et al.*, 2008).

Composition and structure of the fusion peptide

Like most Class I VFPs, the influenza A HA fusion peptide is located at the N-terminus of a fusion subunit (HA2) that has been

Figure 5.7 (opposite) Ribbon diagram of HA monomers of the neutral pH (left) and low-pH (right) structures of HA. The colours are utilized to follow the structural changes and relocation of domains that occur during membrane fusion. Included among these changes are: (1) The membrane distal head domains (light brown) detrimerize. The receptor binding domains and many antigenic regions retain their native structure following detrimerization and remain tethered to the stalk domain by a 'linking peptide' of unknown structure (Bizebard *et al.*, 1995). (2) The fusion peptide is extruded from the trimer interior and relocated towards the target membrane. (3) The fusion peptide is directed to the target membrane because the extended chain region (orange, between HA2 residues 58 and 75) that links the long (yellow) and short (red) helices of HA2, itself becomes helical and extends the central coiled coil in the N-terminal direction. The newly formed α-helices of each monomer extend N-terminally from HA2 residue 106 to 37, where an N-cap structure terminates the central coiled coil. (4) A peptide segment of the long helix located near the fusion peptide in neutral pH HA (green) undergoes a helix-to-loop transition beginning at HA2 residues 106, and the creation of this 'new loop' reorients residues C-terminal to it by 180 degrees. This includes residues shown in purple that pack along a groove against the central coiled coil as the polypeptide chain traces towards the viral transmembrane domain. This converts the molecule into a rod-like structure that brings the fusion peptide and the membrane anchor domain into close proximity with one another.

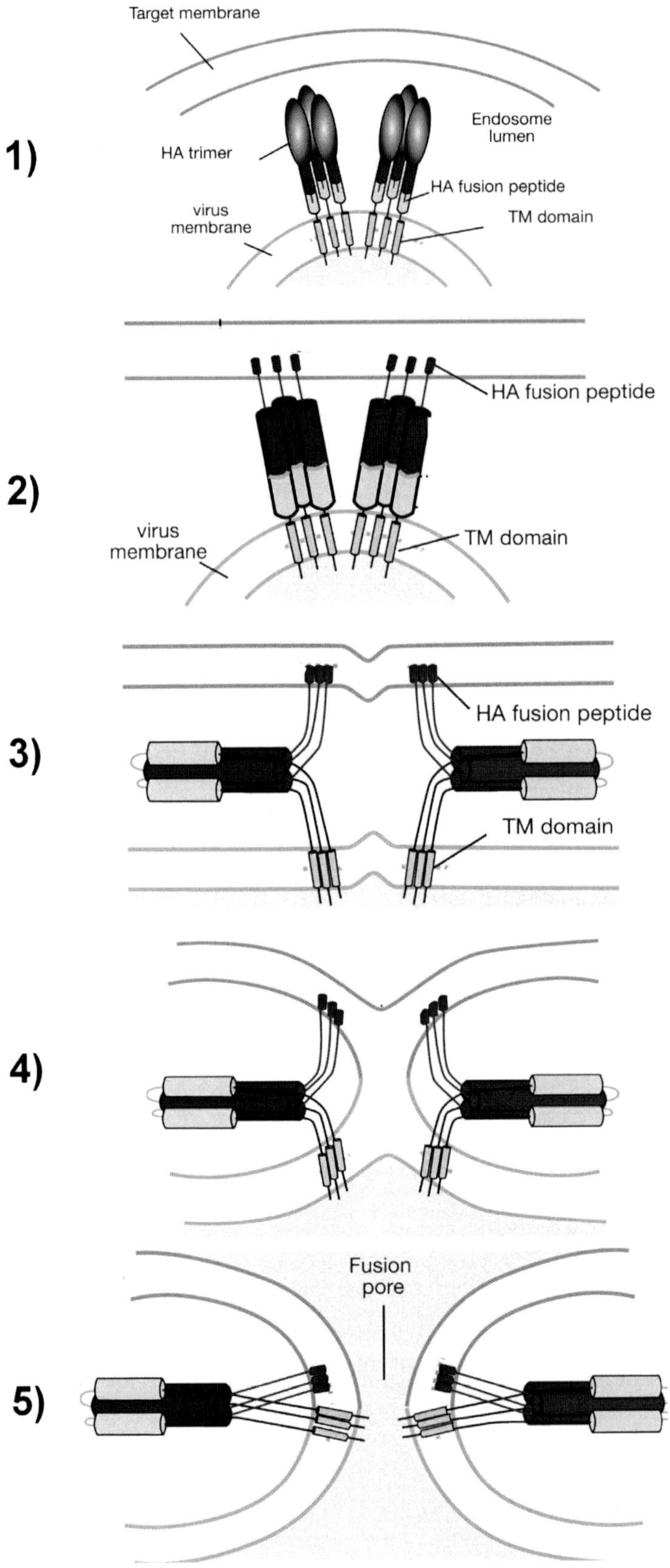
Target membrane
1)
HA trimer
Endosome lumen
HA fusion peptide
virus membrane
TM domain
2)
HA fusion peptide
virus membrane
TM domain
3)
HA fusion peptide
TM domain
4)
Fusion pore
5)

generated by proteolysis of a precursor polypeptide. A comparison of fusion peptide sequences from representative class I VFPs of several viruses demonstrate that although there is little or no direct identity among the more distantly related taxonomic groups, they share some common features (Fig. 5.9). As might be expected, these domains are rich in hydrophobic residues such as Phe, Leu, Ile, and Val, and most fusion peptides also contain a number of alanine residues. Concordantly, very few ionizable residues are represented among these sequences. It is also notable that all of these fusion peptides contain a number of interspersed glycine residues. These are often spaced at intervals that suggest a structural significance, but they do not align directly from one virus family to the next, and do not adhere to definitive motifs.

For the most part, the lengths of fusion peptides are arbitrarily defined based on features such as hydrophobicity and sequence conservation. For influenza A viruses the fusion peptide is often considered as the N-terminal 23 residues of HA2, and an alignment of these positions for representatives of the 16 recognized subtypes is shown in Fig. 5.10. These display a high degree of sequence identity, particularly for HA2 residues 1–11, amino acids with large hydrophobic side chains, and all but one of the glycine residues (Nobusawa *et al.*, 1991). In fact, the HA fusion peptide domains are by far the most highly conserved domains of this notably variable glycoprotein. As discussed above, this conservation is likely to reflect structural roles for fusion peptide residues in each of the three major conformations of HA, and their location in each structure is highlighted in Figs. 5.3 and 5.4.

Numerous mutagenesis studies have addressed the HA fusion peptide domain, and considering their sequence conservation, a surprisingly high percentage of single residue substitution mutants are capable of transporting to the cell surface and responding to acidification with the characteristic structural changes (reviewed in Cross *et al.*, 2009). However, a significant number of these fusion peptide mutants are inhibited for fusion activity, even though they retain the capacity to associate with target membranes. This indicates that the structure adopted by the fusion peptides in association with target membranes is critical for function. The hydrophobic nature of fusion peptides presents a formidable obstacle for obtaining structural information on these domains. For example, in all X-ray crystal structures of low-pH HAs determined to date, the fusion peptides have been removed in order to generate soluble proteolytic fragments capable of forming crystals. Therefore, our current knowledge on fusion peptide structures is largely based on alternative structural and biophysical techniques, but these have yet to provide definitive solutions for the biological structure of fusion peptides in membranes.

Hydrophobic photolabelling studies indicate that only the first 22 residues of HA2 interact with target membranes, and it was suggested by the periodicity of labelling that this segment adopts an amphipathic α-helical structure (Durrer *et al.*, 1996; Harter *et al.*, 1989). Biophysical studies using circular dichroism and FTIR spectroscopy suggest that membrane-bound fusion peptide analogues contain 40% to 60% α-helical structure, with some studies also showing evidence of β-structure (Grey *et al.*, 1996; Han

---

Figure 5.8 (opposite) Schematic model of some of the basic stages of the membrane fusion process. (1) This depicts the target membrane and virus surface with two representative trimers prior to the initiation of conformational changes. The trimeric membrane distal head domains are represented only in this panel. The HA2 stalk region is red and yellow. TM is the viral transmembrane domain and the fusion peptide is buried in the trimer interior at this stage. (2) This shows that following acid-induced conformational changes, the fusion peptide is extruded and relocated to interact with the target membrane due to the extension of the coiled coil. (3) The jack-knife action due to the refolding of the C-terminal portion of the long helices of neutral pH HA draw the TM and fusion peptide domains to the same end of the low pH structure. This brings the two membranes into proximity with one another. (4) Fusion proceeds through a hemifusion intermediate, in which the outer leaflets of the two membranes merge and lipid mixing of the membranes can be detected. (5) This is followed by the formation of a fusion pore and full fusion, which allows for content mixing (or transfer of the viral genome into the cytoplasm).

Influenza A virus	GLFGAIAGFIENGWEGMIDGWYG
Influenza B virus	GFFGAIAGFLEGGWEGMIAGWHG
Influenza C virus	IFGIDDLIIGLLFVAIVEAGIGGYLLGS
Parainfluenza 1 virus	FFGAVIGTIALGVATAAQITAGIALA
Sendai virus	FFGAVIGTIALGVATSAQITAGIALA
Simian virus 5	FAGVVIGLAALGVATAAQVTAAVALVK
Respiratory syncitial virus	FLGFLLGVGSAIASGIAVSKVLHL
Measles virus	FAGVVLAGAALGVATAAQITAGIALH
Ebola virus	GAAIGLAWIPYFGPAAE
Human immunodeficiency virus I	AVGIVGAMFLGFLGAAGSTMGAVSLTLTVQA
Human immunodeficiency virus II	GVFVLGFLGFLATAGSAMGARSLTLSA
Simian immunodeficiency virus	VPFVLGFLGFLGAAGTAMGAAATALTV
Human T-cell leukemia virus	AVPVAVWLVSALAMGAGVAGGITGSMSLASG

Figure 5.9 Fusion peptide sequences of examples from representatives of Class I viral fusion proteins (VFPs). These all reside at the N-terminus of fusion subunits derived following proteolytic activation of a VFP precursor protein. The glycine residues are highlighted in bold type.

	1	2	3	4	5	6	7	8	9	10	11	12	13	14	15	16	17	18	19	20	21	22	23
H1	**G**	L	F	**G**	A	I	A	**G**	F	I	E	**G**	**G**	W	T	**G**	M	I	D	**G**	W	Y	**G**
H2	-	-	-	-	-	-	-	-	-	-	-	-	-	-	Q	-	-	V	-	-	-	-	-
H3	-	-	-	-	-	-	-	-	-	-	-	N	-	-	E	-	-	-	-	-	-	-	-
H4	-	-	-	-	-	-	-	-	-	-	-	N	-	-	Q	-	L	-	-	-	-	-	-
H5	-	-	-	-	-	-	-	-	-	-	-	-	-	-	Q	-	-	V	-	-	-	-	-
H6	-	-	-	-	-	-	-	-	-	-	-	-	-	-	-	-	-	-	-	-	-	-	-
H7	-	-	-	-	-	-	-	-	-	-	-	N	-	-	E	-	L	V	-	-	-	-	-
H8	-	-	-	-	-	-	-	-	-	-	-	-	-	-	S	-	-	-	-	-	-	-	-
H9	-	-	-	-	-	-	-	-	-	-	-	-	-	-	P	-	L	V	A	-	-	-	-
H10	-	-	-	-	-	-	-	-	-	-	-	N	-	-	E	-	-	V	-	-	-	-	-
H11	-	-	-	-	-	-	-	-	-	-	-	-	-	-	P	-	L	-	N	-	-	-	-
H12	-	-	-	-	-	-	-	-	-	-	-	-	-	-	P	-	L	V	A	-	-	-	-
H13	-	-	-	-	-	-	-	-	-	-	-	-	-	-	P	-	L	-	N	-	-	-	-
H14	-	-	-	-	-	-	-	-	-	-	-	N	-	-	Q	-	L	-	-	-	-	-	-
H15	-	-	-	-	-	-	-	-	-	-	-	N	-	-	E	-	L	-	-	-	-	-	-
H16	-	-	-	-	-	-	-	-	-	-	-	-	-	-	P	-	L	-	N	-	-	-	-

Figure 5.10 Fusion peptides of representatives from each of the 16 antigenic subtypes of influenza A viruses. The sequence of the H1 subtype has glycine residues highlighted in bold, and is used as the reference sequence for the purpose of this figure. Dashes indicate sequence identity at the given position.

et al., 1999; Lear and DeGrado, 1987; Murata *et al.*, 1987; Rafalski *et al.*, 1991; Wharton *et al.*, 1988). In some studies of mutant HA peptide analogues, a loose correlation between helical content and the capacity to mediate liposome fusion is demonstrated (Burger *et al.*, 1991; Lear and DeGrado, 1987; Wharton *et al.*, 1988). HA and other viral fusion proteins, as well as apolipoproteins, proteins of neurodegenerative diseases (β-amyloid, human PrP), signal peptides, toxins, and fusion proteins of spermatozoids, can be modelled as oblique oriented or tilted helices with an asymmetric gradient of hydrophobicity that are hypothesized to cause membrane destabilization by perturbing the regular lipid acyl chain packing (Brasseur, 2000).

HA fusion peptides have been modelled as both oligomeric and monomeric structures in association with membranes. Tamm and colleagues have used a combination of NMR structural data and EPR depth analysis to examine the structure of synthetic influenza fusion peptide analogues in DPC micelles (Han *et al.*, 2001; Han and Tamm, 2000). This work involved the addition of the seven residue peptide sequence GCGKKKK to the C-terminal end of a 20-residue HA2 N-terminal domain to mitigate against solubility problems, and these 27-residue peptides were shown to possess pH-dependent activity in lipid mixing and erythrocyte haemolysis assays. Analysis of the peptide structure by NMR showed it to adopt a monomeric inverted V shaped structure in detergent micelles at both neutral pH and pH 5.0, with the residue 12 asparagine forming the apex of the V at the aqueous interface (Han *et al.*, 2001). In both structures the N-terminal portion was found to be helical, with the glycine at positions 1, 4, and 8 aligned to orient towards the aqueous interface. In the neutral pH structure the N-terminal helix was formed by residues L2 to F9 followed by a turn, and a C-terminal segment that was less ordered, with residues W14–G20 forming an extended structure containing a bend between residues G16 and M17. At pH 5, the N-terminal helix was extended to include I10, and following the turn, the C-terminal structure was characterized by a short 3_{10} helix (Han *et al.*, 2001). This allowed for the alignment of non-polar residues at the bottom face of the kinked peptide, creating a hydrophobic pocket and giving the peptide a more closed and deeply inserted structure within the membrane relative to the neutral pH structure. Studies on mutant fusion peptide analogues with either reduced or negative fusogenic activity suggested a structural significance for the hinge region which lies in the middle of the fusion peptide sequence (Hsu *et al.*, 2002; Lai *et al.*, 2005, 2006; Lai and Tamm, 2007). Studies such as these will allow for the continued assessment of models and the correlation of structure and fusion activity, but they are dependent on the assumption that fusion peptides are active in monomeric form, and the oligomeric state of fusion peptide domains during the fusion process remains unknown at present. The effects of pH on the structure of fusion peptides and their relevance for membrane destabilization during fusion are also unclear.

In other studies, tryptophan fluorescence quenching and NMR spectroscopy analyses of a 25 residue HA fusion peptide analogue in SDS micelles indicated that residues 2–14 exist predominately as an α-helix with residues 16–18 residing at the micelle water interface and no observable changes in insertion depth at low pH (Chang *et al.*, 2000). Their evidence suggests that residue E11 is located in the interior of the SDS micelle, and the observation that the peptides tend to self-associate during SDS gel electrophoresis is consistent with models of a peptide trimer in which residues E11, E15 and D19 form a hydrophilic face that lies within the oligomeric interior, while hydrophobic residues I6, F9, W14 and M17 are exposed to the bilayer. Studies using infectious viruses with mutations within the first 10 fusion peptide residues (Cross *et al.*, 2001) are also compatible with a model in which helical fusion peptides orient such that the relatively polar glycine residues form a trimeric interface and the large hydrophobic side chains of residues 2, 6, and 10 reside on the surface to interact with membrane lipids (Skehel *et al.*, 2001). Furthermore, in experiments with HA fusion peptide analogues linked to coiled coil domains in order to promote their trimerization, DeGrado and co-workers demonstrated that trimeric peptides promoted a greater degree of liposomal content leakage and membrane mixing than the monomeric form (Lau *et al.*, 2004).

The effects of pH on fusion peptide structure, and the relevance of this for fusion function, are also unclear. Studies with fusion peptide analogues show that fusion activity is often higher at low pH compared to neutral pH (Han and Tamm, 2000; Murata *et al.*, 1987; Rafalski *et al.*, 1991; Wharton *et al.*, 1988). However, most of these comparative studies are carried out at specific pH such as 7.4 and 5.0, and are not extended to include intermediate pH. The analysis at intermediate pH may be critical for interpreting the relevance of fusion peptide structures if they are considered as separate entities from intact HA. For example, HA mutants that mediate fusion at pH as high as 6.5 have been characterized, and fusion function correlates with the pH of HA ectodomain conformational changes (Daniels *et al.*, 1985; Wharton *et al.*, 1986). It is not clear whether the structure of fusion peptide domains at pH 6.5 more closely resembles that at lower or neutral pH, but this should be taken into consideration, as should the effects of pH and temperature on the membrane. HA-mediated membrane fusion can be promoted at neutral pH by increasing temperature or by the addition of chemical denaturants such as urea (Carr *et al.*, 1997; Ruigrok *et al.*, 1986). In these studies the conditions under which fusion activity was detected again correlated with those at which conformational changes of HA ectodomains were observed, although it was not possible to assess structural changes that may have taken place in the fusion peptide domains.

Studies of mutant HAs with changes in the fusion peptide domain provide some guidelines regarding the requirements for particular classes of amino acids at given positions for folding and fusion function. One constraint for fusion appears to involve the length of the fusion peptide, as deletion of the N-terminal glycine residue results in a fusion negative phenotype, as does deletion of the leucine at position 2 (Steinhauer *et al.*, 1995). This is consistent with results on cleavage activation mutants selected for growth in the presence of the protease thermolysin (Orlich and Rott, 1994), which cleaves HA between HA2 Gly1 and Leu2. In effect, viruses grown in the presence of thermolysin would contain a cleaved HA with a fusion peptide truncated by one residue. However, mutants selected for growth in the presence of thermolysin were found to contain single residue insertions C-terminal to the new N-terminal leucine, which functioned to restore the length of the fusion peptide sequence following cleavage. Despite the observation that fusion peptides with an N-terminal leucine are active, most studies suggest that the glycine residue at the HA2 N-terminus of WT HAs appears to be particularly important for optimal fusion function. In fact, with the exception of alanine, all mutants with single residue substitutions for the N-terminal glycine have been shown to be negative or significantly impaired for fusion activity (Cross *et al.*, 2001; Orlich and Rott, 1994; Qiao *et al.*, 1999; Steinhauer *et al.*, 1995; Walker and Kawaoka, 1993). However, despite the low levels of full fusion activity displayed by the serine substitution mutant at HA2 position 1, it is quite efficient at promoting lipid mixing, a characteristic of the hemifusion phenotype (Cross *et al.*, 2001; Qiao *et al.*, 1999). The conserved glycine at position 4 appears to accept changes reasonably well while maintaining fusion function, although the mutants that have been characterized lead to a significant increase in the pH at which this takes place (Gething *et al.*, 1986; Lin *et al.*, 1997; Steinhauer *et al.*, 1995). Among expressed HAs and mutant viruses, fusion activity is evident for HAs with alanine substitutions or conservative changes for the large hydrophobic residues at positions 2, 3, 6, 9, and 10 (Cross *et al.*, 2001; Daniels *et al.*, 1985; Gething *et al.*, 1986; Lai and Tamm, 2007; Lin *et al.*, 1997; Steinhauer *et al.*, 1995; Verhoeyen *et al.*, 1980; Yewdell *et al.*, 1993). However, alanine is not functional when substituted for the conserved glycine at position 8 (Steinhauer *et al.*, 1995). On the other hand, expressed HAs with glycine substitutions for these large hydrophobic residues are significantly inhibited for fusion activity.

Not surprisingly, the less conserved positions between residues 12 and 23 seem reasonably tolerant of substitutions, whereas the highly conserved hydrophobic tryptophan residues at HA2 positions 14 and 21, and the conserved tyrosine at HA2 22 appear more critical for function (Cross *et al.*, 2009). Our data show that alanine and glycine substitution mutants at these positions were consistently negative for polykaryon formation, and this has been confirmed by others for the position 14 alanine mutant (Lai and Tamm, 2007).

Some noteworthy insights on the requirements and functional constraints of fusion peptide residues derive from studies on attempts to generate infectious viruses containing specific mutations using reverse genetics. Studies with these mutants provide us with additional data to consider from a biological perspective, as a number of them initially replicate quite poorly, and subsequently select for reversions or pseudoreversions upon passage. For example, the G1S and G1L substitutions at the HA2 N-terminus were found to promote heterokaryon formation very poorly (Steinhauer *et al.*, 1995) and using assays for both lipid and content mixing display only hemifusion activity (Cross *et al.*, 2001;

Qiao *et al.*, 1999). The G1S virus reverted to WT almost immediately in one rescue experiment, but was still the predominant genotype following three passages in another experiment before subsequent reversion. Thus, the hemifusion phenotype ascribed to this mutant might be 'leaky' enough for a low level of biological activity that is readily selected against once WT reversions are generated. Similarly, the G1L mutant was observed to mutate readily upon passage, but in this case the change involved a pseudoreversion to phenylalanine.

It was also possible to rescue viruses with phenylalanine to glycine substitutions at either position 3 or 9, even though fusion activity was not detected with these expressed HAs. In each case the viruses replicated poorly and pseudoreversions to valine and alanine respectively, were selected upon passage. The viruses in which large hydrophobic residues at positions 2, 3, 6, 9, and 10 were changed to alanine also proved interesting. When expressed, these mutants were positive for heterokaryon formation and infectious viruses containing these were rescued, but upon passage, pseudoreversion mutations to alternative large hydrophobic residues were selected (true reversion to WT would have required two separate changes at the RNA level for most of these). This was observed for the alanine mutants at L2, F3, I6, and I10, all of which mutated to valine after several rounds of replication.

The major inferences derived from these mutagenesis studies relate to the 10 N-terminal fusion peptide residues, for which glycine appears to be operative only at the positions 1, 4, and 8, where they reside in the WT sequence, but not at other positions. Similarly, the large hydrophobic residues at positions 2, 3, 6, 9, and 10 confer a clear selective advantage. These results are consistent with models for the structure of the fusion peptide in membranes with the glycines oriented on one face of the helix and the large hydrophobic side chains interacting with the membrane bilayer, and such properties can be incorporated into either monomeric or trimeric models for fusion peptide structure.

The functional structure of HA fusion peptides and their oligomeric form within biological membranes remain to be elucidated. The viral transmembrane domain of HA should also be taken into consideration, as in the final fusion state following membrane merger this domain is available to interact with fusion peptides, and such interactions may play a role in the final stages of the fusion process (Chang *et al.*, 2008; Melikyan *et al.*, 2000). Additionally, the presence of the complete HA molecule may impose constraints on the structure that the fusion peptide adopts in membrane that may not be apparent based on studies of peptide analogues. The determination of complete HA structures including transmembrane and/or fusion peptides associated with membranes will be required to fully appreciate fusion mechanism. Indeed, the question of how many complete HA trimers participate in biologically relevant membrane fusion has been addressed but not defined with precision (Bentz, 2000; Blumenthal *et al.*, 1996; Danieli *et al.*, 1996). If and when a consensus emerges, the structural data will have to reconcile with the large quantity of biological information that has accumulated regarding HA fusion peptide mutants.

Comparison of the HA structures of different subtypes

X-ray crystal structures have now been reported for the neutral pH HAs of several influenza A subtype viruses including H1, H3, H5, H7, and H9 (Gamblin *et al.*, 2004; Ha *et al.*, 2001, 2002; Russell *et al.*, 2004; Stevens *et al.*, 2004, 2006; Wilson *et al.*, 1981). In general, the shape and architecture of these HAs appear quite similar, but a more detailed comparison of these shows that they differ in structural regions and domains that are refolded or relocated during membrane fusion. Based on sequence and structural considerations, the 16 influenza A HA subtypes can be separated into five clades, which can be segregated into two distinct groups that are designated as group 1 and group 2. A phylogenetic tree of the HA subtypes is included as part of Fig. 5.11, and shows that the group 1 HAs are composed of three clades and that group 2 contains two clades of three subtypes each. For purposes of discussion, group 1 HAs are represented here by the H1 subtype and group 2 by the H3 subtype HA. The monomer in the centre of the figure is the Aichi H3 subtype HA, and the differences among subtypes at particular regions of HA are highlighted to the left and

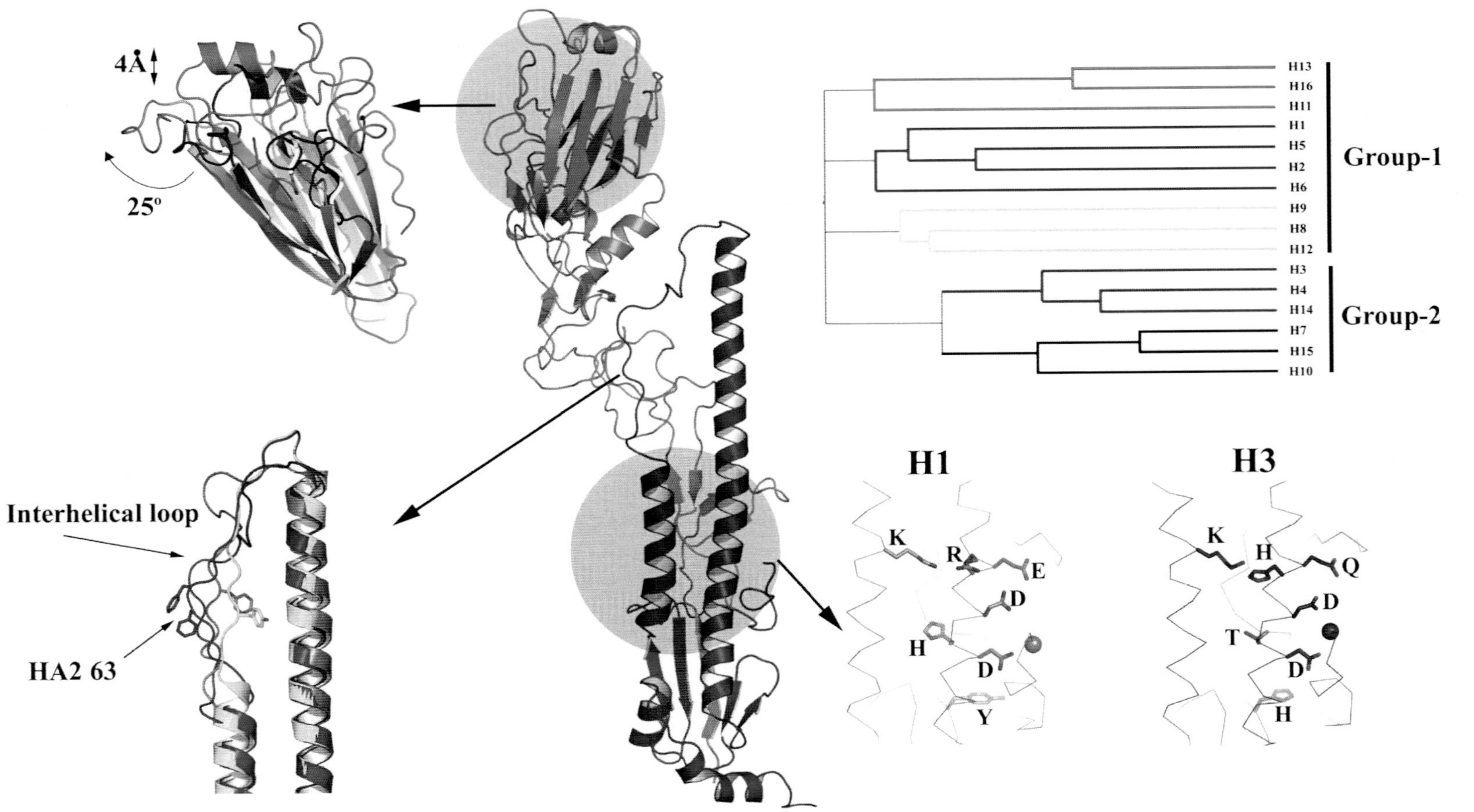
4Å
25°
Interhelical loop
HA2 63
H13
H16
H11
H1
H5
H2
H6
H9
H8
H12
H3
H4
H14
H7
H15
H10
Group-1
Group-2
H1
K
R
E
D
H
D
Y
H3
K
H
Q
D
T
D
H

right of the monomer structure as directed by the arrows.

One of the structural features that differ among the clades involves the membrane distal head domains that detrimerize during fusion. These are situated differently on their stalk domains depending on HA subtype. They vary both in 'height' relative to the stalk domain, and also with respect to the rotational symmetry of head domain monomers along the threefold axis (Russell *et al.*, 2004). The main reason for this appears to involve differences in the structures of the HA2 polypeptide chain formed by residues 58–75, which link the long and short helices of neutral pH HA, and interactions that these residues make with a short helix at the base of the HA1 head domains. HA2 residues 58–75 become helical during fusion to extend the coiled coil, and in the native HA the height and shape of this peptide near the top of the hairpin loop structure seems to dictate the positioning of HA1 head domains relative to the stem. The lower left panel of Fig. 5.11 shows that the structures of the two helices of the HA2 hairpin loop are largely superimposable, but the connecting peptide is quite variable among subtypes. The hairpin loop of H1 subtype HA forms a taller loop with a sharper turn by comparison to H3 HA. This causes the H1 subtype head domains to 'ride' approximately 4 Å higher on the stem, and rotate about 25° in a clockwise direction (viewed from above) relative to the H3 subtype head domains. Despite these differences, the basic architecture of the receptor binding domains is generally quite similar among the subtypes.

Structural comparisons of the region of the neutral pH HA in the vicinity of the fusion peptide are also quite intriguing. When precursor HA is cleaved into HA1 and HA2, only six residues at the C-terminus of HA1 and 12 residues at the N-terminus of HA2 are relocated. Thus, the conformational changes that accompany cleavage alter the accessibility to solvent of only a limited number of ionizable residues in the entire trimer, and these are in the fusion peptide region. The lower right panels of Fig. 5.11 depict the fusion peptide region for H1 and H3 subtype HAs, and shows that some of the ionizable residues that are buried by the fusion peptide after cleavage are completely conserved, while others vary strictly along group-specific lines. In all HAs, HA2 residues K51, D109, and D112 are completely conserved, whereas the residues at positions HA1 17, and HA2 106 and 111 are group-specific. Among H3 group HAs, such as the Aichi HA analysed in this study, HA1 17 and HA2 106 are histidine, and HA2 111 is threonine. In the H1 group HAs, HA1 17 is tyrosine, HA2 111 is histidine, and HA2 106 is a basic residue, either lysine or arginine (Fouchier *et al.*, 2005; Kawaoka *et al.*, 1990; Nobusawa *et al.*, 1991; Rohm *et al.*, 1996b). As described above, these ionizable residues near the fusion peptide may be relevant for protonation events that provide the initial triggers for the conformational changes that lead to fusion. Several of these ionizable residues are known to be significant for HA stability based on

---

Figure 5.11 (opposite) Comparison of structural features of HA subtypes. The phylogenetic tree at the upper right segregates the 16 HA subtypes into five clades. Each of these clades fall into one of two groups based on structural characteristics, which are highlighted here by comparing H1 subtype for group 1 HAs, and the H3 subtype for group 2 HAs. The monomer in the centre of the figure is the H3 HA and is used as a reference for the three regions where notable structural differences are observed among subtypes. The upper left panel shows the H1 HA head domain in blue and the H3 in red, and illustrates that the H1 head domains are 4 Å 'taller' than those of H3, and that they are rotate by 25° along the three-fold axis by comparison. This is due in part, to the structural differences observed for the polypeptide chains that link the helices of the hairpin loop of HA2, as shown in the lower left panel. The five colours in this panel are coordinated with the colours in the branches of the phylogenetic tree. They show that the helices are relatively equivalent in structure among the subtypes, except at the top of the short helix, but the connecting peptides vary considerably in structure. The two panels at the lower right compare the H1 and H3 HAs in the region in the vicinity of the fusion peptide (shaded in the monomer structure). These panels indicate that some of the ionizable residues in this region are highly conserved, whereas others segregate in group-specific fashion. The amino acid positions of the residues are as follows going clockwise from HA2 K51 at the upper left of the H3 structure: HA2 K51, HA2 H106, HA2 Q105, HA2 D109, HA2 D112, HA1 H17, HA2 T111.

observations of mutants that differ with respect to fusion pH. The observation that the kinetics of membrane fusion and HA inactivation can differ among subtypes might relate to differences in the fusion peptide region that are group or subtype specific, and such differences could play a role in the ecology of influenza viruses.

The group-specific differences among subtypes in the region of the fusion peptide may also have significance with regard to antiviral compounds designed to inhibit membrane fusion function. Several compounds have been developed to bind in this region and have been shown to inhibit, or in some cases facilitate HA conformational changes (Bodian *et al.*, 1993; Cianci *et al.*, 1999; Combrink *et al.*, 2000; Hoffman *et al.*, 1997; Luo *et al.*, 1997; Yu *et al.*, 1999). The observation that these compounds often select for changes in the fusion peptide region support the hypothesis that they bind locally, and interestingly, some of the drugs that have been characterized are subtype specific. However, the binding site for one of the drugs that was initially identified, TBHQ, has recently been localized to a region in the vicinity of HA2 residue 57 near the top of the short helix based on the crystal structure of HAs in complex with the compound (Russell *et al.*, 2008). This compound is specific for group 2 HAs, which contain an open cavity in this region making it available for binding. By contrast, the differences in the structure near the top of the helix and bottom of the connecting peptide (Fig. 5.11) appear to be responsible for the lack of anti-viral activity observed for group 1 viruses with TBHQ. These results suggest that a greater understanding of the initiation of conformational changes, the refolding of HA, and the involvement of structural intermediates may aid in the development of new generations of anti-influenza drugs that are designed to inhibit membrane fusion function, and comparative structural analyses among HAs of different subtypes could feature prominently in the progress of these studies.

Functions of HA post-translational modifications

The Aichi HA ectodomain contains six sites at which N-linked carbohydrates are attached to asparagine residues of the HA1 subunit (positions 8, 22, 38, 81, 165, and 285), and one site in HA2 at residue 154 (Ward and Dopheide, 1981; Wilson *et al.*, 1981). With the exception of those attached to 81 and 165, all are processed complex oligosaccharides. HA carbohydrates attached at sites proximal to the N-terminus, positions 8, 22, and 38 for Aichi HA, are involved in the folding of HA following synthesis. In the endoplasmic reticulum (ER) oligosaccharides at these sites are bound by the membrane bound and soluble lectins, calnexin and calreticulin respectively. These chaperone proteins in complex with the co-chaperone thiol oxidoreductase, ERp57, help mediate proper folding and correct disulphide bond formation of HA (Daniels *et al.*, 2003; Helenius and Aebi, 2001; Molinari and Helenius, 2000).

Individual carbohydrates on the membrane distal domains have not been assigned a direct function per se, but clearly have implications for antigenicity as well as receptor binding. Such carbohydrates are known to be capable of modifying or masking sites of antibody binding (Abe *et al.*, 2004; Caton *et al.*, 1982; Skehel *et al.*, 1984; Tsuchiya *et al.*, 2002), and it may be significant that during the process of antigenic drift, the number of carbohydrates attached to H3 subtype HAs has increased from 7 in 1968 to as many as 12 in currently circulating strains (Blackburne *et al.*, 2008).

There are numerous reports showing that the gain or loss of carbohydrates near the receptor binding site on the HA membrane distal domain can influence binding properties and/or virus adaptation, and a few examples are cited here (Gambaryan *et al.*, 1998,1999; Hartley *et al.*, 1992; Iwatsuki-Horimoto *et al.*, 2004; Ohuchi *et al.*, 1997; Robertson *et al.*, 1987). In some cases oligosaccharides originating from the neighbouring HA monomer can alter access to the binding pocket, as is the case with the glycan attached at HA1 position 165 of Aichi HA, which comes into close proximity to the 220 loop of the adjacent head domain. When oligosaccharides such as this come into play, mutations at distal monomer–monomer interfaces that alter the positioning of individual monomeric head domains relative to one another can effect binding as well (Meisner *et al.*, 2008). The role of carbohydrates attached to HA, neuraminidase stalk length, and the functional balance between the two glycoproteins, can play inter-related roles in virus replication and adaptation (Matrosovich *et al.*, 1999; Mitnaul *et al.*, 2000; Wagner *et al.*, 2002; Wagner *et al.*, 2000). For example, the H5N1 viruses that were responsible for the 1997 Hong Kong outbreak in humans that resulted in 18 hospitalizations and 6 deaths, may have evolved in chickens prior to the outbreak. Viruses that caused disease of poultry on chicken farms on the outskirts of Hong Kong during this period, had additional carbohydrates and shorter NA stalk lengths relative to related viruses circulating in wild birds, and that these chicken strains were related to some that subsequently transmitted to humans (Matrosovich *et al.*, 1999). This provides provocative evidence that chickens, and species other than the pig, can serve as a stepping-stone for interspecies transmission of viruses to humans, and that HA glycosylation can be a contributing factor in such events. It should also be noted that there is recent evidence that the elimination of a glycosylation site at position 158 of circulating pathogenic H5N1 strains may play a role in the acquisition of enhanced binding to human-like receptors (Ilyushina *et al.*, 2008; Stevens *et al.*, 2008).

Another type of post-translational modification of influenza HA involves the addition of acyl chains to three conserved cysteine residues, two in the short cytoplasmic tail of HA and one at the interface between the cytoplasmic tail and the transmembrane domain (Schmidt, 1982). Mass spectrometry analysis shows that the two cysteines near the C-terminus of the protein are modified by palmitic acid, whereas the cysteine at the transmembrane interface 12 residues from the C-terminus contains predominantly stearic acid modifications (Kordyukova *et al.*, 2008). Numerous studies have addressed the membrane fusion properties of acylation mutants of HA, and there are several examples in which fusion activity has been demonstrated with acylation negative mutants; however, a direct correlation between fatty acid modification of HA and membrane fusion or virus infectivity has not been established for all subtypes (Chen *et al.*, 2005; Jin *et al.*, 1996; Lin *et al.*, 1997; Melikyan *et al.*, 1997; Naeve and Williams, 1990; Naim *et al.*, 1992; Philipp *et al.*, 1995; Sakai *et al.*, 2002; Simpson and Lamb, 1992; Steinhauer *et al.*, 1991b; Veit *et al.*, 1991; Wagner *et al.*, 2005). This may reflect subtype specific differences, or alternative assay systems employed for these studies. HA acylation also appears to play a role in virus assembly and virion morphology (Chen *et al.*, 2005; Jin *et al.*, 1997), and the cytoplasmic tail residues, as well as the modified cysteine at the transmembrane interface are involved in directing HA to the lipid raft domains from which viruses bud (Chen *et al.*, 2005; Scheiffele *et al.*, 1999; Takeda *et al.*, 2003).

The influenza A HA, NA, and M2 membrane proteins are all targeted to the apical surface of polarized epithelial cells where budding occurs (Hughey *et al.*, 1992; Jones *et al.*, 1985; Rodriquez-Boulan and Sabatini, 1978; Roth *et al.*, 1983; Stephens *et al.*, 1986), and this has a profound influence on the biology of the virus. The assembly of viral components at apical surfaces where budding occurs is one of the factors that restricts influenza virus to the respiratory tract of human hosts, where they both infect from, and bud back into, the luminal side of the epithelial layer. This is critical for their mechanism of transmission among humans in respiratory droplets, and contrasts with viruses such as vesicular stomatitis virus (VSV), which generally causes systemic infection of its animal hosts. VSV targets its glycoprotein to the basolateral surface of polarized cells (Pfeiffer *et al.*, 1985; Rodriquez-Boulan and Sabatini, 1978; Stephens *et al.*, 1986), buds into the vascular system, spreads systemically, and can be transmitted by insect bites.

HA cleavage properties as a determinant of pathogenicity

Influenza A virus pathogenicity is clearly a multifactorial trait (see Brown *et al.*, 2001; Chen *et al.*, 2008; Pappas *et al.*, 2008; Rott *et al.*, 1979; Scholtissek *et al.*, 1977, for examples). However, factors involving HA0 cleavage activation of virus infectivity are critically important when assessing the pathogenicity, or potential pathogenicity of influenza A virus strains. In this regard, the size and amino acid composition of the HA0 cleavage loop have been demonstrated as a significant determinant (Bosch *et al.*, 1981; Bosch *et al.*, 1979; Chen *et al.*, 1998a; Klenk and

Garten, 1994; Steinhauer, 1999; Webster and Rott, 1987). The HA cleavage site of human and most non-pathogenic avian virus strains usually contains only a single arginine residue, which is cleaved extracellularly by trypsin-like proteases after the HA has been transported to the plasma membrane. For human influenza viruses replicating in the respiratory tract, proteases such as those secreted by bronchiolar epithelial Clara cells have been implicated for their role in proteolytic activation of infectivity (Kido *et al.*, 1992). In addition, there are numerous reports citing the capacity of secreted bacterial proteases for activating influenza virus infectivity (Steinhauer, 1999), and this may, in part, play a role in the increased severity of disease when secondary bacterial infections occur during cases of influenza. In most cases, the limited distribution of extracellular activating proteases to customary sites of infection in the host, such as the respiratory tract of humans, may limit the generalized spread of virus and its pathogenic effects within the host.

By contrast, a number of avian strains have been identified and characterized for their unusual cleavage properties and particularly high degree of pathogenicity. These are most notably represented by subsets of the H5 and H7 serotype viruses that contain insertions of polybasic sequences in the cleavage loop that can be recognized by ubiquitous furin-like proteases that reside in the trans-Golgi of most cell types (Stieneke-Grober *et al.*, 1992; Walker *et al.*, 1994). As such, proteolytic activation of the HAs of these viruses can occur intracellularly in a variety of cell and tissue types, which allows the viruses to spread systemically in the host. Fig. 5.12 illustrates structural features of the HA0 precursor that facilitate cleavage by ubiquitously occurring proteases. In many cases pathogenic strains contain an insertion of additional basic residues, providing greater accessibility to the larger cleavage loop by proteases. These insertions often generate the amino acid motif R-X-R/K-R, or derivatives of this, which are classical recognition sequences for furin-like proteases. In addition, the presence or absence of glycosylation sites near the cleavage loop can also have relevance for access of proteases to the cleavage site and HA0 activation, and the site homologous to HA1 position 22 has come into play in this regard for some strains that have acquired pathogenicity (Fig. 5.12).

Pathogenicity of influenza A viruses is generally defined based on mortality rates of experimentally infected chickens, but among mammalian species mice and ferrets have been used most extensively as animal models for disease (Belser *et al.*, 2007; Brown *et al.*, 2001; Hatta *et al.*, 2001; Jiao *et al.*, 2008; Katz *et al.*, 2000; Maines *et al.*, 2008; Tumpey *et al.*, 2007; Ward, 1997). The presence of polybasic HA cleavage sites as determinants of pathogenicity can be demonstrated in each of these model animals, but is not always the sole determinant – it can often be considered necessary, but not sufficient to convey the pathogenic phenotype. Even within the H5 and H7 subtypes, non-pathogenic strains are historically the most prevalent in nature, and for many of the examples that have been documented, pathogenic viruses appear to have derived directly from related non-pathogenic strains due to the direct insertion of a couple or several basic residues at the HA cleavage site (Bashiruddin, 1991; Garcia *et al.*, 1996; Horimoto *et al.*, 1995; Rohm *et al.*, 1996a; Swayne, 1997). Interestingly, pathogenic H5N2 isolates that appeared in the eastern USA in 1983 evolved from nonpathogenic strains due to the loss of a nearby carbohydrate at the site equivalent to HA1 22 highlighted in Fig. 5.12, even though the cleavage site sequences were unchanged (Deshpande *et al.*, 1987; Kawaoka *et al.*, 1984). Although it is not clear how polybasic insertions are generated or why H5 and H7

Figure 5.12 (opposite) Features of the cleavage loop that can facilitate cleavage and influence virus pathogenicity. The left panel indicates the smaller cleavage loop characteristic of most human and nonpathogenic avian viruses in yellow and blue. The larger loop shown in yellow and magenta represents the structure of a cleavage loop following insertions at this site. Such insertions often contain polybasic recognition motifs for ubiquitous furin-like proteases. These larger loops protrude further from the surface of the trimer and are more accessible for activating proteases. The presence or absence of nearby carbohydrates can also influence access by proteases, and the glycosylation state of HA1 position 22 (indicated) has been implicated for its role in the pathogenicity of some virus strains.

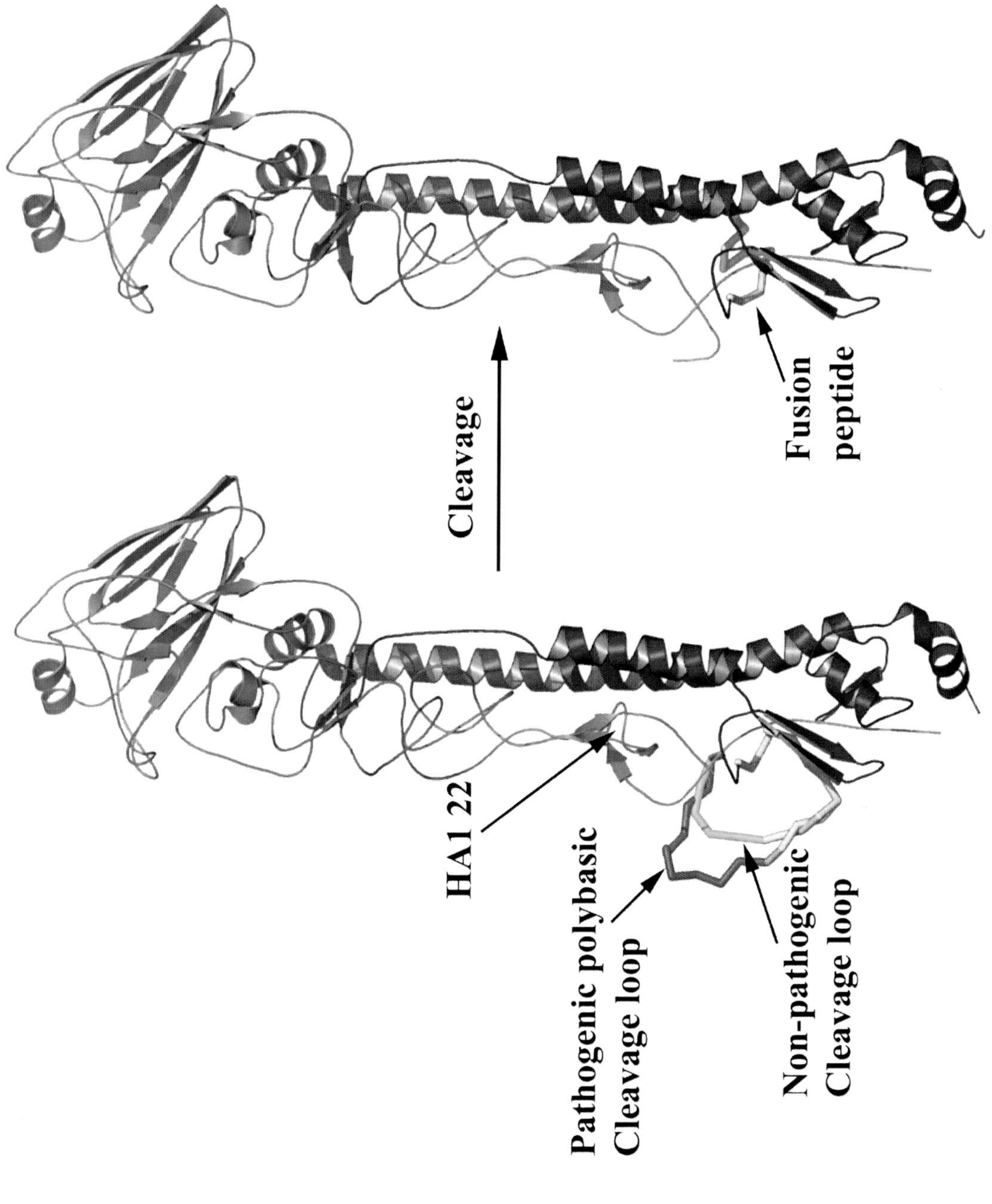
Cleavage
Fusion peptide
HA1 22
Pathogenic polybasic Cleavage loop
Non-pathogenic Cleavage loop

subtypes are particularly prone to this, many H5 and H7 strains are particularly abundant in purines and stretches of poly(A) at the cleavage site coding sequences. Duplication of the purine-rich codons AAA or AAG for lysine or AGA or AGG for arginine due to polymerase slippage could account for the insertion of additional basic residues (Perdue *et al.*, 1997).

Until recently there have been few documented examples of avian isolates that are pathogenic to their natural waterfowl hosts, with the H5N3 A/tern/South Africa/61 virus providing one of the rare exceptions (Becker, 1966). However, some of the H5N1 virus strains that have been circulating widely in recent years are also capable of causing pathogenic disease in their natural hosts. These contain polybasic insertion sequences typical of pathogenic strains, and have caused widespread concern since their emergence as a human threat over the past five years (Guan *et al.*, 2004; Li *et al.*, 2004). The emergence of such strains highlights the unpredictability we face in attempting to understand the ecology, evolution, and transmission of influenza A viruses.

Antigenicity and the mechanism of influenza A virus neutralization

In a process known as antigenic drift, viruses of a single subtype that circulate between pandemics accumulate gradual changes in the antigenic structure of HA from one year to the next. The capacity for closely related influenza A viruses of the same subtype to re-infect individuals and cause widespread illness presents ongoing problems for the design and manufacture of effective vaccines, and poses rather unique challenges for the understanding of the interplay between virus evolution and human immune responses. Currently there are three types of influenza virus circulating in humans, two subtypes of influenza A (H3N2 and H1N1) and influenza B virus. The design of components that constitute annual trivalent vaccines involves predictions of which strains will predominate during the following flu season, and this is often dictated by late-emerging strains from the current season when decisions on the vaccine components need to be made. Some considerations of the problems evoked by antigenic drift can be illustrated using the example of H3N2 subtype 'Hong Kong flu' viruses that emerged in humans in 1968, and continue to this day to be responsible for most of the seasonal influenza epidemics that occur annually. When the X-ray crystal structure of the HA derived from 1968 H3N2 viruses was first determined, it was clear that mutations associated with antigenic drift are located primarily on the surface of the membrane distal head domains of the HA spike (Gerhard *et al.*, 1981; Wiley *et al.*, 1981; Wilson *et al.*, 1981). This appears to result directly from selection pressure due to neutralizing antibody responses in the human population, as illustrated in Fig. 5.13 (Knossow and Skehel, 2006). This figure shows the structures of the 1968 H3 subtype HA with green residues highlighted to show the locations of residues where mutations have been documented. The left panel shows, cumulatively, all amino acid substitutions that have been documented based on sequence analysis of H3 isolates since 1968. The centre panel shows only those that have been fixed in the virus population and carried over to subsequent years. In essence these help constitute the trunk of the phylogenetic tree. The panel on the right shows the locations of amino acid substitutions that have been selected in the laboratory using panels of neutralizing antibodies. The conserved residues involved in receptor binding are represented in yellow. The HA2 subunits, which are responsible for fusion function, are also relatively conserved and are shown in red. As the left panel includes all mutations, several of these are likely to be sporadic changes that were documented only once or a few

Figure 5.13 (opposite) Locations of sequence changes documented for H3 subtype HAs from 1968 through 2005. HA1 is shown in blue and HA2 in red. The yellow spheres represent conserved residues of the receptor binding site and the green represents amino acids that have been substituted. The left panel shows all mutants that have been observed since 1968, while the centre panel represents substitutions that are retained in subsequent years. The right panel shows the sites where substitutions have been detected in laboratory variants that escape neutralization by panels of monoclonal antibodies. Reproduced from Knossow and Skehel (2006), with permission of authors and Blackwell Publishing Ltd.

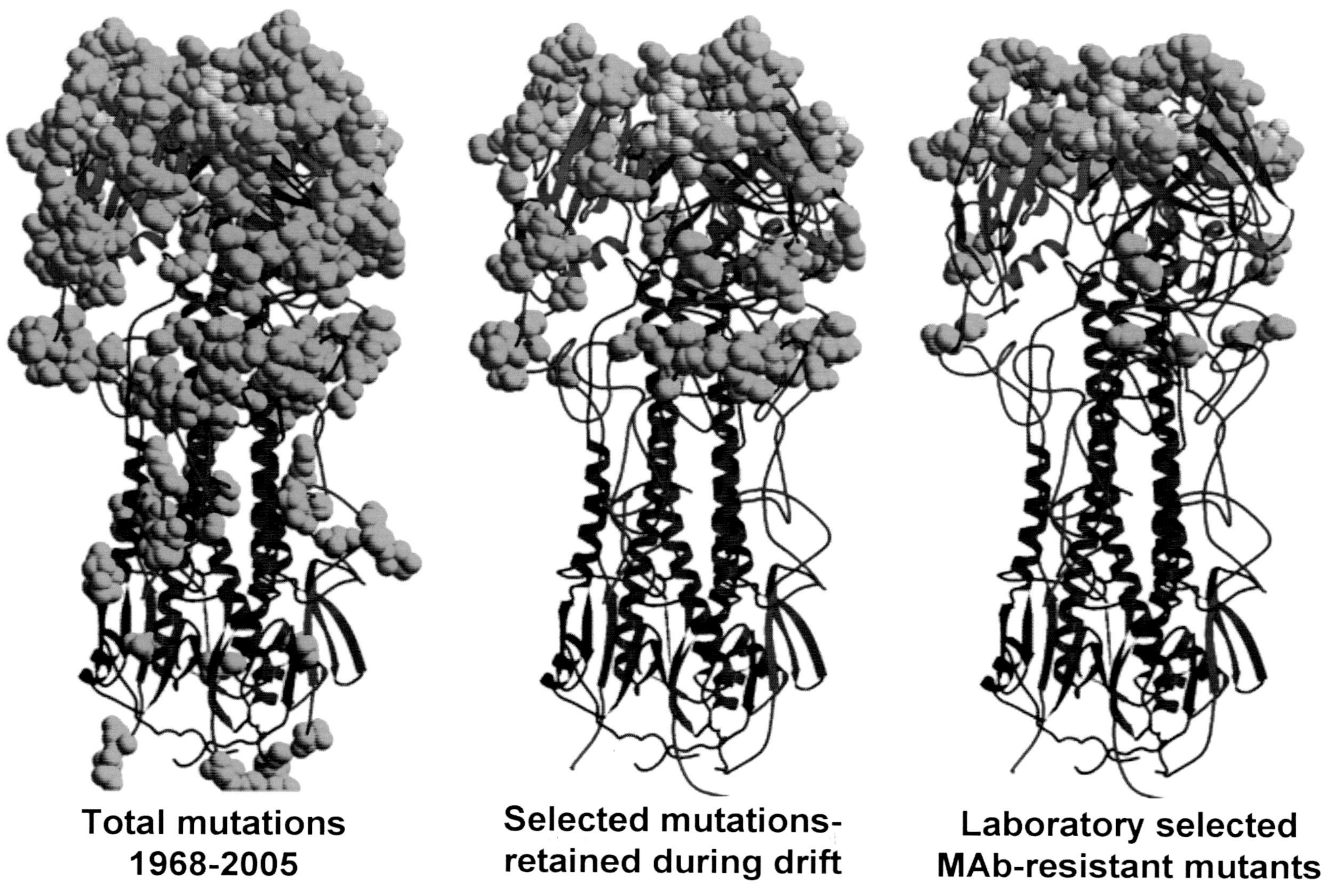
Total mutations
1968-2005
Selected mutations-
retained during drift
Laboratory selected
MAb-resistant mutants

times and disappeared as viral strains evolved in subsequent years. Such evolutionary 'dead end' changes are unlikely to have much significance. On the other hand, the substitutions illustrated in the centre panel are likely to be maintained by immune selective pressure in the human population. As can be noted by comparing the centre and right hand panels, the structural locations of residues selected by human immunity during antigenic drift overlap with the sites at which well documented neutralizing antibodies select for mutations in laboratory-derived antibody resistant variants. Changes associated with antigenic drift are not observed in critical residues at the base of the receptor binding pocket, nor in the relatively conserved residues of the HA stem, which are involved in membrane fusion function. Furthermore, the mutations that are fixed in the evolving virus population during antigenic drift are predominantly restricted to surface regions of the membrane distal HA head domains that are accessible for antibodies, as opposed to residues that are buried in the trimer interior (Wiley and Skehel, 1987; Wilson and Cox, 1990).

An analysis of co-crystals of HA complexed with neutralizing antibodies (Fig. 5.14) by Knossow, Skehel, and colleagues provides a structural rationale for the location of residues involved in drift (Bizebard *et al.*, 1995; Fleury *et al.*, 1999; Knossow *et al.*, 2002; Knossow and Skehel, 2006). For antibodies 1 and 2, shown in the left and centre panels, it is clear that the antibodies bind to epitopes that overlap with, or are in close proximity to the HA receptor binding site. The red spheres indicate residues at which the antibodies select for changes in neutralization-resistant mutant viruses, and for antibodies 1 and 2 these are located within the antibody footprint at positions near the receptor binding pocket. In the HA–antibody complex shown in the right panel, the antibody binds more distal to the receptor site, but the orientation of bound antibody suggests that stearic interference with virus attachment to receptors can still be invoked as the mechanism of neutralization. Significantly, antibodies that bind more distal to the receptor site, such as the one in the right hand panel, are less efficient at neutralizing virus infectivity than those that bind more proximal to the receptor binding site (Knossow *et al.*, 2002). Taken together, the analysis of the structural locations of drift mutations, and antibody binding and neutralization results, indicate that protective immunity may largely result from neutralizing antibodies that block receptor binding. Antibodies that bind non-critical residues in the loop regions that surround the binding site may be most effective. In some cases mutations selected for their effects on HA binding properties can modify antigenicity, or antigenic variants can display altered binding characteristics (Burnet and Bull, 1943; Daniels *et al.*, 1987; Yang *et al.*, 2007). This does not rule out other potential mechanisms of antibody neutralization, including examples mediated by complement pathways, or those involving fusion inhibition (Kida *et al.*, 1985; Reading and Dimmock, 2007; Sui *et al.*, 2009). In fact, a recent study provides evidence that specific antibodies derived from phage display libraries, which target a conserved pocket in the HA stem region, can inhibit fusion of several group 1 HAs and have broad-spectrum neutralizing activity (Ekiert *et al.*, 2009; Sui *et al.*, 2009). However, antibodies to conserved HA2 regions may not be induced efficiently by circulating viruses or current vaccines, as neither provides broadly protective immunity. Overall, it seems that the HA is essentially designed to provoke the immune system into targeting the antigenic loop regions that can tolerate change without compromising its essential functions.

Although vaccine strategies and immunity mediated by T cells are not addressed in this chapter, the locations of drift mutations and binding inhibition as a major mechanism of virus

Figure 5.14 (opposite) Structures of HA complexed with neutralizing monoclonal antibody Fab fragments. HA1 is shown in blue, HA2 in red, receptor site residues in yellow, and antibodies in green. The red spheres indicate the locations of residues that lead to escape from neutralization by the particular antibody shown in that complex. Antibody 1 on the left, and antibody 2 in the centre, bind sites in close proximity to the receptor binding site. Antibody 3 binds more distal to the HA receptor binding pocket and is less efficient at virus neutralization. Reproduced from Knossow and Skehel (2006), with permission of authors and Blackwell Publishing Ltd.

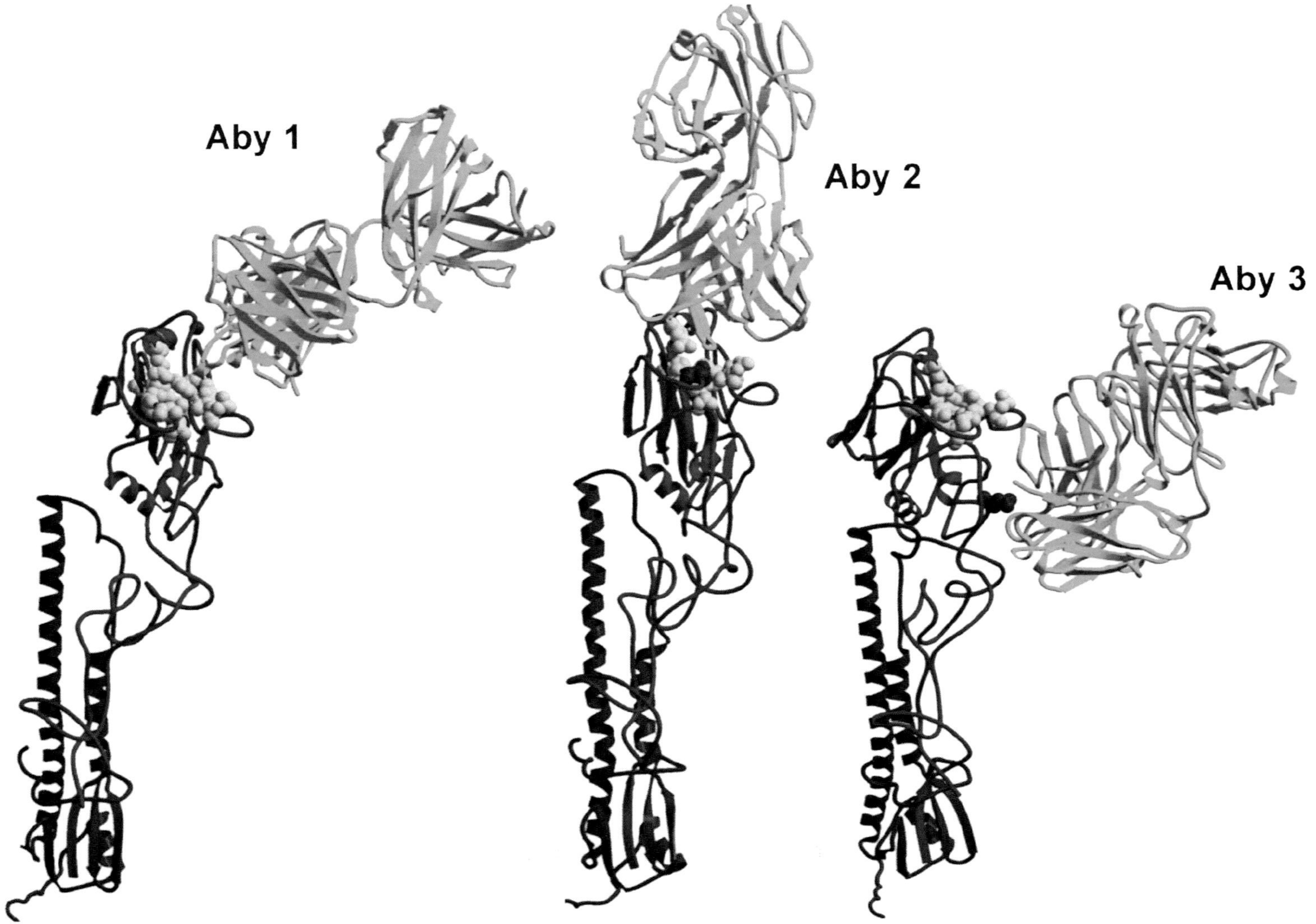
Aby 1
Aby 2
Aby 3

neutralization, provide formidable obstacles with regard to the development of vaccines designed to provide broad spectrum or heterosubtypic immunity. The observation that human individuals become re-infected by influenza A viruses that contain minor antigenic changes to non-essential loop components of the HA head domains suggest that neutralizing antibody responses are the key component of protective immunity. Broad spectrum vaccines that illicit effective T cell responses or suboptimal antibody induction may prove useful for the amelioration of disease symptoms if novel influenza A strains or subtypes emerge in humans. However, vaccines strategies that target conserved regions of HA or other viral antigens will have to be designed to function better than natural infection if the goal is to provide sterilizing immunity.

Acknowledgements

I thank Rupert Russell for comments and generous assistance with figures, and Summer Galloway, Trey Langley, Konrad Bradley, and Dick Compans for critical comments on the manuscript. Support for the work has been provided by NIH Public Service Grant AI66870 and NIH/NIAID contract HHSN266200700006C.

References

Abe, Y., Takashita, E., Sugawara, K., Matsuzaki, Y., Muraki, Y., and Hongo, S. (2004). Effect of the addition of oligosaccharides on the biological activities and antigenicity of influenza A/H3N2 virus hemagglutinin. J. Virol. *78*, 9605–9611.

Ahmed, R., Oldstone, M.B., and Palese, P. (2007). Protective immunity and susceptibility to infectious diseases: lessons from the 1918 influenza pandemic. Nat. Immunol. *8*, 1188–1193.

Appleyard, G., and Maber, H.B. (1974). Plaque formation by influenza viruses in the presence of trypsin. J. Gen. Virol. *25*, 351–357.

Baker, K.A., Dutch, R.E., Lamb, R.A., and Jardetzky, T.S. (1999). Structural basis for paramyxovirus-mediated membrane fusion. Mol. Cell *3*, 309–319.

Bashiruddin, J., Gould, A.R., and Westbury, H.A. (1991). Molecular pathotyping of two avian influenza viruses isolated during the Victoria 1976 outbreak. Aust. Vet. J. *69*, 140–142.

Beare, A.S., and Webster, R.G. (1991). Replication of avian influenza viruses in humans. Arch. Virol. *119*, 37–42.

Becker, W.B. (1966). The isolation and classification of Tern virus: influenza A-Tern South Africa–1961. J Hyg (Lond) *64*, 309–320.

Belser, J.A., Lu, X., Maines, T.R., Smith, C., Li, Y., Donis, R.O., Katz, J.M., and Tumpey, T.M. (2007). Pathogenesis of avian influenza (H7) virus infection in mice and ferrets: enhanced virulence of Eurasian H7N7 viruses isolated from humans. J. Virol. *81*, 11139–11147.

Bender, C., Hall, H., Huang, J., Klimov, A., Cox, N., Hay, A., Gregory, V., Cameron, K., Lim, W., and Subbarao, K. (1999). Characterization of the surface proteins of influenza A (H5N1) viruses isolated from humans in 1997–1998. Virology *254*, 115–123.

Bentz, J. (2000). Minimal aggregate size and minimal fusion unit for the first fusion pore of influenza hemagglutinin-mediated membrane fusion. Biophys. J. *78*, 227–245.

Bizebard, T., Gigant, B., Rigolet, P., Rasmussen, B., Diat, O., Bosecke, P., Wharton, S.A., Skehel, J.J., and Knossow, M. (1995). Structure of influenza virus haemagglutinin complexed with a neutralizing antibody. Nature *376*, 92–94.

Blackburne, B.P., Hay, A.J., and Goldstein, R.A. (2008). Changing selective pressure during antigenic changes in human influenza H3. PLoS Pathog. *4*, e1000058.

Blumenthal, R., Sarkar, D.P., Durell, S., Howard, D.E., and Morris, S.J. (1996). Dilation of the influenza hemagglutinin fusion pore revealed by the kinetics of individual cell–cell fusion events. J. Cell Biol. *135*, 63–71.

Bodian, D.L., Yamasaki, R.B., Buswell, R.L., Stearns, J.F., White, J.M., and Kuntz, I.D. (1993). Inhibition of the fusion-inducing conformational change of influenza hemagglutinin by benzoquinones and hydroquinones. Biochemistry *32*, 2967–2978.

Borrego-Diaz, E., Peeples, M.E., Markosyan, R.M., Melikyan, G.B., and Cohen, F.S. (2003). Completion of trimeric hairpin formation of influenza virus hemagglutinin promotes fusion pore opening and enlargement. Virology *316*, 234–244.

Bosch, F.X., Garten, W., Klenk, H.D., and Rott, R. (1981). Proteolytic cleavage of influenza virus hemagglutinins: primary structure of the connecting peptide between HA1 and HA2 determines proteolytic cleavability and pathogenicity of Avian influenza viruses. Virology *113*, 725–735.

Bosch, F.X., Orlich, M., Klenk, H.D., and Rott, R. (1979). The structure of the hemagglutinin, a determinant for the pathogenicity of influenza viruses. Virology *95*, 197–207.

Brasseur, R. (2000). Tilted peptides: a motif for membrane destabilization (hypothesis). Mol. Membr. Biol. *17*, 31–40.

Brown, E.G. Liu, H., Kit, L.C., Baird, S., and Nesrallah, M. (2001). Pattern of mutation in the genome of influenza A virus on adaptation to increased virulence in the mouse lung: identification of functional themes. Proc. Natl. Acad. Sci. U.S.A. *98*, 6883–6888.

Bullough, P.A., Hughson, F.M., Skehel, J.J., and Wiley, D.C. (1994). Structure of influenza haemagglutinin at the pH of membrane fusion. Nature *371*, 37–43.

Burger, K.N., Wharton, S.A., Demel, R.A., and Verkleij, A.J. (1991). The interaction of synthetic analogs of the N-terminal fusion sequence of influenza virus with a lipid monolayer. Comparison of fusion-active and fusion-defective analogs. Biochim. Biophys. Acta. *1065*, 121–129.

Burnett, F.M., and Bull, D.R. (1943). Changes in influenza virus associated with adaptation to passage in chick embryos. Aust J Exp Med Sci *21*, 55–69.

Carr, C.M., Chaudhry, C., and Kim, P.S. (1997). Influenza hemagglutinin is spring-loaded by a metastable native conformation. Proc. Natl. Acad. Sci. U.S.A. *94*, 14306–14313.

Caton, A.J., Brownlee, G.G., Yewdell, J.W., and Gerhard, W. (1982). The antigenic structure of the influenza virus A/PR/8/34 hemagglutinin (H1 subtype). Cell *31*, 417–427.

Chang, D.K., Cheng, S.F., Deo Trivedi, V., and Yang, S.H. (2000). The amino-terminal region of the fusion peptide of influenza virus hemagglutinin HA2 inserts into sodium dodecyl sulfate micelle with residues 16–18 at the aqueous boundary at acidic pH. Oligomerization and the conformational flexibility. J. Biol. Chem. *275*, 19150–19158.

Chang, D.K., Cheng, S.F., Kantchev, E.A., Lin, C.H., and Liu, Y.T. (2008). Membrane interaction and structure of the transmembrane domain of influenza hemagglutinin and its fusion peptide complex. BMC Biol. *6*, 2.

Chen, B.J., Takeda, M., and Lamb, R.A. (2005). Influenza virus hemagglutinin (H3 subtype) requires palmitoylation of its cytoplasmic tail for assembly: M1 proteins of two subtypes differ in their ability to support assembly. J. Virol. *79*, 13673–13684.

Chen, J., Lee, K.H., Steinhauer, D.A., Stevens, D.J., Skehel, J.J., and Wiley, D.C. (1998a). Structure of the hemagglutinin precursor cleavage site, a determinant of influenza pathogenicity and the origin of the labile conformation. Cell 95, 409–417.

Chen, J., Skehel, J.J., and Wiley, D.C. (1998b). A polar octapeptide fused to the N-terminal fusion peptide solubilizes the influenza virus HA2 subunit ectodomain. Biochemistry 37, 13643–13649.

Chen, J., Skehel, J.J., and Wiley, D.C. (1999). N- and C-terminal residues combine in the fusion-pH influenza hemagglutinin HA(2) subunit to form an N cap that terminates the triple-stranded coiled coil. Proc. Natl. Acad. Sci. U.S.A. *96*, 8967–8972.

Chen, J., Wharton, S.A., Weissenhorn, W., Calder, L.J., Hughson, F.M., Skehel, J.J., and Wiley, D.C. (1995). A soluble domain of the membrane-anchoring chain of influenza virus hemagglutinin (HA2) folds in Escherichia coli into the low-pH-induced conformation. Proc. Natl. Acad. Sci. U.S.A. *92*, 12205–12209.

Chen, L.M., Davis, C.T., Zhou, H., Cox, N.J., and Donis, R.O. (2008). Genetic compatibility and virulence of reassortants derived from contemporary avian H5N1 and human H3N2 influenza A viruses. PLoS Pathog. *4*, e1000072.

Chernomordik, L.V., and Kozlov, M.M. (2008). Mechanics of membrane fusion. Nat. Struct. Mol. Biol. *15*, 675–683.

Choppin, P.W., Murphy, J.S., and Tamm, I. (1960). Studies of two kinds of virus particles which comprise influenza A2 virus strains. III. Morphological characteristics: independence to morphological and functional traits. J. Exp. Med. *112*, 945–952.

Chu, C.M., Dawson, I.M., and Elford, W.J. (1949). Filamentous forms associated with newly isolated influenza virus. Lancet *1*, 602.

Cianci, C., Yu, K.L., Dischino, D.D., Harte, W., Deshpande, M., Luo, G., Colonno, R.J., Meanwell, N.A., and Krystal, M. (1999). pH-dependent changes in photoaffinity labeling patterns of the H1 influenza virus hemagglutinin by using an inhibitor of viral fusion. J. Virol. 73, 1785–1794.

Claas, E.C., Osterhaus, A.D., van Beek, R., De Jong, J.C., Rimmelzwaan, G.F., Senne, D.A., Krauss, S., Shortridge, K.F., and Webster, R.G. (1998). Human influenza A H5N1 virus related to a highly pathogenic avian influenza virus. Lancet *351*, 472–477.

Colman, P.M., and Lawrence, M.C. (2003). The structural biology of type I viral membrane fusion. Nat. Rev. Mol. Cell Biol. *4*, 309–319.

Combrink, K.D., Gulgeze, H.B., Yu, K.L., Pearce, B.C., Trehan, A.K., Wei, J., Deshpande, M., Krystal, M., Torri, A., Luo, G., *et al.* (2000). Salicylamide inhibitors of influenza virus fusion. Bioorg. Med. Chem. Lett. *10*, 1649–1652.

Connor, R.J., Kawaoka, Y., Webster, R.G., and Paulson, J.C. (1994). Receptor specificity in human, avian, and equine H2 and H3 influenza virus isolates. Virology *205*, 17–23.

Cross, K.J., Wharton, S.A., Skehel, J.J., Wiley, D.C., and Steinhauer, D.A. (2001). Studies on influenza haemagglutinin fusion peptide mutants generated by reverse genetics. EMBO J. *20*, 4432–4442.

Cross, K.J., Langley, W.A., Russell, R.J., Skehel, J.J., and Steinhauer, D.A. (2009). Composition and functions of the influenza fusion peptide. Protein Pept. Lett. *16*, 766–778.

Danieli, T., Pelletier, S.L., Henis, Y.I., and White, J.M. (1996). Membrane fusion mediated by the influenza virus hemagglutinin requires the concerted action of at least three hemagglutinin trimers. J. Cell Biol. *133*, 559–569.

Daniels, R.S., Jeffries, S., Yates, P., Schild, G.C., Rogers, G.N., Paulson, J.C., Wharton, S.A., Douglas, A.R., Skehel, J.J., and Wiley, D.C. (1987). The receptor-binding and membrane-fusion properties of influenza virus variants selected using anti-haemagglutinin monoclonal antibodies. EMBO J. *6*, 1459–1465.

Daniels, R., Kurowski, B., Johnson, A.E., and Hebert, D.N. (2003). N-linked glycans direct the cotranslational folding pathway of influenza hemagglutinin. Mol. Cell *11*, 79–90.

Daniels, R.S., Downie, J.C., Hay, A.J., Knossow, M., Skehel, J.J., Wang, M.L., and Wiley, D.C. (1985). Fusion mutants of the influenza virus hemagglutinin glycoprotein. Cell *40*, 431–439.

Deshpande, K.L., Fried, V.A., Ando, M., and Webster, R.G. (1987). Glycosylation affects cleavage of an H5N2 influenza virus hemagglutinin and regulates virulence. Proc. Natl. Acad. Sci. U.S.A. *84*, 36–40.

Dowdle, W.R. (1999). Influenza A virus recycling revisited. Bull. World. Health. Organ. *77*, 820–828.

Durrer, P., Galli, C., Hoenke, S., Corti, C., Gluck, R., Vorherr, T., and Brunner, J. (1996). H+-induced membrane insertion of influenza virus hemagglutinin involves the HA2 amino-terminal fusion peptide but not the coiled coil region. J. Biol. Chem. *271*, 13417–13421.

Eckert, D.M., and Kim, P.S. (2001). Mechanisms of viral membrane fusion and its inhibition. Annu. Rev. Biochem. *70*, 777–810.

Eisen, M.B., Sabesan, S., Skehel, J.J., and Wiley, D.C. (1997). Binding of the influenza A virus to cell-surface receptors: structures of five hemagglutinin–sialyloligosaccharide complexes determined by X-ray crystallography. Virology *232*, 19–31.

Ekiert, D.C., Bhabha, G., Elsliger, M.A., Friesen, R.H., Jongeneelen, M., Throsby, M., Goudsmit, J., and Wilson, I.A. (2009). Antibody recognition of a highly conserved influenza virus epitope. Science *324*, 246–251.

Fleury, D., Barrere, B., Bizebard, T., Daniels, R.S., Skehel, J.J., and Knossow, M. (1999). A complex of influenza hemagglutinin with a neutralizing antibody that binds outside the virus receptor binding site. Nat. Struct. Biol. *6*, 530–534.

Fouchier, R.A., Munster, V., Wallensten, A., Bestebroer, T.M., Herfst, S., Smith, D., Rimmelzwaan, G.F., Olsen, B., and Osterhaus, A.D. (2005). Characterization of a novel influenza A virus hemagglutinin subtype (H16) obtained from black-headed gulls. J. Virol. *79*, 2814–2822.

Fouchier, R.A., Schneeberger, P.M., Rozendaal, F.W., Broekman, J.M., Kemink, S.A., Munster, V., Kuiken, T., Rimmelzwaan, G.F., Schutten, M., Van Doornum, G.J., *et al.* (2004). Avian influenza A virus (H7N7) associated with human conjunctivitis and a fatal case of acute respiratory distress syndrome. Proc. Natl. Acad. Sci. U.S.A. *101*, 1356–1361.

Gambaryan, A.S., Marinina, V.P., Tuzikov, A.B., Bovin, N.V., Rudneva, I.A., Sinitsyn, B.V., Shilov, A.A., and Matrosovich, M.N. (1998). Effects of host-dependent glycosylation of hemagglutinin on receptor-binding properties on H1N1 human influenza A virus grown in MDCK cells and in embryonated eggs. Virology *247*, 170–177.

Gambaryan, A.S., Robertson, J.S., and Matrosovich, M.N. (1999). Effects of egg-adaptation on the receptor-binding properties of human influenza A and B viruses. Virology *258*, 232–239.

Gambaryan, A.S., Tuzikov, A.B., Piskarev, V.E., Yamnikova, S.S., Lvov, D.K., Robertson, J.S., Bovin, N.V., and Matrosovich, M.N. (1997). Specification of receptor-binding phenotypes of influenza virus isolates from different hosts using synthetic sialylglycopolymers: non-egg-adapted human H1 and H3 influenza A and influenza B viruses share a common high binding affinity for 6′-sialyl(N-acetyllactosamine). Virology *232*, 345–350.

Gamblin, S.J., Haire, L.F., Russell, R.J., Stevens, D.J., Xiao, B., Ha, Y., Vasisht, N., Steinhauer, D.A., Daniels, R.S., Elliot, A., *et al.* (2004). The structure and receptor binding properties of the 1918 influenza hemagglutinin. Science *303*, 1838–1842.

Garcia, M., Crawford, J.M., Latimer, J.W., Rivera-Cruz, E., and Perdue, M.L. (1996). Heterogeneity in the haemagglutinin gene and emergence of the highly pathogenic phenotype among recent H5N2 avian influenza viruses from Mexico. J. Gen. Virol. 77 (Pt 7), 1493–1504.

Garten, W., and Klenk, H.D. (1983). Characterization of the carboxypeptidase involved in the proteolytic cleavage of the influenza haemagglutinin. J. Gen. Virol. 64 (Pt 10), 2127–2137.

Gerhard, W., Yewdell, J., Frankel, M.E., and Webster, R. (1981). Antigenic structure of influenza virus haemagglutinin defined by hybridoma antibodies. Nature *290*, 713–717.

Gething, M.J., Doms, R.W., York, D., and White, J. (1986). Studies on the mechanism of membrane fusion: site-specific mutagenesis of the hemagglutinin of influenza virus. J. Cell Biol. *102*, 11–23.

Gottschalk, A., and Lind, P.E. (1949). Product of interaction between influenza virus enzyme and ovomucin. Nature *164*, 232.

Gray, C., Tatulian, S.A., Wharton, S.A., and Tamm, L.K. (1996). Effect of the N-terminal glycine on the secondary structure, orientation, and interaction of the influenza hemagglutinin fusion peptide with lipid bilayers. Biophys. J. *70*, 2275–2286.

Guan, Y., Poon, L.L., Cheung, C.Y., Ellis, T.M., Lim, W., Lipatov, A.S., Chan, K.H., Sturm-Ramirez, K.M., Cheung, C.L., Leung, Y.H., *et al.* (2004). H5N1 influenza: a protean pandemic threat. Proc. Natl. Acad. Sci. U.S.A. *101*, 8156–8161.

Ha, Y., Stevens, D.J., Skehel, J.J., and Wiley, D.C. (2001). X-ray structures of H5 avian and H9 swine influenza virus hemagglutinins bound to avian and human receptor analogs. Proc. Natl. Acad. Sci. U.S.A. *98*, 11181–11186.

Ha, Y., Stevens, D.J., Skehel, J.J., and Wiley, D.C. (2002). H5 avian and H9 swine influenza virus haemagglutinin structures: possible origin of influenza subtypes. EMBO J. *21*, 865–875.

Han, X., Bushweller, J.H., Cafiso, D.S., and Tamm, L.K. (2001). Membrane structure and fusion-triggering conformational change of the fusion domain from influenza hemagglutinin. Nat. Struct. Biol. *8*, 715–720.

Han, X., Steinhauer, D.A., Wharton, S.A., and Tamm, L.K. (1999). Interaction of mutant influenza virus hemagglutinin fusion peptides with lipid bilayers: probing the role of hydrophobic residue size in the central region of the fusion peptide. Biochemistry *38*, 15052–15059.

Han, X., and Tamm, L.K. (2000). A host-guest system to study structure–function relationships of membrane fusion peptides. Proc. Natl. Acad. Sci. U.S.A. *97*, 13097–13102.

Hanson, J.E., Sauter, N.K., Skehel, J.J., and Wiley, D.C. (1992). Proton nuclear magnetic resonance studies of the binding of sialosides to intact influenza virus. Virology *189*, 525–533.

Harrison, S.C. (2008). Viral membrane fusion. Nat. Struct. Mol. Biol. *15*, 690–698.

Harter, C., James, P., Bachi, T., Semenza, G., and Brunner, J. (1989). Hydrophobic binding of the ectodomain of influenza hemagglutinin to membranes occurs through the 'fusion peptide'. J. Biol. Chem. *264*, 6459–6464.

Hartley, C.A., Jackson, D.C., and Anders, E.M. (1992). Two distinct serum mannose-binding lectins function as beta inhibitors of influenza virus: identification of bovine serum beta inhibitor as conglutinin. J. Virol. *66*, 4358–4363.

Hatta, M., Gao, P., Halfmann, P., and Kawaoka, Y. (2001). Molecular basis for high virulence of Hong Kong H5N1 influenza A viruses. Science *293*, 1840–1842.

Helenius, A., and Aebi, M. (2001). Intracellular functions of N-linked glycans. Science *291*, 2364–2369.

Hinshaw, V.S., Webster, R.G., Easterday, B.C., and Bean, W.J., Jr. (1981). Replication of avian influenza A viruses in mammals. Infect. Immun. *34*, 354–361.

Hinshaw, V.S., Webster, R.G., Naeve, C.W., and Murphy, B.R. (1983). Altered tissue tropism of human-avian reassortant influenza viruses. Virology *128*, 260–263.

Hoffman, L.R., Kuntz, I.D., and White, J.M. (1997). Structure-based identification of an inducer of the low-pH conformational change in the influenza virus hemagglutinin: irreversible inhibition of infectivity. J. Virol. *71*, 8808–8820.

Horimoto, T., Rivera, E., Pearson, J., Senne, D., Krauss, S., Kawaoka, Y., and Webster, R.G. (1995). Origin and molecular changes associated with emergence of a highly pathogenic H5N2 influenza virus in Mexico. Virology *213*, 223–230.

Hsu, C.H., Wu, S.H., Chang, D.K., and Chen, C. (2002). Structural characterizations of fusion peptide analogs of influenza virus hemagglutinin. Implication of the necessity of a helix-hinge-helix motif in fusion activity. J. Biol. Chem. *277*, 22725–22733.

Hughey, P.G., Compans, R.W., Zebedee, S.L., and Lamb, R.A. (1992). Expression of the influenza A virus M2 protein is restricted to apical surfaces of polarized epithelial cells. J. Virol. *66*, 5542–5552.

Ibricevic, A., Pekosz, A., Walter, M.J., Newby, C., Battaile, J.T., Brown, E.G. Holtzman, M.J., and Brody, S.L. (2006). Influenza virus receptor specificity and cell tropism in mouse and human airway epithelial cells. J. Virol. *80*, 7469–7480.

Ilyushina, N.A., Govorkova, E.A., Gray, T.E., Bovin, N.V., and Webster, R.G. (2008). Human-like receptor specificity does not affect the neuraminidase-inhibitor susceptibility of H5N1 influenza viruses. PLoS Pathog. *4*, e1000043.

Ito, T., Couceiro, J.N., Kelm, S., Baum, L.G., Krauss, S., Castrucci, M.R., Donatelli, I., Kida, H., Paulson, J.C., Webster, R.G., and Kawaoka, Y. (1998). Molecular basis for the generation in pigs of influenza A viruses with pandemic potential. J. Virol. *72*, 7367–7373.

Ito, T., Suzuki, Y., Suzuki, T., Takada, A., Horimoto, T., Wells, K., Kida, H., Otsuki, K., Kiso, M., Ishida, H., and Kawaoka, Y. (2000). Recognition of N-glycolylneuraminic acid linked to galactose by the alpha2,3 linkage is associated with intestinal replication of influenza A virus in ducks. J. Virol. *74*, 9300–9305.

Iwatsuki-Horimoto, K., Kanazawa, R., Sugii, S., Kawaoka, Y., and Horimoto, T. (2004). The index influenza A virus subtype H5N1 isolated from a human in 1997 differs in its receptor-binding properties from a virulent avian influenza virus. J. Gen. Virol. *85*, 1001–1005.

Jiao, P., Tian, G., Li, Y., Deng, G., Jiang, Y., Liu, C., Liu, W., Bu, Z., Kawaoka, Y., and Chen, H. (2008). A single-amino-acid substitution in the NS1 protein changes the pathogenicity of H5N1 avian influenza viruses in mice. J. Virol. *82*, 1146–1154.

Jin, H., Leser, G.P., Zhang, J., and Lamb, R.A. (1997). Influenza virus hemagglutinin and neuraminidase cytoplasmic tails control particle shape. EMBO J. *16*, 1236–1247.

Jin, H., Subbarao, K., Bagai, S., Leser, G.P., Murphy, B.R., and Lamb, R.A. (1996). Palmitylation of the influenza virus hemagglutinin (H3) is not essential for virus assembly or infectivity. J. Virol. *70*, 1406–1414.

Jones, L.V., Compans, R.W., Davis, A.R., Bos, T.J., and Nayak, D.P. (1985). Surface expression of influenza virus neuraminidase, an amino-terminally anchored viral membrane glycoprotein, in polarized epithelial cells. Mol. Cell. Biol. *5*, 2181–2189.

Katz, J.M., Lu, X., Tumpey, T.M., Smith, C.B., Shaw, M.W., and Subbarao, K. (2000). Molecular correlates of influenza A H5N1 virus pathogenesis in mice. J. Virol. *74*, 10807–10810.

Katz, J.M., Naeve, C.W., and Webster, R.G. (1987). Host cell-mediated variation in H3N2 influenza viruses. Virology *156*, 386–395.

Kawaoka, Y., Krauss, S., and Webster, R.G. (1989). Avian-to-human transmission of the PB1 gene of influenza A viruses in the 1957 and 1968 pandemics. J. Virol. *63*, 4603–4608.

Kawaoka, Y., Naeve, C.W., and Webster, R.G. (1984). Is virulence of H5N2 influenza viruses in chickens associated with loss of carbohydrate from the hemagglutinin? Virology *139*, 303–316.

Kawaoka, Y., Yamnikova, S., Chambers, T.M., Lvov, D.K., and Webster, R.G. (1990). Molecular characterization of a new hemagglutinin, subtype H*14*, of influenza A virus. Virology *179*, 759–767.

Kemble, G.W., Danieli, T., and White, J.M. (1994). Lipid-anchored influenza hemagglutinin promotes hemifusion, not complete fusion. Cell *76*, 383–391.

Kida, H., Yoden, S., Kuwabara, M., and Yanagawa, R. (1985). Interference with a conformational change in the haemagglutinin molecule of influenza virus by antibodies as a possible neutralization mechanism. Vaccine 3, 219–222.

Kido, H., Yokogoshi, Y., Sakai, K., Tashiro, M., Kishino, Y., Fukutomi, A., and Katunuma, N. (1992). Isolation and characterization of a novel trypsin-like protease found in rat bronchiolar epithelial Clara cells. A possible activator of the viral fusion glycoprotein. J. Biol. Chem. *267*, 13573–13579.

Kielian, M., and Rey, F.A. (2006). Virus membrane-fusion proteins: more than one way to make a hairpin. Nat. Rev. Microbiol. *4*, 67–76.

Klenk, E., Faillard, H., and Lempfrid, H. (1955). Uber die enzymatische wirkung von influenzaviren. Hoppe-Seylerias Z Physiol Chem *301*, 235–246.

Klenk, H.D., and Garten, W. (1994). Host cell proteases controlling virus pathogenicity. Trends. Microbiol. *2*, 39–43.

Klenk, H.D., Rott, R., Orlich, M., and Blodorn, J. (1975). Activation of influenza A viruses by trypsin treatment. Virology *68*, 426–439.

Knossow, M., Gaudier, M., Douglas, A., Barrere, B., Bizebard, T., Barbey, C., Gigant, B., and Skehel, J.J. (2002). Mechanism of neutralization of influenza virus infectivity by antibodies. Virology *302*, 294–298.

Knossow, M., and Skehel, J.J. (2006). Variation and infectivity neutralization in influenza. Immunology *119*, 1–7.

Kogure, T., Suzuki, T., Takahashi, T., Miyamoto, D., Hidari, K.I., Guo, C.T., Ito, T., Kawaoka, Y., and Suzuki, Y. (2006). Human trachea primary epithelial cells express both sialyl(alpha2–3)Gal receptor for human parainfluenza virus type 1 and avian influenza viruses, and sialyl(alpha2–6)Gal receptor for human influenza viruses. Glycoconj. J. *23*, 101–106.

Kordyukova, L.V., Serebryakova, M.V., Baratova, L.A., and Veit, M. (2008). S acylation of the hemagglutinin of influenza viruses: mass spectrometry reveals site-specific attachment of stearic acid to a transmembrane cysteine. J. Virol. *82*, 9288–9292.

Korte, T., Ludwig, K., Booy, F.P., Blumenthal, R., and Herrmann, A. (1999). Conformational intermediates and fusion activity of influenza virus hemagglutinin. J. Virol. *73*, 4567–4574.

Lai, A.L., Park, H., White, J.M., and Tamm, L.K. (2006). Fusion peptide of influenza hemagglutinin requires a fixed angle boomerang structure for activity. J. Biol. Chem. *281*, 5760–5770.

Lai, A.L., and Tamm, L.K. (2007). Locking the kink in the influenza hemagglutinin fusion domain structure. J. Biol. Chem. *282*, 23946–23956.

Lamb, R.A., and Jardetzky, T.S. (2007). Structural basis of viral invasion: lessons from paramyxovirus F. Curr. Opin. Struct. Biol. *17*, 427–436.

Lau, W.L., Ege, D.S., Lear, J.D., Hammer, D.A., and DeGrado, W.F. (2004). Oligomerization of fusogenic peptides promotes membrane fusion by enhancing membrane destabilization. Biophys. J. *86*, 272–284.

Lazarowitz, S.G., and Choppin, P.W. (1975). Enhancement of the infectivity of influenza A and B viruses by proteolytic cleavage of the hemagglutinin polypeptide. Virology *68*, 440–454.

Lear, J.D., and DeGrado, W.F. (1987). Membrane binding and conformational properties of peptides representing the NH2 terminus of influenza HA-2. J. Biol. Chem. *262*, 6500–6505.

Li, K.S., Guan, Y., Wang, J., Smith, G.J., Xu, K.M., Duan, L., Rahardjo, A.P., Puthavathana, P., Buranathai, C., Nguyen, T.D., *et al.* (2004). Genesis of a highly pathogenic and potentially pandemic H5N1 influenza virus in eastern Asia. Nature *430*, 209–213.

Li, Y., Han, X., Lai, A.L., Bushweller, J.H., Cafiso, D.S., and Tamm, L.K. (2005). Membrane structures of the hemifusion-inducing fusion peptide mutant G1S and the fusion-blocking mutant G1V of influenza virus hemagglutinin suggest a mechanism for pore opening in membrane fusion. J. Virol. *79*, 12065–12076.

Li, Z.N., Lee, B.J., Langley, W.A., Bradley, K.C., Russell, R.J., and Steinhauer, D.A. (2008). Length requirements for membrane fusion of influenza virus hemagglutinin peptide linkers to transmembrane or fusion peptide domains. J. Virol. *82*, 6337–6348.

Lin, Y.P., Shaw, M., Gregory, V., Cameron, K., Lim, W., Klimov, A., Subbarao, K., Guan, Y., Krauss, S., Shortridge, K., *et al.* (2000). Avian-to-human transmission of H9N2 subtype influenza A viruses: relationship between H9N2 and H5N1 human isolates. Proc. Natl. Acad. Sci. U.S.A. *97*, 9654–9658.

Lin, Y.P., Wharton, S.A., Martin, J., Skehel, J.J., Wiley, D.C., and Steinhauer, D.A. (1997). Adaptation of egg-grown and transfectant influenza viruses for growth in mammalian cells: selection of hemagglutinin mutants with elevated pH of membrane fusion. Virology *233*, 402–410.

Lipatov, A.S., Govorkova, E.A., Webby, R.J., Ozaki, H., Peiris, M., Guan, Y., Poon, L., and Webster, R.G. (2004). Influenza: emergence and control. J. Virol. *78*, 8951–8959.

Luo, G., Torri, A., Harte, W.E., Danetz, S., Cianci, C., Tiley, L., Day, S., Mullaney, D., Yu, K.L., Ouellet, C., *et al.* (1997). Molecular mechanism underlying the action of a novel fusion inhibitor of influenza A virus. J. Virol. *71*, 4062–4070.

Maines, T.R., Szretter, K.J., Perrone, L., Belser, J.A., Bright, R.A., Zeng, H., Tumpey, T.M., and Katz, J.M. (2008). Pathogenesis of emerging avian influenza viruses in mammals and the host innate immune response. Immunol. Rev. *225*, 68–84.

Matrosovich, M., Gao, P., and Kawaoka, Y. (1998). Molecular mechanisms of serum resistance of human influenza H3N2 virus and their involvement in virus adaptation in a new host. J. Virol. *72*, 6373–6380.

Matrosovich, M., Tuzikov, A., Bovin, N., Gambaryan, A., Klimov, A., Castrucci, M.R., Donatelli, I., and Kawaoka, Y. (2000). Early alterations of the receptor-binding properties of H1, H2, and H3 avian influenza virus hemagglutinins after their introduction into mammals. J. Virol. *74*, 8502–8512.

Matrosovich, M., Zhou, N., Kawaoka, Y., and Webster, R. (1999). The surface glycoproteins of H5 influenza viruses isolated from humans, chickens, and wild aquatic birds have distinguishable properties. J. Virol. *73*, 1146–1155.

Matrosovich, M.N., Gambaryan, A.S., Teneberg, S., Piskarev, V.E., Yamnikova, S.S., Lvov, D.K., Robertson, J.S., and Karlsson, K.A. (1997). Avian influenza A viruses differ from human viruses by recognition of sialyloligosaccharides and gangliosides and by a higher conservation of the HA receptor-binding site. Virology *233*, 224–234.

Matrosovich, M.N., Matrosovich, T.Y., Gray, T., Roberts, N.A., and Klenk, H.D. (2004). Human and avian influenza viruses target different cell types in cultures of human airway epithelium. Proc. Natl. Acad. Sci. U.S.A. *101*, 4620–4624.

Meisner, J., Szretter, K.J., Bradley, K.C., Langley, W.A., Li, Z.N., Lee, B.J., Thoennes, S., Martin, J., Skehel, J.J., Russell, R.J., *et al.* (2008). Infectivity studies of influenza virus hemagglutinin receptor binding site mutants in mice. J. Virol. *82*, 5079–5083.

Melikyan, G.B., Jin, H., Lamb, R.A., and Cohen, F.S. (1997). The role of the cytoplasmic tail region of influenza virus hemagglutinin in formation and growth of fusion pores. Virology *235*, 118–128.

Melikyan, G.B., Lin, S., Roth, M.G., and Cohen, F.S. (1999). Amino acid sequence requirements of the transmembrane and cytoplasmic domains of influenza virus hemagglutinin for viable membrane fusion. Mol. Biol. Cell *10*, 1821–1836.

Melikyan, G.B., Markosyan, R.M., Roth, M.G., and Cohen, F.S. (2000). A point mutation in the transmembrane domain of the hemagglutinin of influenza virus stabilizes a hemifusion intermediate that can transit to fusion. Mol. Biol. Cell *11*, 3765–3775.

Mitnaul, L.J., Matrosovich, M.N., Castrucci, M.R., Tuzikov, A.B., Bovin, N.V., Kobasa, D., and Kawaoka, Y. (2000). Balanced hemagglutinin and neuraminidase activities are critical for efficient replication of influenza A virus. J. Virol. *74*, 6015–6020.

Molinari, M., and Helenius, A. (2000). Chaperone selection during glycoprotein translocation into the endoplasmic reticulum. Science *288*, 331–333.

Moscona, A. (2005). Entry of parainfluenza virus into cells as a target for interrupting childhood respiratory disease. J. Clin. Invest. *115*, 1688–1698.

Mosley, V., and Wyckoff, R. (1946). Electron micrography of the virus of influenza. Nature *157*, 263.

Murata, M., Sugahara, Y., Takahashi, S., and Ohnishi, S. (1987). pH-dependent membrane fusion activity of a synthetic twenty amino acid peptide with the same sequence as that of the hydrophobic segment of influenza virus hemagglutinin. J. Biochem. *102*, 957–962.

Naeve, C.W., Hinshaw, V.S., and Webster, R.G. (1984). Mutations in the hemagglutinin receptor-binding site can change the biological properties of an influenza virus. J. Virol. *51*, 567–569.

Naeve, C.W., and Williams, D. (1990). Fatty acids on the A/Japan/305/57 influenza virus hemagglutinin have a role in membrane fusion. EMBO J. *9*, 3857–3866.

Naim, H.Y., Amarneh, B., Ktistakis, N.T., and Roth, M.G. (1992). Effects of altering palmitylation sites on biosynthesis and function of the influenza virus hemagglutinin. J. Virol. *66*, 7585–7588.

Nicholls, J.M., Chan, M.C., Chan, W.Y., Wong, H.K., Cheung, C.Y., Kwong, D.L., Wong, M.P., Chui, W.H., Poon, L.L., Tsao, S.W., *et al.* (2007). Tropism of avian influenza A (H5N1) in the upper and lower respiratory tract. Nat. Med. *13*, 147–149.

Nicholls, J.M., Chan, R.W., Russell, R.J., Air, G.M., and Peiris, J.S. (2008). Evolving complexities of influenza virus and its receptors. Trends. Microbiol. *16*, 149–157.

Nobusawa, E., Aoyama, T., Kato, H., Suzuki, Y., Tateno, Y., and Nakajima, K. (1991). Comparison of complete amino acid sequences and receptor-binding properties among 13 serotypes of hemagglutinins of influenza A viruses. Virology *182*, 475–485.

Ohuchi, M., Ohuchi, R., Feldmann, A., and Klenk, H.D. (1997). Regulation of receptor binding affinity of influenza virus hemagglutinin by its carbohydrate moiety. J. Virol. *71*, 8377–8384.

Orlich, M., and Rott, R. (1994). Thermolysin activation mutants with changes in the fusogenic region of an influenza virus hemagglutinin. J. Virol. *68*, 7537–7539.

Pappas, C., Aguilar, P.V., Basler, C.F., Solorzano, A., Zeng, H., Perrone, L.A., Palese, P., Garcia-Sastre, A., Katz, J.M., and Tumpey, T.M. (2008). Single gene reassortants identify a critical role for PB*1*, HA, and NA in the high virulence of the 1918 pandemic influenza virus. Proc. Natl. Acad. Sci. U.S.A. *105*, 3064–3069.

Park, H.E., Gruenke, J.A., and White, J.M. (2003). Leash in the groove mechanism of membrane fusion. Nat. Struct. Biol. *10*, 1048–1053.

Peiris, J.S., Yu, W.C., Leung, C.W., Cheung, C.Y., Ng, W.F., Nicholls, J.M., Ng, T.K., Chan, K.H., Lai, S.T., Lim, W.L., *et al.* (2004). Re-emergence of fatal human influenza A subtype H5N1 disease. Lancet *363*, 617–619.

Peiris, M., Yuen, K.Y., Leung, C.W., Chan, K.H., Ip, P.L., Lai, R.W., Orr, W.K., and Shortridge, K.F. (1999). Human infection with influenza H9N2. Lancet *354*, 916–917.

Perdue, M.L., Garcia, M., Senne, D., and Fraire, M. (1997). Virulence-associated sequence duplication at the hemagglutinin cleavage site of avian influenza viruses. Virus. Res. *49*, 173–186.

Pfeiffer, S., Fuller, S.D., and Simons, K. (1985). Intracellular sorting and basolateral appearance of the G protein of vesicular stomatitis virus in Madin-Darby canine kidney cells. J. Cell Biol. *101*, 470–476.

Philipp, H.C., Schroth, B., Veit, M., Krumbiegel, M., Herrmann, A., and Schmidt, M.F. (1995). Assessment of fusogenic properties of influenza virus hemagglutinin deacylated by site-directed mutagenesis and hydroxylamine treatment. Virology *210*, 20–28.

Puri, A., Booy, F.P., Doms, R.W., White, J.M., and Blumenthal, R. (1990). Conformational changes and fusion activity of influenza virus hemagglutinin of the H2 and H3 subtypes: effects of acid pretreatment. J. Virol. *64*, 3824–3832.

Qiao, H., Armstrong, R.T., Melikyan, G.B., Cohen, F.S., and White, J.M. (1999). A specific point mutant at position 1 of the influenza hemagglutinin fusion peptide displays a hemifusion phenotype. Mol. Biol. Cell *10*, 2759–2769.

Rafalski, M., Ortiz, A., Rockwell, A., van Ginkel, L.C., Lear, J.D., DeGrado, W.F., and Wilschut, J. (1991). Membrane fusion activity of the influenza virus hemagglutinin: interaction of HA2 N-terminal peptides with phospholipid vesicles. Biochemistry *30*, 10211–10220.

Rapoport, E.M., Mochalova, L.V., Gabius, H.J., Romanova, J., and Bovin, N.V. (2006). Search for additional influenza virus to cell interactions. Glycoconj. J. 23, 115–125.

Ray, N., and Doms, R.W. (2006). HIV-1 coreceptors and their inhibitors. Curr. Top. Microbiol. Immunol. *303*, 97–120.

Reading, S.A., and Dimmock, N.J. (2007). Neutralization of animal virus infectivity by antibody. Arch. Virol. *152*, 1047–1059.

Robertson, J.S. (1993). Clinical influenza virus and the embryonated hen's egg. Rev. Med. Virol. 3, 97–106.

Robertson, J.S., Bootman, J.S., Newman, R., Oxford, J.S., Daniels, R.S., Webster, R.G., and Schild, G.C. (1987). Structural changes in the haemagglutinin which accompany egg adaptation of an influenza A(H1N1) virus. Virology *160*, 31–37.

Rodriquez-Boulan, E., and Sabatini, D.D. (1978). Asymmetric budding of viruses in epithelial monlayers: a model system for study of epithelial polarity. Proc. Natl. Acad. Sci. U.S.A. *75*, 5071–5075.

Rogers, G.N., and D'Souza, B.L. (1989). Receptor binding properties of human and animal H1 influenza virus isolates. Virology *173*, 317–322.

Rogers, G.N., Daniels, R.S., Skehel, J.J., Wiley, D.C., Wang, X.F., Higa, H.H., and Paulson, J.C. (1985). Host-mediated selection of influenza virus receptor variants. Sialic acid-alpha 2,6Gal-specific clones of A/duck/Ukraine/1/63 revert to sialic acid-alpha 2,3Gal-specific wild type in ovo. J. Biol. Chem. *260*, 7362–7367.

Rogers, G.N., and Paulson, J.C. (1983). Receptor determinants of human and animal influenza virus isolates: differences in receptor specificity of the H3 hemagglutinin based on species of origin. Virology *127*, 361–373.

Rogers, G.N., Paulson, J.C., Daniels, R.S., Skehel, J.J., Wilson, I.A., and Wiley, D.C. (1983a). Single amino acid substitutions in influenza haemagglutinin change receptor binding specificity. Nature *304*, 76–78.

Rogers, G.N., Pritchett, T.J., Lane, J.L., and Paulson, J.C. (1983b). Differential sensitivity of human, avian, and equine influenza A viruses to a glycoprotein inhibitor of infection: selection of receptor specific variants. Virology *131*, 394–408.

Rohm, C., Suss, J., Pohle, V., and Webster, R.G. (1996a). Different hemagglutinin cleavage site variants of H7N7 in an influenza outbreak in chickens in Leipzig, Germany. Virology *218*, 253–257.

Rohm, C., Zhou, N., Suss, J., Mackenzie, J., and Webster, R.G. (1996b). Characterization of a novel influenza hemagglutinin, H15: criteria for determination of influenza A subtypes. Virology *217*, 508–516.

Roth, M.G., Compans, R.W., Giusti, L., Davis, A.R., Nayak, D.P., Gething, M.J., and Sambrook, J. (1983). Influenza virus hemagglutinin expression is polarized in cells infected with recombinant SV40 viruses carrying cloned hemagglutinin DNA. Cell 33, 435–443.

Rott, R., Orlich, M., and Scholtissek, C. (1979). Correlation of pathogenicity and gene constellation of influenza A viruses. III. Non-pathogenic recombinants derived from highly pathogenic parent strains. J. Gen. Virol. *44*, 471–477.

Ruigrok, R.W., Andree, P.J., Hooft van Huysduynen, R.A., and Mellema, J.E. (1984). Characterization of three highly purified influenza virus strains by electron microscopy. J. Gen. Virol. 65 (Pt 4), 799–802.

Ruigrok, R.W., Martin, S.R., Wharton, S.A., Skehel, J.J., Bayley, P.M., and Wiley, D.C. (1986). Conformational changes in the hemagglutinin of influenza virus which accompany heat-induced fusion of virus with liposomes. Virology *155*, 484–497.

Russell, R.J., Gamblin, S.J., Haire, L.F., Stevens, D.J., Xiao, B., Ha, Y., and Skehel, J.J. (2004). H1 and H7 influenza haemagglutinin structures extend a structural classification of haemagglutinin subtypes. Virology *325*, 287–296.

Russell, R.J., Stevens, D.J., Haire, L.F., Gamblin, S.J., and Skehel, J.J. (2006). Avian and human receptor binding by hemagglutinins of influenza A viruses. Glycoconj. J. *23*, 85–92.

Russell, R.J., Kerry, P.S., Stevens, D.J., Steinhauer, D.A., Martin, S.R., Gamblin, S.J., and Skehel, J.J. (2008). Structure of influenza hemagglutinin in complex with an inhibitor of membrane fusion. Proc. Natl. Acad. Sci. U.S.A. *105*, 17736–17741.

Ryan-Poirier, K., Suzuki, Y., Bean, W.J., Kobasa, D., Takada, A., Ito, T., and Kawaoka, Y. (1998). Changes in H3 influenza A virus receptor specificity during replication in humans. Virus. Res. *56*, 169–176.

Sakai, T., Ohuchi, R., and Ohuchi, M. (2002). Fatty acids on the A/USSR/77 influenza virus hemagglutinin facilitate the transition from hemifusion to fusion pore formation. J. Virol. *76*, 4603–4611.

Sauter, N.K., Bednarski, M.D., Wurzburg, B.A., Hanson, J.E., Whitesides, G.M., Skehel, J.J., and Wiley, D.C. (1989). Hemagglutinins from two influenza virus variants bind to sialic acid derivatives with millimolar dissociation constants: a 500-MHz proton nuclear magnetic resonance study. Biochemistry *28*, 8388–8396.

Sauter, N.K., Hanson, J.E., Glick, G.D., Brown, J.H., Crowther, R.L., Park, S.J., Skehel, J.J., and Wiley, D.C. (1992). Binding of influenza virus hemagglutinin to analogs of its cell-surface receptor, sialic acid: analysis by proton nuclear magnetic resonance spectroscopy and X-ray crystallography. Biochemistry *31*, 9609–9621.

Scheiffele, P., Rietveld, A., Wilk, T., and Simons, K. (1999). Influenza viruses select ordered lipid domains during budding from the plasma membrane. J. Biol. Chem. *274*, 2038–2044.

Schmidt, M.F. (1982). Acylation of viral spike glycoproteins: a feature of enveloped RNA viruses. Virology *116*, 327–338.

Scholtissek, C. (1994). Source for influenza pandemics. Eur. J. Epidemiol. *10*, 455–458.

Scholtissek, C., Rott, R., Orlich, M., Harms, E., and Rohde, W. (1977). Correlation of pathogenicity and gene constellation of an influenza A virus (fowl plague). I. Exchange of a single gene. Virology *81*, 74–80.

Scholtissek, C., Rohde, W., Von Hoyningen, V., and Rott, R. (1978). On the origin of the human influenza virus subtypes H2N2 and H3N2. Virology *87*, 13–20.

Scholtissek, C., Burger, H., Kistner, O., and Shortridge, K.F. (1985). The nucleoprotein as a possible major factor in determining host specificity of influenza H3N2 viruses. Virology *147*, 287–294.

Schulze, I.T. (1972). The structure of influenza virus. II. A model based on the morphology and composition of subviral particles. Virology *47*, 181–196.

Shinya, K., Ebina, M., Yamada, S., Ono, M., Kasai, N., and Kawaoka, Y. (2006). Avian flu: influenza virus receptors in the human airway. Nature *440*, 435–436.

Shortridge, K.F., Zhou, N.N., Guan, Y., Gao, P., Ito, T., Kawaoka, Y., Kodihalli, S., Krauss, S., Markwell, D., Murti, K.G., *et al.* (1998). Characterization of avian

H5N1 influenza viruses from poultry in Hong Kong. Virology *252*, 331–342.

Sieczkarski, S.B., and Whittaker, G.R. (2005). Viral entry. Curr. Top. Microbiol. Immunol. *285*, 1–23.

Simpson, D.A., and Lamb, R.A. (1992). Alterations to influenza virus hemagglutinin cytoplasmic tail modulate virus infectivity. J. Virol. *66*, 790–803.

Skehel, J.J., Cross, K., Steinhauer, D., and Wiley, D.C. (2001). Influenza fusion peptides. Biochem. Soc. Trans. *29*, 623–626.

Skehel, J.J., Stevens, D.J., Daniels, R.S., Douglas, A.R., Knossow, M., Wilson, I.A., and Wiley, D.C. (1984). A carbohydrate side chain on hemagglutinins of Hong Kong influenza viruses inhibits recognition by a monoclonal antibody. Proc. Natl. Acad. Sci. U.S.A. *81*, 1779–1783.

Skehel, J.J., and Wiley, D.C. (1998). Coiled coils in both intracellular vesicle and viral membrane fusion. Cell *95*, 871–874.

Skehel, J.J., and Wiley, D.C. (2000). Receptor binding and membrane fusion in virus entry: the influenza hemagglutinin. Annu. Rev. Biochem. *69*, 531–569.

Smith, A.E., and Helenius, A. (2004). How viruses enter animal cells. Science *304*, 237–242.

Steinhauer, D.A. (1999). Role of hemagglutinin cleavage for the pathogenicity of influenza virus. Virology *258*, 1–20.

Steinhauer, D.A., Wharton, S.A., Skehel, J.J., and Wiley, D.C. (1995). Studies of the membrane fusion activities of fusion peptide mutants of influenza virus hemagglutinin. J. Virol. *69*, 6643–6651.

Steinhauer, D.A., Wharton, S.A., Skehel, J.J., Wiley, D.C., and Hay, A.J. (1991a). Amantadine selection of a mutant influenza virus containing an acid-stable hemagglutinin glycoprotein: evidence for virus-specific regulation of the pH of glycoprotein transport vesicles. Proc. Natl. Acad. Sci. U.S.A. *88*, 11525–11529.

Steinhauer, D.A., Wharton, S.A., Wiley, D.C., and Skehel, J.J. (1991b). Deacylation of the hemagglutinin of influenza A/Aichi/2/68 has no effect on membrane fusion properties. Virology *184*, 445–448.

Stephens, E.B., Compans, R.W., Earl, P., and Moss, B. (1986). Surface expression of viral glycoproteins is polarized in epithelial cells infected with recombinant vaccinia viral vectors. EMBO J. *5*, 237–245.

Stevens, J., Blixt, O., Chen, L.M., Donis, R.O., Paulson, J.C., and Wilson, I.A. (2008). Recent avian H5N1 viruses exhibit increased propensity for acquiring human receptor specificity. J. Mol. Biol. *381*, 1382–1394.

Stevens, J., Blixt, O., Tumpey, T.M., Taubenberger, J.K., Paulson, J.C., and Wilson, I.A. (2006). Structure and receptor specificity of the hemagglutinin from an H5N1 influenza virus. Science *312*, 404–410.

Stevens, J., Corper, A.L., Basler, C.F., Taubenberger, J.K., Palese, P., and Wilson, I.A. (2004). Structure of the uncleaved human H1 hemagglutinin from the extinct 1918 influenza virus. Science *303*, 1866–1870.

Stieneke-Grober, A., Vey, M., Angliker, H., Shaw, E., Thomas, G., Roberts, C., Klenk, H.D., and Garten, W. (1992). Influenza virus hemagglutinin with multibasic cleavage site is activated by furin, a subtilisin-like endoprotease. EMBO J. *11*, 2407–2414.

Stray, S.J., Cummings, R.D., and Air, G.M. (2000). Influenza virus infection of desialylated cells. Glycobiology *10*, 649–658.

Suarez, D.L., Perdue, M.L., Cox, N., Rowe, T., Bender, C., Huang, J., and Swayne, D.E. (1998). Comparisons of highly virulent H5N1 influenza A viruses isolated from humans and chickens from Hong Kong. J. Virol. *72*, 6678–6688.

Subbarao, K., Klimov, A., Katz, J., Regnery, H., Lim, W., Hall, H., Perdue, M., Swayne, D., Bender, C., Huang, J., *et al.* (1998). Characterization of an avian influenza A (H5N1) virus isolated from a child with a fatal respiratory illness. Science *279*, 393–396.

Sui, J., Hwang, W.C., Perez, S., Wei, G., Aird, D., Chen, L.M., Santelli, E., Stec, B., Cadwell, G., Ali, M., *et al.* (2009). Structural and functional bases for broad-spectrum neutralization of avian and human influenza A viruses. Nat. Struct. Mol. Biol. *16*, 265–273.

Swayne, D.E. (1997). Pathobiology of H5N2 Mexican avian influenza virus infections of chickens. Vet. Pathol. *34*, 557–567.

Takeda, M., Leser, G.P., Russell, C.J., and Lamb, R.A. (2003). Influenza virus hemagglutinin concentrates in lipid raft microdomains for efficient viral fusion. Proc. Natl. Acad. Sci. U.S.A. *100*, 14610–14617.

Taubenberger, J.K., Reid, A.H., Lourens, R.M., Wang, R., Jin, G., and Fanning, T.G. (2005). Characterization of the 1918 influenza virus polymerase genes. Nature *437*, 889–893.

Thoennes, S., Li, Z.N., Lee, B.J., Langley, W.A., Skehel, J.J., Russell, R.J., and Steinhauer, D.A. (2008). Analysis of residues near the fusion peptide in the influenza hemagglutinin structure for roles in triggering membrane fusion. Virology *370*, 403–414.

Tiffany, J.M., and Blough, H.A. (1970). Estimation of the number of surface projections on myxo- and paramyxoviruses. Virology *41*, 392–394.

Tsuchiya, E., Sugawara, K., Hongo, S., Matsuzaki, Y., Muraki, Y., Li, Z.N., and Nakamura, K. (2002). Effect of addition of new oligosaccharide chains to the globular head of influenza A/H2N2 virus haemagglutinin on the intracellular transport and biological activities of the molecule. J. Gen. Virol. *83*, 1137–1146.

Tumpey, T.M., Maines, T.R., Van Hoeven, N., Glaser, L., Solorzano, A., Pappas, C., Cox, N.J., Swayne, D.E., Palese, P., Katz, J.M., and Garcia-Sastre, A. (2007). A two-amino acid change in the hemagglutinin of the 1918 influenza virus abolishes transmission. Science *315*, 655–659.

van Riel, D., Munster, V.J., de Wit, E., Rimmelzwaan, G.F., Fouchier, R.A., Osterhaus, A.D., and Kuiken, T. (2006). H5N1 Virus Attachment to Lower Respiratory Tract. Science *312*, 399.

Veit, M., Kretzschmar, E., Kuroda, K., Garten, W., Schmidt, M.F., Klenk, H.D., and Rott, R. (1991). Site-specific mutagenesis identifies three cysteine residues in the cytoplasmic tail as acylation sites of influenza virus hemagglutinin. J. Virol. *65*, 2491–2500.

Verhoeyen, M., Fang, R., Jou, W.M., Devos, R., Huylebroeck, D., Saman, E., and Fiers, W. (1980).

Antigenic drift between the haemagglutinin of the Hong Kong influenza strains A/Aichi/2/68 and A/Victoria/3/75. Nature *286*, 771–776.

Vines, A., Wells, K., Matrosovich, M., Castrucci, M.R., Ito, T., and Kawaoka, Y. (1998). The role of influenza A virus hemagglutinin residues 226 and 228 in receptor specificity and host range restriction. J. Virol. *72*, 7626–7631.

Wagner, R., Herwig, A., Azzouz, N., and Klenk, H.D. (2005). Acylation-mediated membrane anchoring of avian influenza virus hemagglutinin is essential for fusion pore formation and virus infectivity. J. Virol. *79*, 6449–6458.

Wagner, R., Matrosovich, M., and Klenk, H.D. (2002). Functional balance between haemagglutinin and neuraminidase in influenza virus infections. Rev. Med. Virol. *12*, 159–166.

Wagner, R., Wolff, T., Herwig, A., Pleschka, S., and Klenk, H.D. (2000). Interdependence of hemagglutinin glycosylation and neuraminidase as regulators of influenza virus growth: a study by reverse genetics. J. Virol. *74*, 6316–6323.

Walker, J.A., and Kawaoka, Y. (1993). Importance of conserved amino acids at the cleavage site of the haemagglutinin of a virulent avian influenza A virus. J. Gen. Virol. 74 (Pt 2), 311–314.

Walker, J.A., Molloy, S.S., Thomas, G., Sakaguchi, T., Yoshida, T., Chambers, T.M., and Kawaoka, Y. (1994). Sequence specificity of furin, a proprotein-processing endoprotease, for the hemagglutinin of a virulent avian influenza virus. J. Virol. *68*, 1213–1218.

Ward, A.C. (1997). Virulence of influenza A virus for mouse lung. Virus Genes *14*, 187–194.

Ward, C.W., and Dopheide, T.A. (1981). Amino acid sequence and oligosaccharide distribution of the haemagglutinin from an early Hong Kong influenza virus variant A/Aichi/2/68 (X-31). Biochem. J. *193*, 953–962.

Watowich, S.J., Skehel, J.J., and Wiley, D.C. (1994). Crystal structures of influenza virus hemagglutinin in complex with high-affinity receptor analogs. Structure *2*, 719–731.

Webster, R.G., and Rott, R. (1987). Influenza virus A pathogenicity: the pivotal role of hemagglutinin. Cell *50*, 665–666.

Weis, W., Brown, J.H., Cusack, S., Paulson, J.C., Skehel, J.J., and Wiley, D.C. (1988). Structure of the influenza virus haemagglutinin complexed with its receptor, sialic acid. Nature 333, 426–431.

Weissenhorn, W., Dessen, A., Harrison, S.C., Skehel, J.J., and Wiley, D.C. (1997). Atomic structure of the ectodomain from HIV-1 gp41. Nature *387*, 426–430.

Weissenhorn, W., Hinz, A., and Gaudin, Y. (2007). Virus membrane fusion. FEBS Lett. *581*, 2150–2155.

Wharton, S.A., Martin, S.R., Ruigrok, R.W., Skehel, J.J., and Wiley, D.C. (1988). Membrane fusion by peptide analogues of influenza virus haemagglutinin. J. Gen. Virol. 69 (Pt 8), 1847–1857.

Wharton, S.A., Skehel, J.J., and Wiley, D.C. (1986). Studies of influenza haemagglutinin-mediated membrane fusion. Virology *149*, 27–35.

White, J.M., Delos, S.E., Brecher, M., and Schornberg, K. (2008). Structures and mechanisms of viral membrane fusion proteins: multiple variations on a common theme. Crit Rev Biochem Mol Biol *43*, 189–219.

Wiley, D.C., and Skehel, J.J. (1987). The structure and function of the hemagglutinin membrane glycoprotein of influenza virus. Annu. Rev. Biochem. *56*, 365–394.

Wiley, D.C., Wilson, I.A., and Skehel, J.J. (1981). Structural identification of the antibody-binding sites of Hong Kong influenza haemagglutinin and their involvement in antigenic variation. Nature *289*, 373–378.

Wilson, I.A., and Cox, N.J. (1990). Structural basis of immune recognition of influenza virus hemagglutinin. Annu. Rev. Immunol. *8*, 737–771.

Wilson, I.A., Skehel, J.J., and Wiley, D.C. (1981). Structure of the haemagglutinin membrane glycoprotein of influenza virus at 3 A resolution. Nature *289*, 366–373.

Yamada, S., Suzuki, Y., Suzuki, T., Le, M.Q., Nidom, C.A., Sakai-Tagawa, Y., Muramoto, Y., Ito, M., Kiso, M., Horimoto, T., *et al.* (2006). Haemagglutinin mutations responsible for the binding of H5N1 influenza A viruses to human-type receptors. Nature *444*, 378–382.

Yang, Z.Y., Wei, C.J., Kong, W.P., Wu, L., Xu, L., Smith, D.F., and Nabel, G.J. (2007). Immunization by avian H5 influenza hemagglutinin mutants with altered receptor binding specificity. Science *317*, 825–828.

Yewdell, J.W., Taylor, A., Yellen, A., Caton, A., Gerhard, W., and Bachi, T. (1993). Mutations in or near the fusion peptide of the influenza virus hemagglutinin affect an antigenic site in the globular region. J. Virol. *67*, 933–942.

Yu, K.L., Ruediger, E., Luo, G., Cianci, C., Danetz, S., Tiley, L., Trehan, A.K., Monkovic, I., Pearce, B., Martel, A., *et al.* (1999). Novel quinolizidine salicylamide influenza fusion inhibitors. Bioorg. Med. Chem. Lett. *9*, 2177–2180.

Zhao, X., Singh, M., Malashkevich, V.N., and Kim, P.S. (2000). Structural characterization of the human respiratory syncytial virus fusion protein core. Proc. Natl. Acad. Sci. U.S.A. *97*, 14172–14177.

Structure and Mechanism of the M2 Channel

6

James J. Chou and Jason R. Schnell

Abstract

Viral ion channels have minimalist architecture. Despite their relatively simple structure, viral channels can achieve highly specific gating and selection of ions, and the particular mechanisms may be different from those of prokaryotes and eukaryotes. The M2 proton channel of influenza A virus is a model viral ion channel. This small channel, whose pore is formed by four equivalent transmembrane helices, is the target of two widely used anti-influenza A drugs, amantadine and rimantadine, both belonging to the adamantane class of compounds. However, resistance of influenza A to adamantane is now widespread. Naturally occurring resistant mutants have been observed in as many as six different positions in the transmembrane segment of M2. How could there be so many different resistance-conferring mutations along a transmembrane helix of 25 amino acids? The recently determined high-resolution structures of M2 in complex with adamantane allow us to begin answering this question.

Introduction

As a proven drug target for treatment of influenza A, M2 has been the subject of intense research. Amantadine, the first effective anti-influenza drug, was discovered in the early 1960's at Du Pont (Davies *et al.*, 1964). It was initially approved for treatment of avian influenza A infection, and was later approved for treatment and prophylaxis of all influenza A strains. In the mid-1980's, mutations that confer resistance to the adamantanes were found in M2, a single-pass membrane protein (Hay *et al.*, 1985), resulting in a flurry of research through which it was discovered that M2 was embedded in the viral lipid envelope (Lamb *et al.*, 1985), exhibited proton-specific ion conductance (Pinto *et al.*, 1992; Sugrue and Hay, 1991), and was activated at low pH (Pinto *et al.*, 1992). The adamantanes, which include amantadine and rimantadine, inhibit M2 proton conductance (Pinto *et al.*, 1992; Wang *et al.*, 1993), and prevent efficient viral replication (Takeda *et al.*, 2002).

An early and late role for M2 in the viral life cycle have been suggested, both of which are related to acidification of host cell vesicles (Hay *et al.*, 1985; Helenius, 1992). The early role during viral entry involves dissociation of matrix proteins from the viral ribonucleoprotein complex, which requires low pH (Martin and Helenius, 1991). After endocytosis, the host cell acidifies the endosome, but M2 is needed to conduct those protons across the viral envelope into the viral interior. Fusion of the viral envelope with the endosomal membrane then releases the uncoated ribonucleoproteins into the cytosol for transport into the nucleus. Membrane fusion is catalysed by haemagglutinin (HA), which is anchored to the viral envelope. Like M2, HA is pH-activated, and undergoes an irreversible conformational rearrangement to a fusion-active form when the endosome becomes sufficiently acidic. Presumably, M2-mediated conduction of protons into the small virion must occur prior to HA-mediated fusion, because once the fusion takes place, leading to the viral interior becoming contiguous with and buffered by the host cytosol, the dissociation of matrix proteins from the ribonucleoprotein complex, which requires low pH, cannot happen.

At a later stage of replication, the host cell also acidifies the trans-Golgi network (TGN), by which newly synthesized viral proteins are exported to the cell surface. The viral membrane proteins are topologically inverted in the TGN membrane, with the eventual extracellular domains exposed to the acidified Golgi lumen. In this acidified environment, HA is susceptible to premature rearrangement to the fusion-active conformation, which would result in the budding of fusion-incompetent virus particles. Thus, M2 must activate and equilibrate the Golgi pH with that of the more basic cytosol before HA rearranges (Fig. 6.1). Thus, in both the early and late role, M2 must activate at a higher pH than HA. This functional relationship between M2 and HA was recognized early on (Hay *et al.*, 1985), and adamantane-resistant mutants are occasionally observed in HA (Scholtissek *et al.*, 1998; Steinhauer *et al.*, 1991). The drug-resistant HA mutants are more acid-stable, perhaps restoring the M2-HA order of activation.

Amantadine and rimantadine have served as first-choice antiviral drugs against community outbreaks of influenza A viruses for several decades. Unfortunately, the majority of the circulating virus strains are now resistant to those drugs (Bright *et al.*, 2005; Bright *et al.*, 2006). High-resolution structural data, which was long overdue before the emergence of widespread resistance, has now been obtained (Schnell and Chou, 2008; Stouffer *et al.*, 2008) and should facilitate development of a new generation of drugs.

Structure of the M2 channel in the closed state

M2 is a 97-residue single-pass membrane protein with its N and C terminus directed towards the outside and inside of the virus, respectively (Lamb *et al.*, 1985). It consists of three segments: an extracellular N-terminal segment (residues 1–23), a transmembrane (TM) segment (residues 24–46), and an intracellular C-terminal segment (residues 47–97). Cross linking experiments showed that the channel is formed by a parallel tetrameric array of M2 monomers, in which the formation of intermonomer disulphide bonds at Cys17 and Cys19 may stabilize, but are not essential, for oligomeric assembly (Holsinger and Lamb, 1991; Sugrue and Hay, 1991). Mutagenesis studies have identified two pore-lining residues responsible for channel gating. The imidazole of His37 serves as the pH sensor (Wang *et al.*, 1995), and the indole of Trp41 is the gate which occludes the C-terminal end of the closed channel (Tang *et al.*, 2002).

Obtaining a high resolution structure of this viral proton channel has been a long struggle because the small membrane protein resisted arduous efforts to crystallize, and it was not until very recently that solution NMR technology demonstrated the capability of tackling membrane protein systems. Before the availability of any high-resolution structures, a number of structural models of the channel region had been constructed by means of computational modelling, biochemistry and solid-state NMR (ssNMR) spectroscopy. One model was built by evaluating the functional perturbation of replacing each transmembrane (TM) residue by a cysteine (Pinto *et al.*, 1997). The model is a left-handed four-helix bundle with a twist angle of ~12° and hydrophilic residues Ser31, Gly34, His37 and Trp41 lining the pore. As we will see later in the article, among low-resolution models, this model has the best agreement with the high resolution NMR structure of the closed channel. However, results from this study are not quantitative enough to provide a more detailed and accurate picture of channel mechanism because (1) the functional perturbation was evaluated by whole cell channel recording, in which the M2 activity could depend on many variables such as protein expression level, membrane incorporation, and low levels of endogenous channels (M2 conductance is relatively low above pH 6), and (2) for a minimalist channel such as M2, a single unnatural mutation (multiplied by four in the tetramer) can significantly alter channel assembly. Another model proposed from site-directed infrared dichroism also yielded a left-handed four-helix bundle, but with a twist angle of ~31° (Kukol *et al.*, 1999). Cross and co-workers used ssNMR PISEMA experiment to measure the backbone NH bond vector orientations of TM residues of various M2 constructs in lipid bilayers, and derived a number of structural models. However, due to the limited resolution of the ssNMR methods employed, combined with the variation in construct length, these structures are significantly different, with

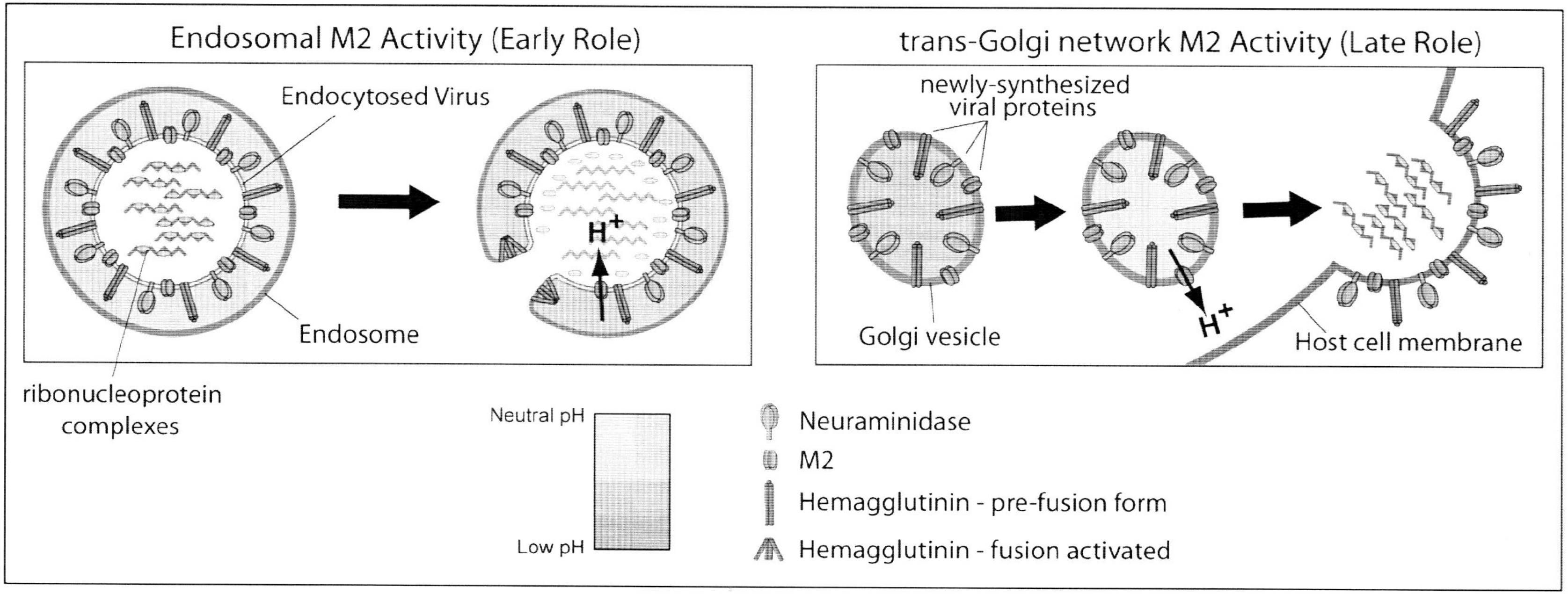

Figure 6.1 Role of M2 in viral replication. An early role in uncoating the viral ribonucleoprotein complex by passing protons from the endosome into the viral interior (left) (Helenius, 1992), and a late role in shunting the acidic pH of the trans-Golgi to prevent conformational rearrangement of haemagglutinin (HA) (right) have been proposed (Hay *et al.*, 1985). In both cases, M2 conducts protons down a pH gradients, and is required to activate before HA converts to its fusion-active form.

packing angles ranging from 25° to 37° (Kovacs *et al.*, 2000; Tian *et al.*, 2003; Wang *et al.*, 2001). Although the above models agree on the overall structural features, they differ in the details essential for understanding the mechanism of channel opening, drug inhibition and drug resistance. Nonetheless, these models have together provided valuable knowledge about the basic structural features of the proton channel that facilitated the eventual high-resolution structure determination.

The solution NMR structure of the closed channel

In 2008, Schnell and Chou succeeded in using solution NMR spectroscopy to provide the first atomic resolution structure of the closed M2 channel in the presence of the inhibitor rimantadine (Schnell and Chou, 2008). The major challenge of solution NMR study of membrane proteins is finding a combination of protein construct and detergent/lipid solubilization system that both support the native function of the protein and yield high quality NMR spectra. A M2 construct that met these requirements includes residues 18 to 60, denoted here as M2 (18–60). This construct, which includes the TM domain as well as 15 residues of the C-terminal extension (Fig. 6.2), forms a very stable tetramer in several detergents, and yielded particularly high-quality NMR spectra in dihexanoyl-phosphocholine (DHPC) detergent micelles. At very low peptide concentrations (~20 μM), chemical cross-linking using dithiobis(succinimidyl)-propionate (DSP) resulted in a homogeneous tetramer. At high protein concentrations used for NMR experiments (0.75 mM monomer), M2 (18–60) runs as a homogeneous tetramer in SDS-PAGE in the absence of cross-linkers, indicating a very stable assembly. It has been confirmed also that when reconstituted into lipid bilayers, M2 (18–60) exhibits specific proton conductance that can be inhibited by low concentrations of both amantadine and rimantadine (Schnell and Chou, unpublished results). Consistent with the pH activated conductance in liposomes, NMR pH titration experiments also showed dramatic changes in spectra as the pH of the sample was lowered, indicating that the channels formed by M2 (18–60) in DHPC micelles could be opened by lowering pH. The above validations of the M2 (18–60) construct approved a full-scale structure determination. Due to the increased stability of the M2 channel at higher pH and thus more homogeneous NMR spectra, the solution structure was determined at pH 7.5. The final NMR sample used for structure determination contained 0.75 mM M2 (18–60) (monomer), ~300 mM DHPC, 40 mM rimantadine, and 30 mM glutamate. Given that DHPC has an aggregation number of 27 (Chou *et al.*, 2004), and rimantadine partitions strongly into phospholipids, locally, there are about four rimantadine molecules per micelle compartment in which the channel resides.

In the closed conformation at pH 7.5 and in the presence of rimantadine, M2 (18–60) is a homotetramer, in which each subunit has an unstructured N-terminus (residues 18–23), a channel-forming TM helix (residues 25–46), a short flexible loop (residues 47–50), and a C-terminal amphipathic (AP) helix (residues 51–59). The TM helices assemble into a four-helix bundle with a left-handed twist angle of ~23° and a well-defined pore (Fig. 6.3A). The N-terminal opening of the channel is very narrow (~3 Å in inner diameter), constricted by a ring of methyl groups from Val27. The pore gradually widens towards the C-terminal end, and becomes the widest at the Gly34 position with an inner diameter of ~ 6 Å. In

```
                                                           TM
M2 Udorn:  MSLLTEVETP IRNEWGCRCN DSSDPLVAA SIIGILHLIL WILDRLFFKC  50
             AP
           IYRFFEHGLK RGPSTEVPE SMREEYRKEQ QSAVDADDSH FVSIELE     97
```

Figure 6.2 Full-length sequence of M2 (A/Udorn/307/1972, H3N2). The transmembrane domain (TM) and the C-terminal amphipathic helix (AP) are boxed and labelled. Sequence covered by the constructs used in solving the solution NMR structure (underlined; residues 18–60) and the x-ray crystal structure (overline, dashed; residues 22–46) are shown.

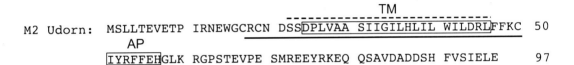

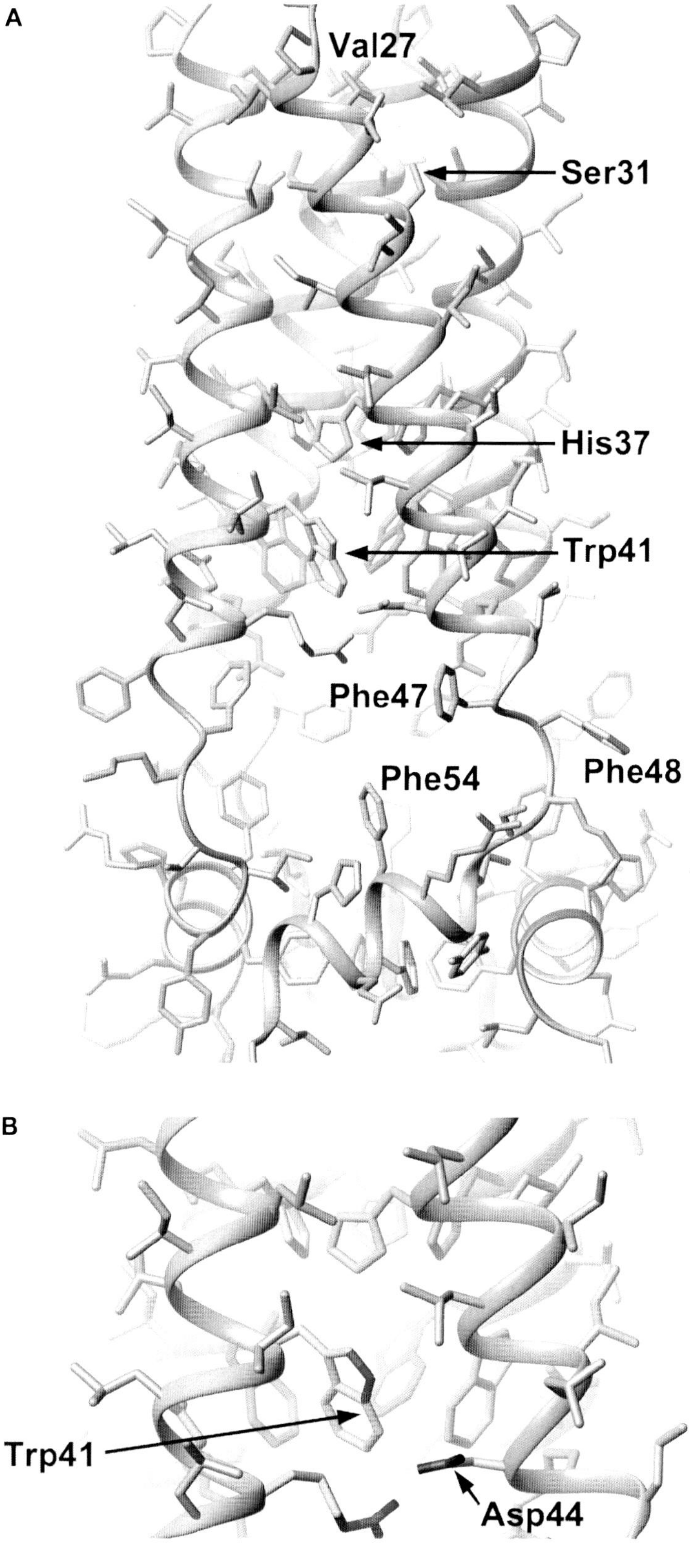

Figure 6.3 NMR structure (2RLF) of the closed channel showing (A) the overall architecture, and (B) the residues involved in channel gating.

the C-terminal half of the channel, the side chains of His37 and Trp41 form the narrowest points, too small to allow anything to pass. Protonation of His37 imidazole is probably the critical event in channel opening, and therefore the His37 side chain conformation is of great interest. According to the three-bond ^{15}N–$^{13}C\gamma$ scalar coupling constant (3JNCγ) of 1.5 Hz, the His37 χ1 rotamer is predominantly trans, but experiences significant rotameric averaging. Therefore, it is unlikely that the four imidazole rings inside the pore are stably interacting with each other, as suggested by the ssNMR studies (Hu *et al.*, 2006). On the other hand, the channel gate, Trp41, is essentially locked in the closed channel, because 3JNCγ of 2.6 Hz indicates that the χ1 rotamer is locked at the trans conformation, and χ2 is also fixed at around −120° based on NOE and residual dipolar coupling data. The restricted motion of Trp41 is partly due to close packing of the four indoles that occludes the C-terminal pore. Moreover, Trp41 indole amine of one subunit is within hydrogen bonding distance to the Asp44 carboxyl carbon of the adjacent subunit. Thus the two residues form a specific inter-subunit contact that stabilizes the Trp41 gate in the closed conformation (Fig. 6.3B). In addition, the Arg45 side chain is also in proximity of forming a salt bridge with Asp44 that may further stabilize the closed gate.

At the C-terminal end of the channel, there is a short loop (residues 47–50) that connects the TM domain to the C-terminal AP helix. The AP helix is roughly perpendicular (~82°) to the TM helix, and its orientation and amphipathic character suggest that it lies on the surface of the membrane (Tian *et al.*, 2003). In addition to the transmembrane four-helix bundle, the AP helices also form a separate tetrameric domain with a head-to-tail assembly and a right-handed packing mode. It is possible that the function of the AP helix tetramer is to further stabilize the M2 tetramer, in particular, when the TM helices open up during activation. The TM construct alone without the AP helix does not form a stable tetramer, which may explain the sample-to-sample variability in helix tilt observed by ssNMR (Li *et al.*, 2007). Inclusion of the AP helix has been shown to limit conformational variation in lipid bilayers (Cady *et al.*, 2007).

A somewhat less satisfying aspect of the solution NMR structure is the lack of NMR signals, thus structural restraints, for residues 47–50 in the detergent micelles. Consequently, in the NMR structure refinement, the AP helix is slightly detached from the C-terminal end of the TM domain because the conformation of the loop (residues 47–50) is poorly defined. However, the loop could adopt a more stable conformation in the viral membrane. In the M2 (18–60) construct used in the NMR study, Cys50 was replaced by serine to avoid potential disulphide formation. Cys50 is palmitoylated *in vivo*, although it is not required for function (Sugrue *et al.*, 1990). Modelling shows that extending the TM helix to Phe48 would place residue 50 facing the membrane, allowing for proper insertion of the palmitoyl acyl chain into the lipid bilayer. We believe the incorporation of the Cys50 palmitoyl chain in lipid bilayer would rigidify the loop and anchor the beginning of the AP helix to the membrane. This minor rearrangement of the loop would also bring the AP helices closer to the TM domain, and possibly form a phenylalanine cage consisting of phenylalanines 47, 48, 54, and 55 on the C-terminal side of the pore.

The X-ray structure of the transmembrane domain

In a separate crystallographic study, DeGrado and co-workers determined the 2.05 Å crystal structure of a shorter construct, M2 (22–46) with the I33M mutation, in β-octylglucoside detergent (Fig. 6.4). Although the peptide was crystallized at pH 7.3, which in principle supports the closed channel conformation, the crystal structure is significantly different from the closed channel structure determined by NMR. In the crystal structure, the four-helix bundle is tightly packed at the N-terminal end of the TM domain as in the NMR structure, but the helices dramatically splay outward at an average angle of ~40° at the C-terminus. As a result of the large inter-helical angle, the C-terminal end of the channel is wide open with no obvious channel gate. The splayed conformation of the C-terminus is stabilized by non-native inter-tetramer crystal contacts. A plausible explanation for the striking difference between the NMR and the crystal structures, both

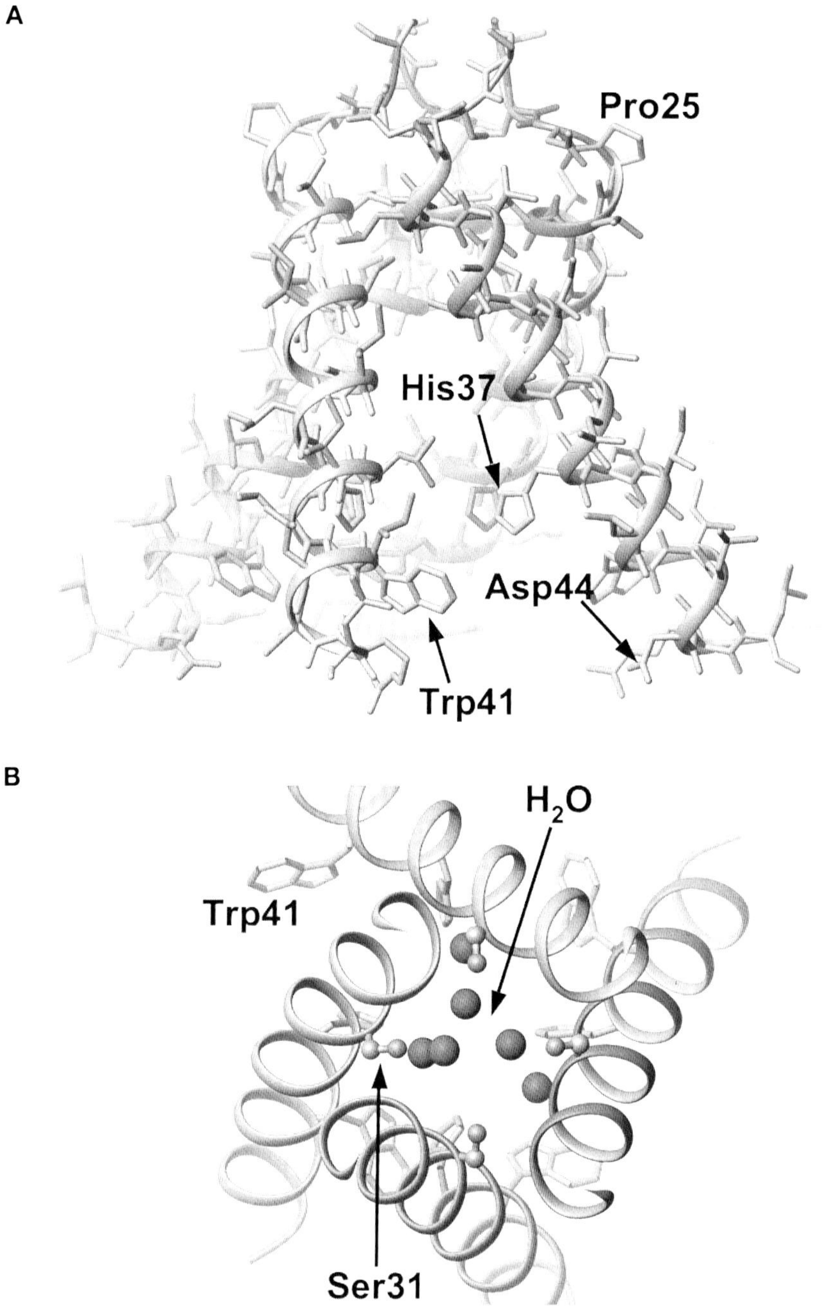

Figure 6.4 X-ray structure (3BKD) of an M2 transmembrane segment (A) from within the membrane, and (B) into the channel pore, with Ser31 in ball-and-stick representation. The several water molecules bound to the Ser31 hydroxyls are shown as spheres.

experimentally determined to high resolution at a similar pH, is the different lengths of constructs used in the two studies. The NMR structure described above shows that inter-molecular interactions between the AP helices are important forces for stabilizing the C-terminal half of the channel assembly. In order to accommodate the bulky Trp41 side chains within the pore, the TM helices splay slightly at the C-terminus of the TM domain, resulting in little interhelical contacts below Leu38. The unfavourable helical packing of this portion of the channel is compensated by the tetramerization of the AP helices. Indeed, truncation of the AP helix results in channels that rapidly lose channel activity (Salom, 2000; Tobler *et al.*, 1999).

Mechanism of proton conductance

It is widely recognized that hydration of the channel pore is required for proton conductance, because the only two proton acceptors above the Trp41 gate, the hydrophilic residue at position 31 and the absolutely conserved His37, are too far apart along the TM helix to relay protons in the absence of water molecules. A popular view of proton translocation in water is the classic Grotthuss mechanism, in which protons hop from one water molecule to the next along a structured water wire (reviewed in Cukierman, 2006). This mechanism provides for highly efficient conductance because protons can bind to one side of a water wire and another proton could leave on the other side by forming and breaking of hydrogen bonds. An example of a protein utilizing a proton-hopping mechanism is the gramicidin proton conductor, a 15 amino acid peptide antibiotic from the soil bacterial species *Bacillus brevis* that can translocate protons across the membrane at an amazing rate of $> 2 \times 10^9$ H^+/s. This peptide forms a short beta-helix in the lipid bilayer, and two helices dimerize in a head-to-head fashion to form a small transmembrane channel. Water molecules are coordinated inside the beta-helices such that protons can be translocated rapidly in the Grotthuss type mechanism. However, channel recordings of M2 in liposomes indicate a conductance rate on the order of 10 H+/s/tetramer (Lin and Schroeder, 2001), a rate that is more compatible with hydronium ion diffusion across the pore. A simple calculation based on the translational diffusion coefficient of water at room temperature ($2 \times 10^9\ m^2/s$), the dimensions of the hydrated pore (40 × 4 Å), and a pH gradient from 6 to 7 across the membrane, gives a rate of ~ 40 H^+/s. Based on the above considerations, it is unlikely that protons are conducted through a structured water wire inside the pore, as in the case of Gramicidin. A more plausible scenario is that when the channel opens, the pH gradient drives the hydroniums from the low pH side to the high pH side by diffusion, while a few hydrophilic residues in the channel pore serve as relay points.

Structure and dynamics of the M2 channel in the open state

The open channel must support a hydrated pore for conducting protons. Although a well-defined structure of the open channel, if it does exist, is not yet available, some features of the open state may be gleaned from the solution NMR and X-ray studies. Channel activation occurs upon protonation of the His37 imidazoles that are closely packed in the channel pore. This must result in electrostatic repulsion that substantially weakens helical packing in the TM domain. Presumably this widens the pore to admit water molecules. The NMR pH titration experiment showed that lowering the pH from 7.5 to 6.0 severely broadens most of the NMR resonances corresponding to the channel-forming TM helix (Schnell and Chou, 2008). Since, the 15N R1 and R2 relaxation rates and self-diffusion coefficients were essentially unchanged between pH 7.5 and 6.0 (Schnell & Chou, unpublished results), the resonance broadening must be caused by exchange between multiple conformations rather than protein aggregation. The open channel is probably not a unique structure like the closed channel, but in dynamic exchange between multiple energetically comparable states. M2 lacks the extensive structural scaffolding observed in larger ion channels that undergo exquisite conformational changes between the open and closed conformation. Thus, M2 appears to have evolved a two-state gating mechanism in which the closed state is structurally rigid, but the open state is dynamic with a loose quaternary structure.

The picture of multiple open channel conformers qualitatively agrees with the 2.05 Å crystal structure of the TM peptide. As reasoned above, in the absence of the AP helix, the TM helix does not assemble into a stable tetramer, and thus the crystal structure could represent an intermediate of channel opening despite the high pH of the buffer. An interesting feature of the crystal structure is that the packing of the tetramer deviates substantially from C_4 rotational symmetry, i.e. the four pairs of helix–helix packing interfaces are significantly different from each other with some being more closely packed than others. Although this could be a result of crystal packing artefact (the surface area of crystal contacts is comparable to that of the intra-tetramer helix–helix interface in the asymmetric unit), the various helix–helix packing arrangements could partially represent the intermediate states of channel opening. The N-terminal portions of the crystal structures are most similar, with divergence occurring at the highly conserved Gly34 in the middle of the TM domain. This results in greater or lesser separation between the helices at the C-terminus, and a large variation in the helix–helix crossing angle. Coarse-grained molecular dynamics simulations, however, suggest that charging of the histidines results in a somewhat different conformation in which the helices become more separated at the C-terminus, but maintain a relatively narrow helix–helix crossing angle (Carpenter *et al.*, 2008).

In one of the four subunit interfaces in the X-ray structure, the N-terminal helices are close enough for side chains to interact. It is interesting to mention that a salt bridge is formed between Asp44 of one helix and Arg45 of the other helix. This is in contrast to the Asp44 to Trp41 interhelical hydrogen bond that was observed in the closed conformation by solution NMR. Although experiments are needed to test this hypothesis, switching between the closed and open conformation may involve swapping the salt-bridge partner of Asp44 from the Trp41 gate to Arg45. This would allow the gate to swing open, as well as provide for a wider pore for hydration.

In contrast to the TM domain, the NMR resonances of the AP helices are essentially unaffected by lowering the pH, indicating that the C-terminal base of the tetramer remains intact and may be needed to preserve the overall tetrameric state of M2 when the TM helices are destabilized in the open state. In addition to the AP helices, a pair of N-terminal cysteines, Cys17 and Cys19, have been shown to form inter-subunit disulphides *in vivo* (Holsinger and Lamb, 1991; Sugrue and Hay, 1991). Although these cysteines are not required for channel function, they are conserved in nature and may play a role in keeping the tetramer together in the open state.

The solution NMR system also permitted the measurement of channel gate dynamics as a function of pH. Since the indole amide resonance of Trp41 remained intense as the pH was lowered from 7.5 to 6.0, it serves as a useful NMR probe for monitoring channel opening. We compared the millisecond timescale dynamics of the Trp41 indole ring between the closed and open states by carrying out relaxation-compensated Carr–Purcell–Meiboom–Gill (CPMG) experiments (Loria *et al.*, 1999) at pH 7.5, 7.0, and 6.0. A two-site exchange model (Allerhand and Thiele, 1966) fits the dependence of ^{15}N relaxation due to chemical shift exchange on the frequency of pulses ($1/\tau cp$) that refocus chemical shift evolution. As the pH was lowered from 7.5 to 6.0, the rate of fluctuation increased by more than four-fold, indicating that the gate was 'unlocked' upon channel activation. Adding rimantadine to the channel at an intermediate pH of 7.0 slowed the timescale of the gate motion to that of a drug-free gate at pH 7.5. Those results provided a direct proof that Trp41 indole is the gate and is unlocked upon channel activation and relocked by drug binding. The exchange rate derived from the CPMG experiment does not reflect the rate of pore breathing between the open and closed states. The exchange rate probably reflects the rate of Trp41 indole flipping because the chemical environment of the indole amine is dominated by side chain $\chi 1$ and $\chi 2$ rotamers. Nevertheless, the increased dynamics of Trp41 indole is a consequence of channel opening.

Events of channel activation

The first event in channel activation is sensing the change of pH outside. The NMR structure of the closed channel shows little room for a water molecule to enter the pore from the N-terminal end: a ring of methyl groups from Val27 constricts the N-terminal end of the pore to ~3.1 Å. However, the static picture does not describe the dynamic aspects of the channel in nature. Even at pH 7.5, the NMR spectra show resonances that are clearly broadened due to conformational/chemical exchange, indicating that the channel undergoes a small amount of conformational exchange. The transient opening of the channel allows for sampling of bulk solution pH even when it is largely closed at high pH.

The first point of hydration, or proton relay, after entering the pore from the N-terminal end is probably Ser31 because a polar residue (Ser, Asn, or Lys) is present at this position in all sequenced variants of M2 (Ito *et al.*, 1991). According to NMR water NOE experiments, the TM region of the reconstituted channel is largely protected from water by the DHPC micelle, and the Val27 ring at the N-terminus and the Trp41 gate at the C-terminus block water from freely diffusing into the pore. However, inside the pore, water NOEs concentrate at Ser31, suggesting the presence of water at that position. A cluster of water molecules coordinated by the hydroxyl groups of Ser31 was also observed in the crystal structure of the TM domain at 2.05 Å (Stouffer *et al.*, 2008).

The force behind channel opening comes mostly from protonation of His37 at low pH. In this case, the His37 imidazole rings become positively charged and the electrostatic repulsion is responsible for destabilizing helix–helix packing and widening of the pore (Hu *et al.*, 2006). The unpacking of the four-helix bundle also breaks the inter-molecular hydrogen bonds between the Trp41 indole amine and the Asp44 carboxyl oxygen, allowing the gate to open and protons or hydroniums to pass. The positively charged imidazoles also have greatly increased affinity for water and thereby facilitate the hydration of the widened pore for proton conductance.

Finally, where do protons exit from the channel? As reasoned above, in a true membrane environment where the Cys50 palmitoyl chain of the native M2 is incorporated, the base formed by AP helices would be closer to C-terminal end of the TM domain. In this conformation, the region immediately below the Trp41 gate is densely packed with hydrophobic phenylalanines, commonly referred to as the phenylalanine cage. If this picture is true, then the proton passage is not as simple as entering from the N-terminal opening and exiting axially at the other side of the pore. In this scenario, after passing the Trp41 gate, the protons or hydroniums most likely exit laterally to the hydrophilic head group region of the membrane. Asp44 and Arg45, two hydrophilic residues in the otherwise hydrophobic portion of the channel, may serve to assist proton exit by accepting protons or hydroniums. Asp44 is highly conserved (it is occasionally replaced by asparagine) whereas Arg45 is completely conserved.

Drug inhibition

Although the NMR structure provides a detailed model for understanding how the M2 channel works, a matter of great biomedical interest (and perhaps even greater debate) is how the adamantane drugs inhibit channel activity. There are two proposed mechanisms of channel inhibition. One mechanism is that the molecule serves as a plug that stearically blocks the channel passage. In this mechanism the drug could block access to the channel, like charybdotoxin, which blocks Ca2+-activated K+ channels (MacKinnon and Miller, 1988), or bind somewhere inside the pore like the tertiary amines (Zhou *et al.*, 2001). Alternatively, the molecule could prevent conformational changes needed for channel opening. This type of mechanism may be allosteric, since the drug-binding site could be located anywhere on the channel that is affected by channel opening. An example of such a mechanism is the blocking of the voltage-gated K^+ channel by spider venom hanatoxin (Lee and MacKinnon, 2004; Swartz and MacKinnon, 1997).

Drug binding site

Before the availability of a high-resolution structure of the channel bound to the drug, the drug binding site has been predicted from the location of mutations that confer resistance. Most models placed the drug inside the pore, in the vicinity of V27, A30 and S31 (for example in Stauffer *et al.*, 2005). However, this does not explain all of the observed resistance mutations. Mutations that confer drug resistance also occur at L26, G34, and L38. Mapping all of these residues onto the structure reveals that Val27, Ala30, and Gly34 are pore-lining, but Leu26, Ser31 and Leu38 are in the helix–helix packing interface (Fig. 6.5). Moreover, these mutations are spread out over more than three turns of the TM helix, covering a distance much larger than the dimensions of amantadine or rimantadine. It is difficult to imagine, from a mechanistic point of view, that a single, specific drug-binding site inside the pore can be responsible for generating all these resistance mutations. Another seemingly intuitive observation, which supports a pore binding mechanism,

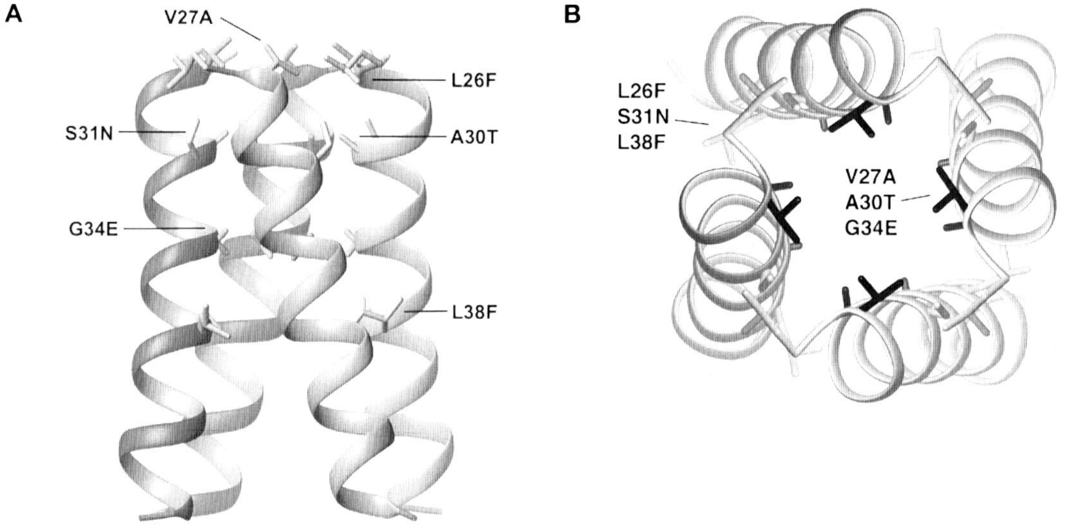

Figure 6.5 Positions of drug-resistant mutants are mapped onto the NMR structure. (A) Resistance mutations can occur over nearly 2/3 of the transmembrane domain, suggesting an allosteric mechanism of resistance. (B) Resistance mutations that increase hydrophilicity point into the pore (shaded black), whereas those that conserve hydropathy are found at the helix–helix interfaces (shaded grey).

is that amantadine only inhibits channel activity when it is applied to the medium bathing the N-terminal ectodomain, and not when applied to the C-terminal cytoplasmic tail; injection of 50 μM rimantadine into cells expressing the A/M2 channel does not inhibit the channel, whereas application of the same concentration to the solution bathing the ectodomain inhibits the channel fully (Chizhmakov *et al.*, 1996). This is a very perplexing result, since the hydrophobic adamantanes can easily pass through membranes (which gives rise to the non-specific antiviral activity observed at high concentrations due to buffering of the virus interior). Thus, one would expect inhibition of the channel even in the case of an N-terminal pore-blocking site. We suspect that the observed lack of inhibition is due to the enormous complexity of the whole cell system. In a Xenopus oocyte, there are a large number of membrane-bound organelles that could absorb the hydrophobic adamantane compound.

Solution NMR spectroscopy of M2 (18–60) in DHPC detergent micelles provided the first definitive observation of a specific drug binding pocket on the closed channel. In this study, the location of rimantadine binding was determined by a number of nuclear Overhauser enhancement (NOE) experiments that observes magnetization transfer between the channel and the drug at specific chemical shifts. The NOEs to rimantadine were observed from the backbone amides of Leu43 and Arg45, and the side chain methyls of Leu40 and Ile42. The drug NOEs were independently validated by a NOE spectrum recorded on a sample containing uniform ^{15}N- and ^{2}H-labelled protein, deuterated detergent, and protonated rimantadine. In this sample, any NOEs observed by the exchangeable amine protons of the channel must arise from the drug. No NOEs from drug were observed in any other region of the protein, including the widely proposed drug-binding site in the channel pore.

The NOEs were also consistent with simple NMR chemical shift perturbation experiments. Upon addition of rimantadine at pH 7.0, the line width of NMR resonances in the ^{1}H–^{15}N correlation spectrum become significantly sharper and more homogeneous, similar to the spectra at pH 7.5. Most dramatically, the resonances of residues 43 – 45 at the C-terminus of the TM helix, which are absent in the drug-free sample, become observable (Schnell and Chou, 2008). Consistent with the NOE experiments at pH 7.5, chemical shift perturbation experiments at both pH 7.5 and 7.0 showed that for the largely closed channel, addition of high concentrations of rimantadine affects neither the chemical shift nor the line width of the pore-lining residues around Gly34, strongly suggesting that rimantadine does not bind the closed channel in the pore.

The binding site that was identified is between adjacent helices at the C-terminal end of the TM domain near the Trp41 gate (Fig. 6.6A). The binding pocket is clearly external – accessible from the membrane side of the channel – and consists of residues 40–45. As expected of a specific drug-binding site, the drug pocket in the closed channel has unique structural properties (Fig. 6.6B). The amine head-group of rimantadine is in contact with the polar side chains of Asp44, Arg45, and the indole amine of Trp41. The side chains of Ile42 from one helix, and Leu40 and Leu43 from another helix form the hydrophobic walls of the binding pocket that interact with the adamantane group of rimantadine. Thus, rimantadine covers a unique polar patch in the otherwise hydrophobic environment of that region of the TM domain. Interactions between rimantadine and the channel are consistent with structure–activity relationships of the adamantane group (Aldrich *et al.*, 1971). In particular, the basic nitrogen group and size limits at the methyl site are critical. These requirements are the result of the interactions with Asp44 and the small hydrophobic pocket formed by Ile42, respectively.

Thus, adamantanes bind at the subunit interface and stabilize the tight assembly of the closed channel. Because channel activation at low pH arises from protonation and destabilization of channel assembly, the action of drug in stabilizing the closed conformation is apparent. Stabilization of the closed conformation also explains the shift in pK_a of His37 that is observed in the presence of drug (Hu *et al.*, 2007): protonation of the imidazole rings when they are closely packed in the drug-stabilized closed conformation is highly unfavourable due to electrostatic repulsion.

The crystallographic study of the M2 transmembrane domain suggests a completely different mode of drug binding. In this study, M2 (22–46) with the G34A mutation was crystallized in β-octylglucoside detergent at pH 5.3 in the presence of amantadine (Stouffer *et al.*, 2008). The 3.5 Å crystallographic map shows electron density inside the pore near Ser31. The authors assigned this density to amantadine and positioned the drug such that its hydrophobic adamantane cage is coordinated by hydroxyl groups from Ser31, and its hydrophilic amine interacts with the methyl of Ala34 (Fig. 6.6C). However, this assignment is not definitive because the dimensions of the roughly spherical adamantane cage are between 2.9 and 3.5 Å, comparable to the resolution of the diffraction data. More importantly, in the same publication, a 2.05 Å crystal structure of M2 (22–46) solved in the absence of drug shows that there is a cluster of four structured water molecules coordinated by the Ser31 hydroxyl groups at exactly the same site where the amantadine was assigned in the 3.5 Å structure (Fig. 6.6D). It was not possible, however, to identify water molecules at the lower resolution at 3.5 Å (we note that in contrast to the coordinates for the 3.5 Å structure, which contains a single amantadine molecule plus M2 peptide, the coordinates for the 2.05 Å structure include three beta-octylglucoside detergent molecules, a chloride ion, four amino groups, three polyethylene glycols, and 19 water molecules per M2 tetramer). Since the TM peptides crystallized in the absence and presence of amantadine have essentially the same structure (backbone pairwise RMSD of 0.9 Å), and the NMR structure bound to the drug has water at the Ser31 position, it is entirely possible that the proposed density of amantadine in the 3.5 Å crystal structure also comes from structured water molecules. Therefore, experimental support for the pore binding site awaits further verification.

Figure 6.6 The adamantane binding site observed by solution NMR (A) showing interactions with the channel gating mechanism of M2. (B) The amine (blue) and adamantyl-cage (orange) of rimantadine interact with the polar (Asp44/Arg45) and hydrophobic (Ile42/Leu40/Leu43) components, respectively, of the NMR drug pocket. In contrast, (C) the proposed drug–M2 complex based on 3.4 Å diffraction data positions the hydrophobic adamantane cage in contact with the Ser31 hydroxyls (3C9J). (D) In a much higher resolution data set (3BKD; 2.05 Å) that allows detergent, ions, water, and polyethylene glycol to be identified, a cluster of water molecules are bound to the Ser31 hydroxyls.

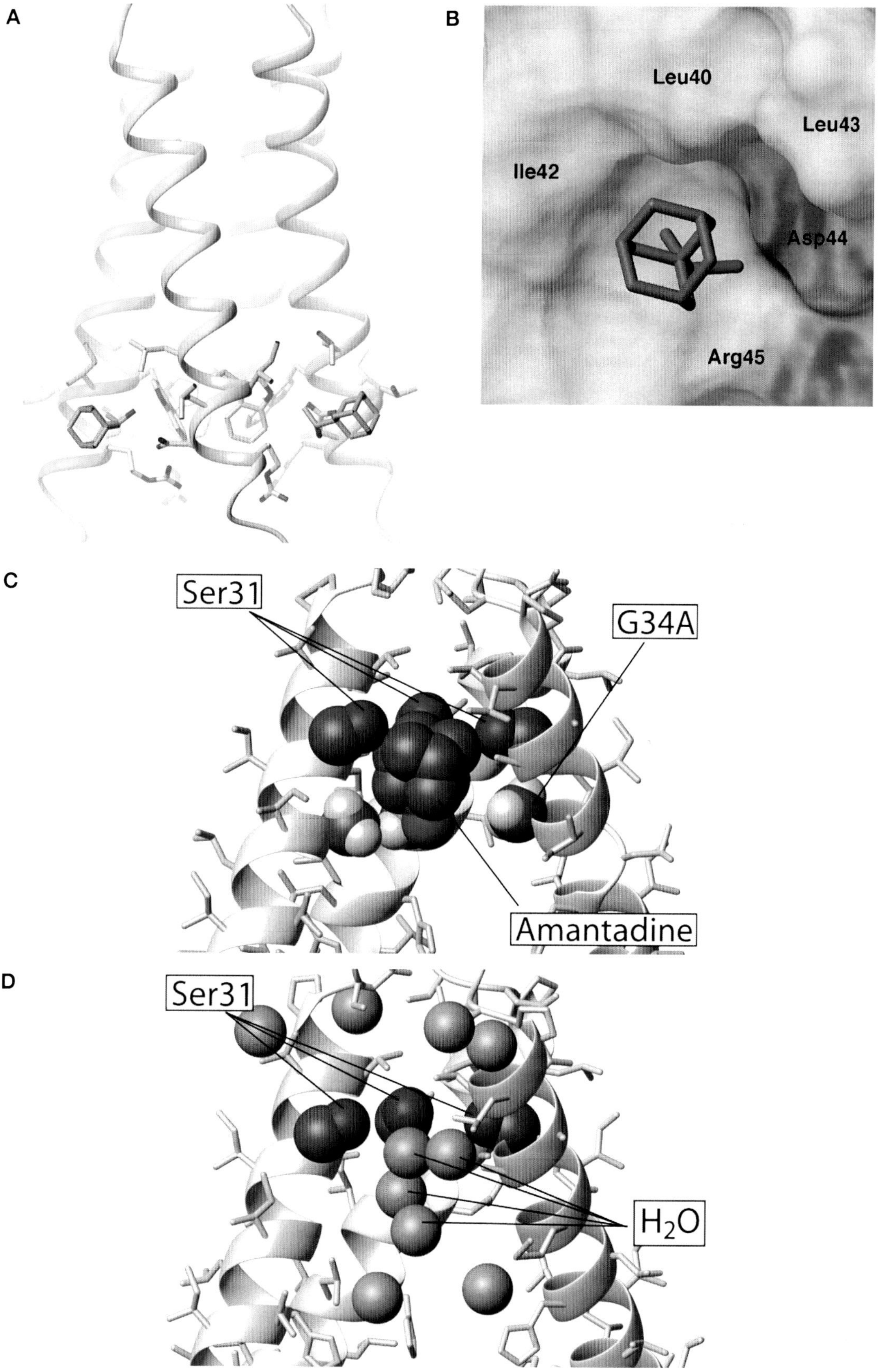
A
B
Leu40
Leu43
Ile42
Asp44
Arg45
C
Ser31
G34A
Amantadine
D
Ser31
H_2O

Drug-binding affinity

There is not a definitive binding constant for the drug. Early channel recording experiments using whole cells measured an isochronic apparent inhibitory constant (K_i) of 9 μM for the wild type Udorn M2 (Wang *et al.*, 1993), although K_i does not necessarily reflect the dissociation constant (K_d) in a complex environment such as a whole cell. In a later study (Czabotar *et al.*, 2004), a K_d of 10 nM was obtained by measuring changes in Trp41 fluorescence of purified Weybridge M2 in LDAO detergent micelles upon addition of rimantadine at pH 5.0. This K_d is stronger than the K_i in Wang *et al.* (1993) by almost three orders of magnitude. However, two concerns with this experimental K_d are that (1) the M2 solubilized in LDAO detergent in this study is not tetrameric (Figure 3B of the original paper), and (2) rimantadine is very lipophilic, e.g. it has very low water solubility, thus changes in tryptophan fluorescence are probably due to the non-specific effect of drug partitioning into the detergent micelles. Finally, a more recent surface plasmon resonance (SPR) study on purified M2 peptide introduced into liposome bilayers determined a K_d of ~0.4 mM for the Udorn M2 (Astrahan *et al.*, 2004). However, this study used a TM-only M2 construct, which is known to be a much less stable tetramer than the native protein, raising the question of whether the K_d is physiologically meaningful. The above numbers differ by as much as 10^4 fold, making it impossible for us to get even a crude idea of the actual strength and specificity of drug binding to the M2 channel. We attribute this lack of understanding to technical difficulties related to obtaining a reproducibly pure and quantitative amount of reconstituted M2 sample that supports the native conformation, and the inability to separate drug–protein and drug–lipid/detergent interactions in a rigorous fashion.

While the apparent inhibitory constant K_i is sufficient for therapeutic purposes, a true measure of K_d is important for understanding the mechanism of drug binding. Solution NMR study of M2 (18–60) in DHPC detergent micelles shows that the NOEs between the channel and the drug rimantadine observed at pH 7.5 correspond to the same chemical shift as that of the free rimantadine in DHPC solution, suggesting that the drug binds in fast exchange within the time regime of NMR chemical shift evolution. Moreover, preliminary results from NMR chemical shift perturbation experiment using the same system shows that the amount of rimantadine required to saturate M2 binding is dependent on the amount of free detergent micelles in the solution, which also have millimolar affinity for the drug. Although more detailed studies are required, the initial NMR characterization suggests that the K_d is in the submillimolar to micromolar range, but certainly not nanomolar.

Drug resistance

Based on the dynamic nature of the open state, the inter-helical position of the adamantane binding site, and the nature of the drug resistance mutations, we have proposed an allosteric inhibition mechanism (Schnell and Chou, 2008). In our model, drug binding stabilizes the tightly packed closed channel conformation, making it harder to initiate conductance, whereas drug-resistance mutations destabilize the closed conformation, allowing activation to proceed as normal upon activation of His37. Observed resistance-conferring mutations in M2 include L26F, V27A, A30T, S31N, G34A, and L38F. Together, these positions span more than two-thirds of the TM domain. However, when these residues are mapped onto the high-resolution structure of the closed channel, they divide into two classes: those at helix–helix packing interfaces and those within the pore (Fig. 6.5). Mutations in pore residues (A30T and G34E) increase the hydrophilicity of the channel, whereas mutations in residues at the helix interface (L26F, S31N, L38F) preserve hydropathy but alter helix–helix packing, and thus the energetics of channel assembly. V27A forms a tight ring at the N-terminus of the channel and could be considered both inter-helical and pore-lining. Thus, all mutations are either positioned at helix–helix interfaces and potentially alter the stability of the closed channel conformation, and/or point into the pore and lower the energy barrier for pore hydration.

There are no resistance mutations near the drug-binding site. This is not surprising given the functional requirements of the residues that are in proximity to the channel pH sensor (His37) and the channel gate (Trp41). In fact, few mutations are ever observed in this region of the channel, drug-resistant or otherwise (Ito *et al.*, 1991). Occasionally, D44N is observed, but this preserves the polar nature of position 44. The residues that form the hydrophobic walls of the binding pocket – Leu40, Ile42, and Leu43 – are on the lipid face of the channel and must be hydrophobic to maintain stable membrane partitioning. Occasionally Leu43 is mutated to a phenylalanine, which is compatible with forming the hydrophobic wall of the adamantane binding pocket.

Evolution seems to have selected exclusively allosteric drug-resistant mutants. No doubt this is partly due to the functional constraints on the actual drug-binding site. But also, the drug pocket consists of residues from two adjacent TM helices, which makes channel destabilization an efficient route to prevent drug binding. Consistent with this conclusion, position 38 is the most C-terminal site in which a resistance mutation has been observed (L38F), and is also the most C-terminal residue besides the critical Trp41-Asp44 gating interaction that makes interhelical contacts. Since the helices splay slightly at the C-terminus of the TM domain to accommodate the bulky Trp41 indole rings within the channel pore, residues beside the gating residues (41, 44, and 45) below Leu38 no longer play important roles in helix–helix packing and are thus not mutated to resist drug binding. We note that destabilization of the channel quaternary structure is necessary, but not sufficient, for channel hydration and conductance. Protonation of His37 and other hydration points along the channel are also required for native activity. Thus, at least for the M2 channel, allosteric mutations are the only mutations that provide an opportunity for the virus to evade drug inhibition without significantly altering channel activity.

References

Aldrich, P.E., Hermann, E.C., Meier, W.E., Paulshock, M., Prichard, W.W., Snyder, J.A., and Watts, J.C. (1971). Antiviral agents. 2. Structure–activity relationships of compounds related to 1-adamantanamine. J. Med. Chem. *14*, 535–543.

Allerhand, A., and Thiele, E. (1966). Analysis of Carr-Purcell spin-echo NMR Experiments on multiple-spin systems. II. The effect of chemical exchange by NMR spectroscopy. J. Am. Chem. Soc. *121*, 2331–2332.

Astrahan, P., Kass, I., Cooper, M.A., and Arkin, I.T. (2004). A novel method of resistance for influenza against a channel-blocking antiviral drug. Proteins *55*, 251–257.

Bright, R.A., Medina, M.J., Xu, X., Perez-Oronoz, G., Wallis, T.R., Davis, X.M., Povinelli, L., Cox, N.J., and Klimov, A.I. (2005). Incidence of adamantane resistance among influenza A (H3N2) viruses isolated worldwide from 1994 to 2005: a cause for concern. Lancet *366*, 1175–1181.

Bright, R.A., Shay, D.K., Shu, B., Cox, N.J., and Klimov, A.I. (2006). Adamantane resistance among influenza A viruses isolated early during the 2005–2006 influenza season in the United States. Jama *295*, 891–894.

Cady, S.D., Goodman, C., Tatko, C.D., DeGrado, W.F., and Hong, M. (2007). Determining the orientation of uniaxially rotating membrane proteins using unoriented samples: a 2H, 13C, AND 15N solid-state NMR investigation of the dynamics and orientation of a transmembrane helical bundle. J. Am. Chem. Soc. *129*, 5719–5729.

Carpenter, T., Bond, P., Khalid, S., and Sansom, M.S. (2008). Self-assembly of a simple membrane protein: coarse-grained molecular dynamics simulations of the influenza M2 channel. Biophys. J. *95*, 3790–3801.

Chizhmakov, I.V., Geraghty, F.M., Ogden, D.C., Hayhurst, A., Antoniou, M., and Hay, A.J. (1996). Selective proton permeability and pH regulation of the influenza virus M2 channel expressed in mouse erythroleukaemia cells. J. Physiol. *494*, 329–336.

Chou, J.J., Baber, J.L., and Bax, A. (2004). Characterization of phospholipid mixed micelles by translational diffusion. J. Biomol. NMR *29*, 299–308.

Cukierman, S. (2006). Et tu, Grotthuss! and other unfinished stories. Biochim. Biophys. Acta *1757*, 876–885.

Czabotar, P.E., Martin, S.R., and Hay, A.J. (2004). Studies of structural changes in the M2 proton channel of influenza A virus by tryptophan fluorescence. Virus. Res. *99*, 57–61.

Davies, W.L., Grunert, R.R., Haff, R.F., McGahen, J.W., Neumayer, E.M., Paulshock, M., Watts, J.C., Wood, T.R., Hermann, E.C., and Hoffmann, C.E. (1964). Antiviral activity of 1-adamantanamine (amantadine). Science *144*, 862–863.

Hay, A.J., Wolstenholme, A.J., Skehel, J.J., and Smith, M.H. (1985). The molecular basis of the specific anti-influenza action of amantadine. EMBO J. *4*, 3021–3024.

Helenius, A. (1992). Unpacking the incoming influenza virus. Cell *69*, 577–578.

Holsinger, L.J., and Lamb, R.A. (1991). Influenza virus M2 integral membrane protein is a homotetramer stabilized by formation of disulfide bonds. Virology *183*, 32–43.

Hu, J., Fu, R., and Cross, T.A. (2007). The chemical and dynamical influence of the anti-viral drug amantadine on the M2 proton channel transmembrane domain. Biophys. J. *93*, 276–283.

Hu, J., Fu, R., Nishimura, K., Zhang, L., Zhou, H.X., Busath, D.D., Vijayvergiya, V., and Cross, T.A. (2006). Histidines, heart of the hydrogen ion channel from influenza A virus: toward an understanding of conductance and proton selectivity. Proc. Natl. Acad. Sci. U.S.A. *103*, 6865–6870.

Ito, T., Gorman, O.T., Kawaoka, Y., Bean, W.J., and Webster, R.G. (1991). Evolutionary analysis of the influenza A virus M gene with comparison of the M1 and M2 proteins. J. Virol. *65*, 5491–5498.

Kovacs, F.A., Denny, J.K., Song, Z., Quine, J.R., and Cross, T.A. (2000). Helix tilt of the M2 transmembrane peptide from influenza A virus: an intrinsic property. J. Mol. Biol. *295*, 117–125.

Kukol, A., Adams, P.D., Rice, L.M., Brunger, A.T., and Arkin, T.I. (1999). Experimentally based orientational refinement of membrane protein models: A structure for the Influenza A M2 H+ channel. J. Mol. Biol. *286*, 951–962.

Lamb, R.A., Zebedee, S.L., and Richardson, C.D. (1985). Influenza virus M2 protein is an integral membrane protein expressed on the infected-cell surface. Cell *40*, 627–633.

Lee, S.Y., and MacKinnon, R. (2004). A membrane-access mechanism of ion channel inhibition by voltage sensor toxins from spider venom. Nature *430*, 232–235.

Li, C., Qin, H., Gao, F.P., and Cross, T.A. (2007). Solid-state NMR characterization of conformational plasticity within the transmembrane domain of the influenza A M2 proton channel. Biochim. Biophys. Acta *1768*, 3162–3170.

Lin, T.I., and Schroeder, C. (2001). Definitive assignment of proton selectivity and attoampere unitary current to the M2 ion channel protein of influenza A virus. J. Virol. *75*, 3647–3656.

Loria, J.P., Rance, M., and Palmer, A.G., 3rd (1999). A relaxation-compensated Carr-Purcell-Meiboom-Gill sequence for characterizing chemical exchange by NMR spectroscopy. J. Am. Chem. Soc. *121*, 2331–2332.

MacKinnon, R., and Miller, C. (1988). Mechanism of charybdotoxin block of the high-conductance, Ca^{2+}-activated K^{+} channel. J. Gen. Physiol. *91*, 335–349.

Martin, K., and Helenius, A. (1991). Nuclear transport of influenza virus ribonucleoproteins: the viral matrix protein (M1) promotes export and inhibits import. Cell *67*, 117–130.

Pinto, L.H., Dieckmann, G.R., Gandhi, C.S., Papworth, C.G., Braman, J., Shaughnessy, M.A., Lear, J.D., Lamb, R.A., and DeGrado, W.F. (1997). A functionally defined model for the M2 proton channel of influenza A virus suggests a mechanism for its ion selectivity. Proc. Natl. Acad. Sci. U.S.A. *94*, 11301–11306.

Pinto, L.H., Holsinger, L.J., and Lamb, R.A. (1992). Influenza virus M2 protein has ion channel activity. Cell *69*, 517–528.

Salom, D., Hill, B.R., Lear, J.D., and DeGrado, W.F. (2000). pH-dependent tetramerization and amantadine binding of the transmembrane helix of M2 from the influenza A virus. Biochemistry *39*, 14160–14170.

Schnell, J.R., and Chou, J.J. (2008). Structure and mechanism of the M2 proton channel of influenza A virus. Nature *451*, 591–595.

Scholtissek, C., Quack, G., Klenk, H.D., and Webster, R.G. (1998). How to overcome resistance of influenza A viruses against adamantane derivatives. Antiviral. Res. *37*, 83–95.

Steinhauer, D.A., Wharton, S.A., Skehel, J.J., Wiley, D.C., and Hay, A.J. (1991). Amantadine selection of a mutant influenza virus containing an acid-stable hemagglutinin glycoprotein: evidence for virus-specific regulation of the pH of glycoprotein transport vesicles. Proc. Natl. Acad. Sci. U.S.A. *88*, 11525–11529.

Stouffer, A.L., Nanda, V., Lear, J.D., and DeGrado, W.F. (2005). Sequence determinants of a transmembrane proton channel: an inverse relationship between stability and function. J. Mol. Biol. *347*, 169–179.

Stouffer, A.L., Acharya, R., Salom, D., Levine, A.S., Di Costanzo, L., Soto, C.S., Tereshko, V., Nanda, V., Stayrook, S., and DeGrado, W.F. (2008). Structural basis for the function and inhibition of an influenza virus proton channel. Nature *451*, 596–599.

Sugrue, R.J., Belshe, R.B., and Hay, A.J. (1990). Palmitoylation of the influenza A virus M2 protein. Virology *179*, 51–56.

Sugrue, R.J., and Hay, A.J. (1991). Structural characteristics of the M2 protein of influenza A viruses: evidence that it forms a tetrameric channel. Virology *180*, 617–624.

Swartz, K.J., and MacKinnon, R. (1997). Hanatoxin modifies the gating of a voltage-dependent K+ channel through multiple binding sites. Neuron *18*, 665–673.

Takeda, M., Pekosz, A., Shuck, K., Pinto, L.H., and Lamb, R.A. (2002). Influenza a virus M2 ion channel activity is essential for efficient replication in tissue culture. J. Virol. *76*, 1391–1399.

Tang, Y., Zaitseva, F., Lamb, R.A., and Pinto, L.H. (2002). The gate of the influenza virus M2 proton channel is formed by a single tryptophan residue. J. Biol. Chem. *277*, 39880–39886.

Tian, C., Gao, P.F., Pinto, L.H., Lamb, R.A., and Cross, T.A. (2003). Initial structural and dynamic characterization of the M2 protein transmembrane and amphipathic helices in lipid bilayers. Protein. Sci. *12*, 2597–2605.

Tobler, K., Kelly, M.L., Pinto, L.H., and Lamb, R.A. (1999). Effect of cytoplasmic tail truncations on the activity of the M(2) ion channel of influenza A virus. J. Virol. *73*, 9695–9701.

Wang, C., Lamb, R.A., and Pinto, L.H. (1995). Activation of the M2 ion channel of influenza virus: a role for the transmembrane domain histidine residue. Biophys. J. *69*, 1363–1371.

Wang, C., Takeuchi, K., Pinto, L.H., and Lamb, R.A. (1993). Ion channel activity of influenza A virus M2 protein: characterization of the amantadine block. J. Virol. *67*, 5585–5594.

Wang, J., Kim, S., Kovacs, F., and Cross, T.A. (2001). Structure of the transmembrane region of the M2 protein H(+) channel. Protein. Sci. *10*, 2241–2250.

Zhou, M., Morais-Cabral, J.H., Mann, S., and MacKinnon, R. (2001). Potassium channel receptor site for the inactivation gate and quaternary amine inhibitors. Nature *411*, 657–661.

Virulence Genes of the 1918 Pandemic Influenza Virus

7

Natalie Pica, Terrence M. Tumpey, Adolfo García-Sastre and Peter Palese

Abstract

The pandemic influenza virus of 1918 was extremely virulent and caused significant morbidity and mortality to millions of people worldwide. The extinct virus caused severe pathology in both the upper and lower respiratory tract, resulting in fatal respiratory complications and bacterial pneumonia. The pathology associated with 1918 influenza virus infections is thought to be the result of the exposure of an immunologically naïve host population to an unusually virulent virus. Using reverse genetics, the 1918 pandemic virus has been studied in different animal models in an attempt to determine which viral genes contribute to the increased virulence. Studies to date point to the role of the haemagglutinin, neuraminidase, and the polymerase basic protein 1 genes as the virulence genes responsible for the high pathogenicity seen with the 1918 influenza virus.

Introduction: a deadly virus

The 1918 influenza pandemic, or 'Spanish influenza', killed more than 50 million people across the globe and is considered the most deadly pandemic in recorded history (Johnson and Mueller, 2002). It is estimated that 675,000 Americans died from this influenza virus (Crosby, 1989), depressing the average life expectancy in the USA by more than 10 years (Arias, 2004). In addition to affecting individuals at both ends of the age spectrum, high rates of mortality were seen in 18- to 30-year-olds (Linder and Grove, 1947), a characteristic specific to pandemics seen in the 20th century (Ahmed *et al.*, 2007; Linder and Grove, 1947). Interestingly, the novel swine influenza virus (H1N1) that emerged in April 2009 also affects young adults (Michaelis *et al.*, 2009). However, it is unlikely that we will see the same virulence of the 1918 influenza virus with this novel virus.

The 1918 influenza virus could not be studied until the end of the twentieth century, as the virus was not isolated at the time of the pandemic. Viral RNA was extracted from formalin-fixed, paraffin-embedded lung samples from the Armed Forces Institute of Pathology, as well as from a frozen lung sample from an Alaskan victim buried in the permafrost. Subsequent reverse transcription PCR and sequencing provided insight into the genetic make-up of the pathogen (Basler *et al.*, 2001; Reid *et al.*, 1999, 2000, 2002, 2004; Taubenberger *et al.*, 1997, 2005). Tumpey and colleagues at the Centers for Disease Control, with the collaboration of the Basler, Garcia-Sastre, and Palese laboratories at Mount Sinai School of Medicine, generated a recombinant virus containing eight RNA segments from the 1918 pandemic strain (Tumpey *et al.*, 2005). The reconstruction of the virus by reverse genetics allowed for the first characterization of its properties and was the basis of further studies in the field (Tumpey *et al.*, 2005). The virus was also studied *in vivo* in mice (Tumpey *et al.*, 2005), ferrets (Tumpey *et al.*, 2007a; Watanabe *et al.*, 2008), guinea pigs (Van Hoeven *et al.*, 2009) and macaques (Kobasa *et al.*, 2007) without previous adaptation.

This chapter reviews the origin of the virus and the pathology of the disease associated its infection, as well as the features that contribute to its extraordinary virulence.

Origin of the 1918 influenza virus

The origin of the 1918 influenza virus remains controversial. It is considered to be a mammalian virus with a haemagglutinin (HA) sequence most similar to the H1N1 classic swine virus A/Swine/Iowa/30 (Swine/30) (Reid *et al.*, 1999). Indeed, immunization with Swine/30 virus conferred protective immunity to mice challenged with lethal recombinant 1918 virus (Tumpey *et al.*, 2004). In addition, monoclonal antibodies isolated from individuals born during or before 1915 were not only protective against the 1918 virus, but were also shown to be cross-reactive and neutralizing against the HA of Swine/30 (Yu *et al.*, 2008). The pandemic strain also shares structural and sequence similarities with avian viruses isolated after 1918 (Reid *et al.*, 2000; Taubenberger *et al.*, 2005) and is the mammalian virus that is most similar to avian viruses in phylogenic analyses (Reid *et al.*, 1999).

Due to these sequence similarities between the 1918 influenza viral genes and avian consensus sequences, it has been postulated that the pandemic strain emerged from a series of adaptations that improved the fitness of a previous avian virus (Taubenberger *et al.*, 2005). Sequences of sufficient avian influenza viruses collected before 1918 are not available for comparison, however. This sampling issue makes it difficult to rely on the analysis of avian viruses to decide the issue.

An alternative theory suggests that the 1918 virus is a reassortant between avian and human strains- like the two other pandemic viruses of the last century that emerged in 1957 and in 1968 (Wright *et al.*, 2006). In these latter events, three and two gene segments were derived from an avian parent virus, respectively. A definitive answer to the question of the origin of the virus will have to await sequence analysis of pre-1918 viruses.

Pathology seen in disease caused by the 1918 influenza virus

Contemporary influenza virus isolates are known to cause acute respiratory distress, high fever, cough, headache, fatigue, and inflammation of the upper respiratory tract and trachea (Wright *et al.*, 2006). People with chronic pulmonary or cardiac diseases or diabetes mellitus have been shown to have an increased risk of haemorrhagic bronchitis, pneumonia and death (Taubenberger and Morens, 2008). In contrast, infection with the 1918 virus caused death in young, healthy adults (Ahmed *et al.*, 2007; Linder and Grove, 1947). On autopsy, severe tracheitis, necrotizing bronchiolitis (Lucke *et al.*, 1919), and oedema (Winternitz *et al.*, 1920) were appreciated, with increased hyaline deposition seen in the membranes of alveoli and alveolar ducts (Goodpasture, 1919). Secondary bacterial pneumonia and respiratory complications are thought to be major causes of adult death (Johnson and Mueller, 2002; Morens *et al.*, 2008; Taubenberger *et al.*, 2000).

The extreme pathology seen in those infected with the 1918 virus is thought to be a result of the virulence genes of the pandemic strain and the immune response of the host. Studies to identify these factors are discussed below.

Non-structural protein 1: less critical than once thought

The non-structural (NS) gene codes for the NS1 protein as well as the nuclear export protein (NEP/NS2), the product of a spliced mRNA (Lamb and Choppin, 1979). The N-terminal domain of the NS1 protein has been shown to be involved in RNA binding and many cellular proteins have been reported to interact with the NS1 protein (Palese and Shaw, 2006).

In the response to early influenza virus infection, the retinoic acid-inducible gene I (RIG-I)-mediated induction of interferon-beta (IFNβ) is thought to be essential (Mibayashi *et al.*, 2007). After recognizing viral single stranded RNA bearing a 5′ phosphate (Pichlmair *et al.*, 2006) or double-stranded RNA (Yoneyama *et al.*, 2004), it is hypothesized that the caspase activation recruitment domain (CARD) of RIG-I signals to the interferon-beta promoter stimulator 1 (IPS1) (Kawai *et al.*, 2005; Kumar *et al.*, 2006) activating transcription factors activator protein 1 (AP1), interferon regulatory factor 3 (IRF3), and nuclear factor-kappa B (NF-κB) via cellular signalling. As a result, upregulation of IFN and several IFN-stimulated genes (ISGs) is seen. The NS1 protein of influenza A viruses (Guo *et al.*, 2007; Mibayashi *et al.*, 2007; Opitz *et al.*, 2007;

Pichlmair *et al.*, 2006), including the NS1 from the 1918 isolate A/New York/1/18/(Kochs *et al.*, 2007), has been shown to complex with RIG-I in infected cells and block the upregulation of IFN and ISGs (Fig. 7.1).

Additionally, NS1 has been shown to interact with several known ISGs, including RNA dependent protein kinase R (PKR) (Bergmann *et al.*, 2000; Hatada *et al.*, 1999; Lu *et al.*, 1995) and 2′–5′ oligo (A) synthetase (Min and Krug, 2006). NS1 blocks the antiviral effects of PKR (Bergmann *et al.*, 2000; Hatada *et al.*, 1999; Lu *et al.*, 1995), a serine threonine protein kinase that inhibits eukaryotic translation initiation factor 2 (eIF-2α) (Gale and Katze, 1998). eIF-2α is needed for translation of both cellular and viral mRNAs, making PKR an important anti-viral factor. It is also proposed that NS1 sequesters double stranded RNA, thereby blocking the activation of 2′–5′ oligo(A) synthetase (2,5-OAS) (Min and Krug, 2006). 2,5-OAS is believed to be an activator of RNaseL, an endonuclease that causes cellular and viral RNA degradation (Dong and Silverman, 1995; Min and Krug, 2006; Moss *et al.*, 1978; Wang *et al.*, 2002; Zhou *et al.*, 1997).

It has also been reported that NS1 interferes with the host's transcriptional activities. The post-translational processing of host mRNA is inhibited through the binding of NS1 to the 30-kDa cleavage and polyadenylation specificity factor (CPSF), (Noah *et al.*, 2003) and poly(A) binding protein (PABII) (Chen *et al.*, 1999). It is postulated that these interactions lead to nuclear accumulation of host pre-mRNA, including those of antiviral proteins (Noah *et al.*, 2003). Moreover, the NS1 directly blocks nuclear export of cellular mRNAs by interacting with components of the nuclear pore machinery (Satterly *et al.*, 2007).

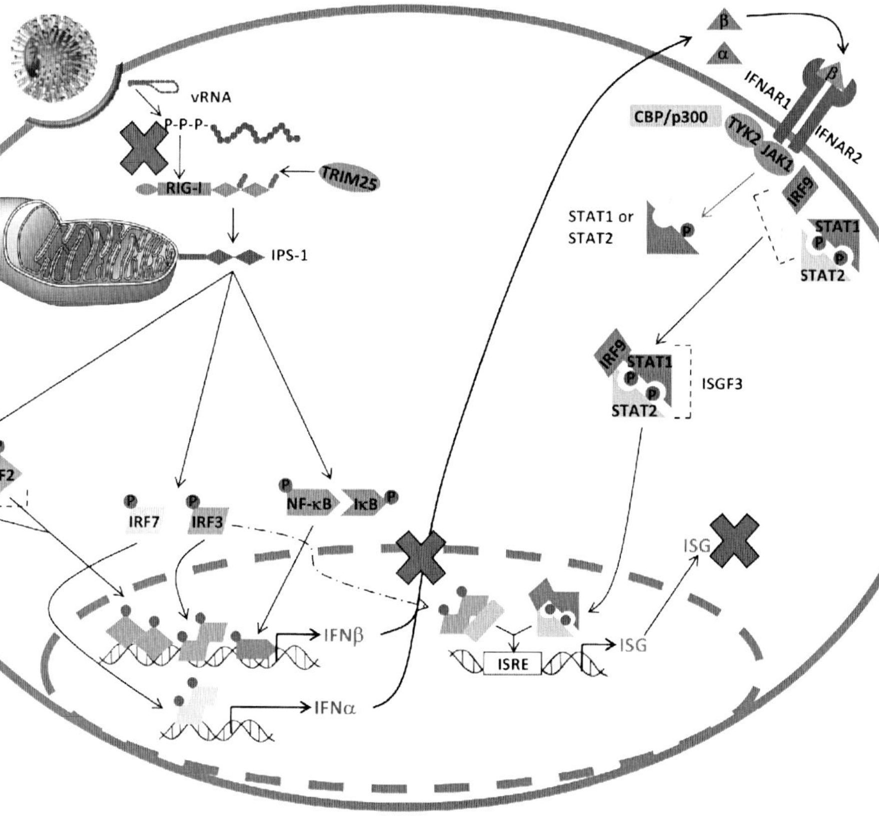

Figure 7.1 Antagonism of the interferon pathway by the influenza virus NS1 protein. NS1 binds to RIG-I and blocks the activation of transcription factors required for IFN expression, AP1, IRF3, and NF-κB. NS1 has also been shown to antagonize specific interferon stimulated genes (ISGs), including 2′–5′ oligo(A)synthetase and protein kinase R (PKR). In addition, NS1 interferes with host transcriptional activities, resulting in the accumulation of host pre-mRNAs in the nucleus. (Kindly provided by Mila B. Ortigoza.)

Viral mRNAs are still able to exit the nucleus for translation (Chen *et al.*, 1999), perhaps contributing to preferential viral transcription and subsequent translation.

NS1's ability to antagonize the host's IFN response and interrupt normal host cell operations made it a candidate virulence factor of the 1918 influenza virus. It was hypothesized that the actions of NS1 were strong enough to allow for the virulence and host damage that is seen with this viral strain. Mice infected with a reassortant virus containing the 1918 NS1 gene segment and seven other RNA segments from A/WSN/33 (WSN) virus, however, did not have higher viral titres than the mice infected with the WSN parent virus (Basler *et al.*, 2001). In further studies, reassortant virus 1918 HA/NA/M/NS:WSN (four segments from 1918 virus: 4 remaining segments from WSN virus) was extremely virulent, suggesting that the additional viral gene segments play a larger role in pathogenicity than NS (Tumpey *et al.*, 2004). Similarly, viral activity was attenuated in a reassortant virus containing the 1918 NS gene segment in the A/Texas/36/91 (Tx/91) virus background (Pappas *et al.*, 2008). The attenuation of these viruses could be the result of the failure of a human virulence factor to be active in a mouse-adapted virus (Basler *et al.*, 2001), but it is more suggestive that NS does not play as large of a role in the virulence of the 1918 virus as was originally hypothesized.

Haemagglutinin and neuraminidase: two proteins that contribute to virulence

It is possible that the virulence of the 1918 influenza virus is due to increased rates of host cell entry, viral RNA replication, and viral progeny release as compared to other less virulent influenza strains (Basler and Aguilar, 2008). Changes in the haemagglutinin (HA) and neuraminidase (NA) genes could therefore contribute to the virulence of the virus.

The HA molecule is an integral membrane glycoprotein that is cleaved by host proteases into HA1 and HA2 (Palese and Shaw, 2006). The HA homotrimer binds virus to the surface of the cell via the low-affinity binding pocket on the HA1 subunit to sialic acid receptors, and promotes viral penetration by mediating viral membrane fusion to the endosome via HA2 fusion peptide exposure (Palese and Shaw, 2006) (Fig. 7.2).

HA receptor specificity is critical to the spread of the 1918 influenza virus. While avian and equine viruses preferentially bind receptors on host cells that end in sialic acid α-2,3-galactose (SAα2,3), human and swine viruses preferentially recognize oligosaccharides ending in sialic acid α-2,6-galactose (SAα2,6) (Glaser *et al.*, 2005; Srinivasan *et al.*, 2008; Stevens *et al.*, 2006). In addition to being found in the bronchi and in terminal and respiratory bronchioles, these receptors are found at the sites of viral transmission: epithelial cells of nasal mucosa, paranasal sinuses, pharynx, and trachea (Shinya *et al.*, 2006). Viral binding and replication in the upper respiratory tract would allow for the spread of the virus via sneezing and coughing and is therefore suggested that the binding specificity to receptors in these locations plays an important role in viral spread from human to human. It has been shown that the 1918 isolate A/South Carolina/1/18 HA binds sialic acid receptors with an α-2,6 linkage while the HA of the A/New York/1/18/isolate can bind both forms of the receptor (Glaser *et al.*, 2005; Srinivasan *et al.*, 2008). When HA affinity is converted from SAα2,6 to SAα2,3 experimentally, the 1918 influenza virus retains lethality and replication efficiency in ferrets inoculated with 10^6 plaque-forming units (pfu) but blocked transmission between ferrets (Tumpey *et al.*, 2007a). In addition, it was shown that viral strains that displayed a mixed affinity for both SAα2,6 and SAα2,3 transmitted less efficiently than strains with only a SAα2,6 preference (Tumpey *et al.*, 2007a). These data further support the hypothesis that the specificity for HA binding plays an important role in the virulence of the 1918 virus and its ability to spread from host to host.

HA is likely to play a large role in the virulence seen with infection of the pandemic strain. Infection with reassortant viruses containing the HA from the 1918 virus in human virus backgrounds that are non-pathogenic in mice (A/Memphis/8/88 and A/Kawasaki/173/2001) resulted in lethal infection (Kobasa *et al.*, 2004). Experiments using reassortants with 1918 viral gene segments in Tx/91 backgrounds or Tx/91 gene segments in a 1918 background corroborated these data (Fig. 7.3). The reassortant virus

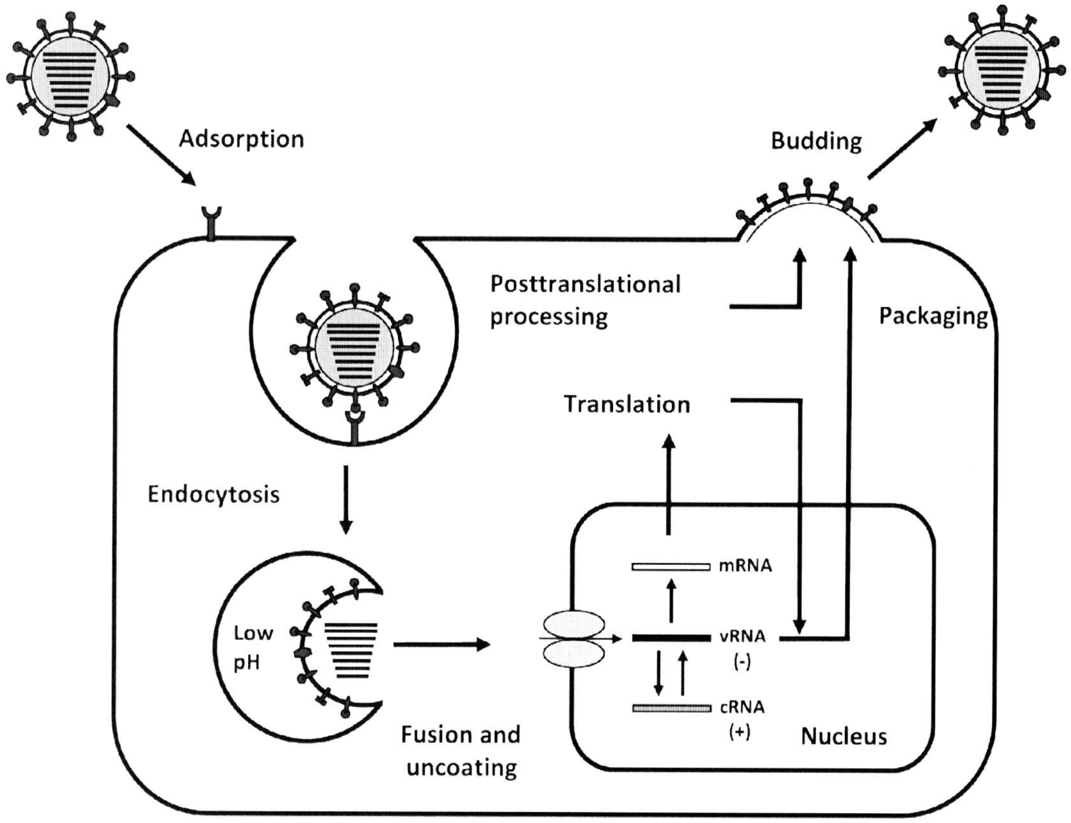

Figure 7.2 Influenza virus replication. The HA molecule of the virus binds to the surface of the host cell and the virus is then endocytosed. The M2 protein channel mediates acidification of the endosome, causing the exposure of the fusion peptide on the HA molecule and dissociation of the RNPs from the M1 protein. After fusion of endosomal and viral membranes, RNPs enter the cytoplasm and are imported to the nucleus. Here, vRNAs serve as templates for transcription and are replicated through a cRNA intermediate. Newly replicated vRNA sequences and viral proteins are assembled as new viruses at the plasma membrane. Viruses bud from the host cell and are released by virtue of NA enzymatic activity. Adapted from Palese and Shaw (2006).

containing the HA from the 1918 virus grew to high titre and caused death in infected mice (Pappas *et al.*, 2008). In contrast, mice infected with a reassortant virus containing the HA from the Tx/91 virus in a 1918 background survived infection. Thus, the HA is probably an important virulence factor to the 1918 virus.

The NA protein is also believed to contribute to the virulence of the 1918 influenza virus. Like HA, NA projects through the viral envelope and interacts with sialic acid molecules on the host. NA allows for the mobility of progeny viruses by removing sialic acid residues from the glycoproteins of virions and infected cells during release (and possibly entry) (Palese and Shaw, 2006) (Fig. 7.2). A more efficient NA could therefore contribute to increased virulence.

Virus particles with low NA enzymatic activity accumulate as aggregates on the cell surface as a result of the normal levels of HA activity (Palese and Schulman, 1974; Palese *et al.*, 1974a; Palese *et al.*, 1974b). It has been shown that HA and NA kinetics must be balanced to ensure proper binding followed by entry and subsequent budding (Mitnaul *et al.*, 2000). While the ratio of HA to NA varies by strain, it is thought that these two molecules complement one another (Mitnaul *et al.*, 2000), possibly working in concert to increase 1918 influenza virulence. The importance of HA and NA in the pathogenicity of the 1918 influenza virus has been demonstrated in experiments using reassortant viruses. Lethality was seen in mice exposed to reassortant 1918 HA/NA:Tx91 virus (two genes from 1918 virus: six remaining

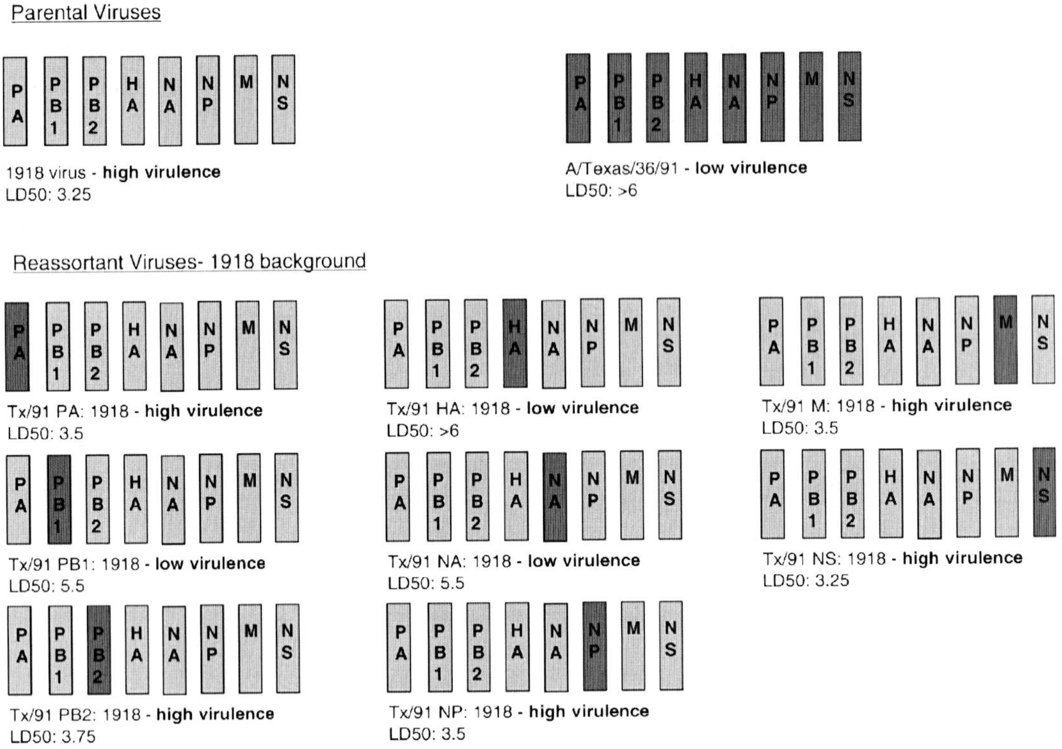

Figure 7.3 Properties of reassortant viruses containing Tx/91 viral gene segments in 1918 virus backgrounds. Pictured above is a schematic of the viral gene segments derived from the parental viruses (1918 virus in light grey; A/Tx/91 virus in dark grey) with their $\log_{10}LD_{50}$ in mice. Reassortants containing the HA, NA, or PB1 Tx/91 gene segments in 1918 virus backgrounds (remaining seven segments) were 100 fold more attenuated than the parental wild type 1918 virus. Adapted from Pappas *et al.* (2008).

genes from the Tx91 virus) (Pappas *et al.*, 2008), further supporting the role of HA and NA in 1918 virulence.

It is of note that the 1918 virus replicates and causes plaques in Madin-Darby canine kidney epithelial cells (MDCK) without trypsin being added to the assay (Tumpey *et al.*, 2005). This phenotype is also seen in reassortant viruses 1918 HA/NA:Tx/91 (two genes from 1918 virus: six genes from the Tx91 virus) and 1918NA:Tx91 (one gene from 1918 virus: seven genes from the Tx91 virus) (Tumpey *et al.*, 2005). It is therefore hypothesized that the 1918 NA plays a role in the cleavage of the HA molecule and exposure of the fusion peptide (Tumpey *et al.*, 2005), an event essential to infection. Interestingly, the sequences of 1918 HA and NA do not point to mechanisms that allow for replication in the absence of trypsin, as is seen in other highly pathogenic viruses. Most mammalian and non-pathogenic avian virus strains have an HA that is cleaved only in the presence of certain host proteases, thereby limiting its spread from the respiratory tract (Steinhauer, 1999). In contrast, the HA molecule of pathogenic avian strains contains a multibasic cleavage site (Bosch *et al.*, 1981) that can be cleaved by ubiquitous proteases found in many tissue types, leading to systemic viral infection (Stieneke-Grober *et al.*, 1992). The HA of the 1918 influenza virus does not possess this multibasic cleavage site (Taubenberger *et al.*, 1997), nor is this site seen in the mouse-pathogenic WSN virus (Steinhauer, 1999). The systemic infection that is seen in mice infected with the WSN virus has been attributed to structural components of the NA protein. It is thought that a C-terminal lysine (Goto and Kawaoka, 1998) or a loss of glycosylation at position 130 (position 146 by N2 numbering) (Li *et al.*, 1993) on the NA protein may be involved in the recruitment of plasminogen to cleave HA molecules (Goto and Kawaoka, 1998; Goto *et al.*, 2001). The NA of the 1918 virus lacks these structural components, however, suggesting a possible novel mechanism of the NA facilitating HA cleavage in the 1918 influenza virus (Goto and Kawaoka, 1998; Tumpey *et al.*, 2005).

The 1918 viral PB1 gene: an alternative reading frame that contributes to virulence

Because high replication rates are frequently associated with high rates of virulence (Grimm *et al.*, 2007), the role of the 1918 viral polymerase as a virulence factor has also been explored. The influenza RNA polymerase is a heterotrimer consisting of polymerase basic protein 1 (PB1), polymerase basic protein 2 (PB2) and polymerase acidic protein (PA) (Palese and Shaw, 2006). The PB1 subunit, encoded by the PB1 gene, is a component of the RNA polymerase responsible for initiation of both replication and transcription (Palese and Shaw, 2006). Its replicative function converts negative sense vRNA to positive sense cRNA, a template for additional vRNA synthesis. PB2 binds the cap of host pre-mRNA molecules to serve as primers for mRNA synthesis (Palese and Shaw, 2006). The PA protein continues to be studied, though it is known to be involved in replication and transcription (Palese and Shaw, 2006) (Fig. 7.2).

Studies in mice by Pappas *et al.* substantiated the role of PB1 in the virulence of the 1918 influenza virus. The 1918 PB1:Tx/91 reassortant virus (one gene from 1918 virus: seven remaining genes from the Tx91 virus) showed increased rates of replication and created larger plaques on MDCK cells, suggesting that PB1 contributes to increased virulence compared to the parent Tx/91 strain (Pappas *et al.*, 2008). In addition, it has been shown that the contemporary H1N1 virus, A/Kawasaki/173/2001 (K173), that normally infects only the upper respiratory tract, can cause infection in the lower respiratory tract when the PB1 gene is replaced with that from the 1918 virus in the ferret animal model (Watanabe *et al.*, 2008)

This increased virulence of the 1918 influenza strain could therefore be the result of a more efficient replication strategy led by the PB1 protein or the result of the viral protein PB1-F2. The PB1-F2 protein is a 90 amino acid peptide generated from a +1 reading frame by virtue of ribosomal scanning (Chen *et al.*, 2001). The protein has a putative role in apoptosis (Chen *et al.*, 2001) and 1918 influenza virulence (Conenello *et al.*, 2007; Gibbs *et al.*, 2003; McAuley *et al.*, 2007).

Chen *et al.* discovered PB1-F2 while searching for antigenic peptides that are presented on MHC I molecules following influenza infection with influenza A/Puerto Rico/8/34 (PR8) virus (Chen *et al.*, 2001). An early study recognized the similarity of the proposed structure of PB1-F2 and the arginine rich domains of Vpr in HIV (Chen *et al.*, 2001), a protein that forms holes in infected host cells (Ferri *et al.*, 2000). Interestingly, PB1-F2 sequence comparisons between high- and low-pathogenicity strains of the Hong Kong/156/97 virus (HK/97) identified a difference at amino acid 66. This amino acid change, N66S, was also found in the PB1-F2 segment of the 1918 virus. Introduction of S66N into the 1918 virus resulted in reduced pathogenicity and decreased mortality and morbidity compared with wild-type 1918 virus, suggesting the importance of this amino acid to its virulence (Conenello *et al.*, 2007).

In vitro studies have supported the role of PB1-F2 as a pro-apoptotic factor. PB1-F2 localized to the mitochondria and induced changes in mitochondrial morphology and membrane potential in MDCK cells infected with PR8 virus as compared to recombinant PR8 viruses where PB1-F2 was not expressed (Chen *et al.*, 2001). Additionally, transfection of PB1-F2 in HeLa cells causes the release of cytochrome c from the mitochondria (Chen *et al.*, 2001). The importance of PB1-F2 in viral pathogenicity has also been supported *in vivo*. Mice inoculated with a virus that did not express PB1-F2 had lowered morbidity and mortality and were able to clear the virus two days earlier than mice inoculated with the parent virus expressing the protein (Zamarin *et al.*, 2006). The role of PB1-F2 in the apoptosis of certain cell types, including those recruited for immune responses such as monocytes (Chanturiya *et al.*, 2004; Chen *et al.*, 2001; Zamarin *et al.*, 2006) continues to be elucidated.

By monitoring caspase activity, loss of mitochondrial membrane potential, and visualization of cells, it has been shown that PB1-F2 sensitizes both the extrinsic and intrinsic apoptotic pathways to cell-death stimuli (Zamarin *et al.*, 2005).

PB1-F2 potentiates the release of cytochrome c from the mitochondria by tBid and interacts with voltage dependent anion channel (VDAC) and adenine nucleotide translocator (ANT3) (Zamarin *et al.*, 2005). It is thought that the interaction of these two proteins, in addition to others, create the permeability transcription pore complex (PTPC), a putative regulator of cell death (Zamzami and Kroemer, 2001). Previous structural studies have shown that the C-terminal end of PB1-F2 is necessary for localization to the inner mitochondrial membrane, by virtue of a positively charged alpha helix (Gibbs *et al.*, 2003). In later studies, it was shown that the C-terminal domain binds to the inner mitochondrial membrane protein ANT3 while both the N-terminal and C-terminal domains have been shown to interact with VDAC1 (Zamarin *et al.*, 2005). It is therefore hypothesized that PB1-F2 can be a mediator of the formation of the PTPC (Zamarin *et al.*, 2005), causing increased permeability of the mitochondrial membranes (Zamzami and Kroemer, 2001).

Consistent with these studies, it has been shown that amino acids 46–75 of the PB1 protein are necessary and sufficient for mitochondrial localization (Yamada *et al.*, 2004). It is of note that while 87% of influenza A viruses in GenBank are at least 78 amino acids long, all of the contemporary H1N1 viruses have truncated PB1-F2s consisting of 57 amino acids (Zell *et al.*, 2007). This truncation results in a PB1-F2 peptide that is most likely non-functional due to the loss of its mitochondrial targeting sequence (Zell *et al.*, 2007) and could therefore explain the decreased pathogenicity and localization of infection to the upper respiratory tract seen in contemporary H1N1 viruses. By the same token, it is proposed that the C-terminal portion of PB1-F2, and specifically serine in position 66, contribute to the virulence of the 1918 influenza virus.

In addition to its apoptotic role, PB1-F2 increases susceptibility to secondary bacterial pneumonia in mice (McAuley *et al.*, 2007). Mice infected with wild-type PR8 compared to infection with a PR8 virus that lacks PB1-F2 expression showed increased weight loss, greater induction of pneumonia, high mortality, increased white blood cell recruitment and epithelial cell damage after viral infection and subsequent bacterial challenge with *Streptococcus pneumoniae* (McAuley *et al.*, 2007). Experiments were also conducted to suggest that the C-terminal portion of PB1-F2 increases host susceptibility to S. pneumoniae infection (McAuley *et al.*, 2007). This is particularly noteworthy in the context of the 1918 influenza pandemic, as secondary bacterial pneumonia was a major cause of death (Johnson and Mueller, 2002; Morens *et al.*, 2008; Taubenberger *et al.*, 2000). Recombinant viruses expressing PB1-F2 of the 1918 influenza virus in a PR8 background displayed enhanced growth as well as larger mean plaque size (McAuley *et al.*, 2007). In animal models, this virus displayed more rapid replication at 2 days post infection, increased recruitment of white blood cells, and diffuse coagulative necrosis compared with the patchy areas of necrosis seen in the lungs of wild type-PR8 infected mice (McAuley *et al.*, 2007). In addition, it was shown that these mice had increased levels of cytokines known to play a role in secondary bacterial pneumonia (McAuley *et al.*, 2007) resulting in increased influenza pathogenesis independent of bacteraemia (Smith *et al.*, 2007). Increased macrophage and neutrophil infiltration, as well as increased cytokine levels, have also been seen in mice infected with the 1918 influenza virus as compared to infection with Tx/91 (Perrone *et al.*, 2008). It is therefore possible that the pro-apoptotic function of PB1-F2's results in the recruitment of cytokines and increases susceptibility to secondary bacterial pneumonia in 1918 influenza virus infection, though the exact mechanism of the protein continues to be studied.

The study of the 1918 pandemic virus in the guinea pig animal model

The guinea pig animal model has been shown to be useful to the study of 1918 influenza virus transmission, as mice do not transmit the virus to one another (Lowen *et al.*, 2006). While transmission of the 1918 pandemic virus has been studied in the ferret model (Tumpey *et al.*, 2007a; Watanabe *et al.*, 2008), these animals are expensive and difficult to work with (Lowen *et al.*, 2006). For these reasons, the guinea pig is seen as an ideal model in which to study 1918 pandemic virus transmission.

While the mouse model has been used to study morbidity and mortality caused by the 1918 virus, standard laboratory mouse strains, such as BALB/c and C57BL/6, have the limitation of carrying deletions in the Mx gene (Staeheli *et al.*, 1988). The Mx GTPase is one of many antiviral proteins induced during the IFN response (Sadler and Williams, 2008), and has been shown to significantly contribute to the control and respiratory restriction of highly pathogenic influenza strains, including the 1918 pandemic influenza virus (Tumpey *et al.*, 2007b). The ferret as well as the guinea pig, however, possesses a functional Mx gene (Horisberger and Gunst, 1991), highlighting these animal models as better suited to study the IFN response to influenza virus infection. Recently, it has been shown that the intranasal administration of human IFN-α induces Mx gene expression in the guinea pig animal model (Van Hoeven *et al.*, 2009). In addition, IFN treatment of guinea pigs before and after viral infection with the 1918 virus (days −1, 0, 1, 3, 5 and 7) reduced viral replication in guinea pig lungs at high and low viral doses (Van Hoeven *et al.*, 2009). Viral infection was completely blocked in 50% of guinea pigs that were treated with a multi-day course of human IFN-α and infected with 10^3 pfu (Van Hoeven *et al.*, 2009). These studies represent possible treatment strategies that could be adapted to humans and underscore the importance of the guinea pig model in the study of the 1918 virus.

Conclusion

The 1918 influenza pandemic took a devastating toll on the human population. While the HA, NA and PB1-F2 (coded for by the PB1 gene) proteins are likely to be crucial components of the 1918 virus causing its high virulence, the remaining genes are also important as only the interplay of all eight segments and their eleven encoded proteins could result in this unique virus. Thus, we have to look at this virus as a pathogen which, by virtue of its multigenic traits and infection of immunologically naïve hosts, caused havoc on a population. The mechanisms of the gene products continue to be studied, with the goal of providing a better understanding of the virulence of the 1918 influenza virus.

Acknowledgements

This work was supported by the National Institutes of Health Grants PO1 AI058113-03, UC19 AI062623-023, T32 AI07647-06, HHSN266200700010C, U01 AI070469-01 and by the Louis and Rachel Rudin Foundation.

References

Ahmed, R., Oldstone, M.B., and Palese, P. (2007). Protective immunity and susceptibility to infectious diseases: lessons from the 1918 influenza pandemic. Nat. Immunol. *8*, 1188–1193.

Arias, E. (2004). United States life tables, 2001. In National Vital Statistics Reports 52, 14.

Basler, C.F., and Aguilar, P.V. (2008). Progress in identifying virulence determinants of the 1918 H1N1 and the Southeast Asian H5N1 influenza A viruses. Antiviral. Res. *79*, 166–178.

Basler, C.F., Reid, A.H., Dybing, J.K., Janczewski, T.A., Fanning, T.G., Zheng, H., Salvatore, M., Perdue, M.L., Swayne, D.E., Garcia-Sastre, A., *et al.* (2001). Sequence of the 1918 pandemic influenza virus nonstructural gene (NS) segment and characterization of recombinant viruses bearing the 1918 NS genes. Proc. Natl. Acad. Sci. U.S.A. *98*, 2746–2751.

Bergmann, M., Garcia-Sastre, A., Carnero, E., Pehamberger, H., Wolff, K., Palese, P., and Muster, T. (2000). Influenza virus NS1 protein counteracts PKR-mediated inhibition of replication. J. Virol. *74*, 6203–6206.

Bosch, F.X., Garten, W., Klenk, H.D., and Rott, R. (1981). Proteolytic cleavage of influenza virus hemagglutinins: primary structure of the connecting peptide between HA1 and HA2 determines proteolytic cleavability and pathogenicity of Avian influenza viruses. Virology *113*, 725–735.

Chanturiya, A.N., Basanez, G., Schubert, U., Henklein, P., Yewdell, J.W., and Zimmerberg, J. (2004). PB1-F2, an influenza A virus-encoded proapoptotic mitochondrial protein, creates variably sized pores in planar lipid membranes. J. Virol. *78*, 6304–6312.

Chen, W., Calvo, P.A., Malide, D., Gibbs, J., Schubert, U., Bacik, I., Basta, S., O'Neill, R., Schickli, J., Palese, P., *et al.* (2001). A novel influenza A virus mitochondrial protein that induces cell death. Nat. Med. *7*, 1306–1312.

Chen, Z., Li, Y., and Krug, R.M. (1999). Influenza A virus NS1 protein targets poly(A)-binding protein II of the cellular 3′-end processing machinery. EMBO J. *18*, 2273–2283.

Conenello, G.M., Zamarin, D., Perrone, L.A., Tumpey, T., and Palese, P. (2007). A single mutation in the PB1-F2 of H5N1 (HK/97) and 1918 influenza A viruses contributes to increased virulence. PLoS Pathog. *3*, 1414–1421.

Crosby, W. (1989). America's Forgotten Pandemic: The Influenza of 1918 (Cambridge, Cambridge University Press).

Dong, B., and Silverman, R.H. (1995). 2–5A-dependent RNase molecules dimerize during activation by 2–5A. J. Biol. Chem. *270*, 4133–4137.

Ferri, K.F., Jacotot, E., Blanco, J., Este, J.A., and Kroemer, G. (2000). Mitochondrial control of cell death induced by HIV-1-encoded proteins. Ann. N.Y. Acad. Sci.*926*, 149–164.

Gale, M., Jr., and Katze, M.G. (1998). Molecular mechanisms of interferon resistance mediated by viral-directed inhibition of PKR, the interferon-induced protein kinase. Pharmacol. Ther. *78*, 29–46.

Gibbs, J.S., Malide, D., Hornung, F., Bennink, J.R., and Yewdell, J.W. (2003). The influenza A virus PB1-F2 protein targets the inner mitochondrial membrane via a predicted basic amphipathic helix that disrupts mitochondrial function. J. Virol. *77*, 7214–7224.

Glaser, L., Stevens, J., Zamarin, D., Wilson, I.A., Garcia-Sastre, A., Tumpey, T.M., Basler, C.F., Taubenberger, J.K., and Palese, P. (2005). A single amino acid substitution in 1918 influenza virus hemagglutinin changes receptor binding specificity. J. Virol. *79*, 11533–11536.

Goodpasture, E. (1919). The significance of certain pulmonary lesion in relation to the etiology of influenza. Am. J. Med. Sci. *158*, 863–870.

Goto, H., and Kawaoka, Y. (1998). A novel mechanism for the acquisition of virulence by a human influenza A virus. Proc. Natl. Acad. Sci. U.S.A. *95*, 10224–10228.

Goto, H., Wells, K., Takada, A., and Kawaoka, Y. (2001). Plasminogen-binding activity of neuraminidase determines the pathogenicity of influenza A virus. J. Virol. *75*, 9297–9301.

Grimm, D., Staeheli, P., Hufbauer, M., Koerner, I., Martinez-Sobrido, L., Solorzano, A., Garcia-Sastre, A., Haller, O., and Kochs, G. (2007). Replication fitness determines high virulence of influenza A virus in mice carrying functional Mx1 resistance gene. Proc. Natl. Acad. Sci. U.S.A. *104*, 6806–6811.

Guo, Z., Chen, L.M., Zeng, H., Gomez, J.A., Plowden, J., Fujita, T., Katz, J.M., Donis, R.O., and Sambhara, S. (2007). NS1 protein of influenza A virus inhibits the function of intracytoplasmic pathogen sensor, RIG-I. Am. J. Respir. Cell. Mol. Biol. *36*, 263–269.

Hatada, E., Saito, S., and Fukuda, R. (1999). Mutant influenza viruses with a defective NS1 protein cannot block the activation of PKR in infected cells. J. Virol. *73*, 2425–2433.

Horisberger, M.A., and Gunst, M.C. (1991). Interferon-induced proteins: identification of Mx proteins in various mammalian species. Virology *180*, 185–190.

Johnson, N.P., and Mueller, J. (2002). Updating the accounts: global mortality of the 1918–1920 'Spanish' influenza pandemic. Bull. Hist. Med. *76*, 105–115.

Kawai, T., Takahashi, K., Sato, S., Coban, C., Kumar, H., Kato, H., Ishii, K.J., Takeuchi, O., and Akira, S. (2005). IPS-*1*, an adaptor triggering RIG-I- and Mda5-mediated type I interferon induction. Nat. Immunol. *6*, 981–988.

Kobasa, D., Jones, S.M., Shinya, K., Kash, J.C., Copps, J., Ebihara, H., Hatta, Y., Kim, J.H., Halfmann, P., Hatta, M., *et al.* (2007). Aberrant innate immune response in lethal infection of macaques with the 1918 influenza virus. Nature *445*, 319–323.

Kobasa, D., Takada, A., Shinya, K., Hatta, M., Halfmann, P., Theriault, S., Suzuki, H., Nishimura, H., Mitamura, K., Sugaya, N., *et al.* (2004). Enhanced virulence of influenza A viruses with the haemagglutinin of the 1918 pandemic virus. Nature *431*, 703–707.

Kochs, G., Garcia-Sastre, A., and Martinez-Sobrido, L. (2007). Multiple anti-interferon actions of the influenza A virus NS1 protein. J. Virol. *81*, 7011–7021.

Kumar, H., Kawai, T., Kato, H., Sato, S., Takahashi, K., Coban, C., Yamamoto, M., Uematsu, S., Ishii, K.J., Takeuchi, O., *et al.* (2006). Essential role of IPS-1 in innate immune responses against RNA viruses. J. Exp. Med. *203*, 1795–1803.

Lamb, R.A., and Choppin, P.W. (1979). Segment 8 of the influenza virus genome is unique in coding for two polypeptides. Proc. Natl. Acad. Sci. U.S.A. *76*, 4908–4912.

Li, S., Schulman, J., Itamura, S., and Palese, P. (1993). Glycosylation of neuraminidase determines the neurovirulence of influenza A/WSN/33 virus. J. Virol. *67*, 6667–6673.

Linder, F.E., and Grove, R.D. (1947). Vital Statistics Rates in the United States: 1900–1940 (Washington, D.C., Government Printing Office).

Lowen, A.C., Mubareka, S., Tumpey, T.M., Garcia-Sastre, A., and Palese, P. (2006). The guinea pig as a transmission model for human influenza viruses. Proc. Natl. Acad. Sci. U.S.A. *103*, 9988–9992.

Lu, Y., Wambach, M., Katze, M.G., and Krug, R.M. (1995). Binding of the influenza virus NS1 protein to double-stranded RNA inhibits the activation of the protein kinase that phosphorylates the elF-2 translation initiation factor. Virology *214*, 222–228.

Lucke, B., T., W., and E., K. (check authors?)(1919). Pathological anatomy and bacteriology of influenza. Archiv. Intern. Med. *24*, 154–237.

McAuley, J.L., Hornung, F., Boyd, K.L., Smith, A.M., McKeon, R., Bennink, J., Yewdell, J.W., and McCullers, J.A. (2007). Expression of the 1918 influenza A virus PB1-F2 enhances the pathogenesis of viral and secondary bacterial pneumonia. Cell Host Microbe *2*, 240–249.

Mibayashi, M., Martinez-Sobrido, L., Loo, Y.M., Cardenas, W.B., Gale, M., Jr., and Garcia-Sastre, A. (2007). Inhibition of retinoic acid-inducible gene I-mediated induction of beta interferon by the NS1 protein of influenza A virus. J. Virol. *81*, 514–524.

Michaelis, M., Doerr, H.W., and Cinatl, J., Jr. (2009). Novel swine-origin influenza A virus in humans: another pandemic knocking at the door. Med. Microbiol. Immunol. *198*, 175–183.

Min, J.Y., and Krug, R.M. (2006). The primary function of RNA binding by the influenza A virus NS1 protein in infected cells: Inhibiting the 2′–5′ oligo (A) synthetase/RNase L pathway. Proc. Natl. Acad. Sci. U.S.A. *103*, 7100–7105.

Mitnaul, L.J., Matrosovich, M.N., Castrucci, M.R., Tuzikov, A.B., Bovin, N.V., Kobasa, D., and Kawaoka, Y. (2000). Balanced hemagglutinin and neuraminidase activities are critical for efficient replication of influenza A virus. J. Virol. *74*, 6015–6020.

Morens, D.M., Taubenberger, J.K., and Fauci, A.S. (2008). Predominant role of bacterial pneumonia as a cause of death in pandemic influenza: implications for pandemic influenza preparedness. J. Infect. Dis. *198*, 962–970.

Moss, B., Keith, J.M., Gershowitz, A., Ritchey, M.B., and Palese, P. (1978). Common sequence at the 5′ ends of the segmented RNA genomes of influenza A and B viruses. J. Virol. *25*, 312–318.

Noah, D.L., Twu, K.Y., and Krug, R.M. (2003). Cellular antiviral responses against influenza A virus are countered at the posttranscriptional level by the viral NS1A protein via its binding to a cellular protein required for the 3′ end processing of cellular pre-mRNAS. Virology *307*, 386–395.

Opitz, B., Rejaibi, A., Dauber, B., Eckhard, J., Vinzing, M., Schmeck, B., Hippenstiel, S., Suttorp, N., and Wolff, T. (2007). IFNbeta induction by influenza A virus is mediated by RIG-I which is regulated by the viral NS1 protein. Cell. Microbiol. *9*, 930–938.

Palese, P., and Schulman, J. (1974). Isolation and characterization of influenza virus recombinants with high and low neuraminidase activity. Use of 2-(3′-methoxyphenyl)-*n*-acetylneuraminic acid to identify cloned populations. Virology *57*, 227–237.

Palese, P., Schulman, J.L., Bodo, G., and Meindl, P. (1974a). Inhibition of influenza and parainfluenza virus replication in tissue culture by 2-deoxy-2,3-dehydro-N-trifluoroacetylneuraminic acid (FANA). Virology *59*, 490–498.

Palese, P., and Shaw, M.L. (2006). Orthomyxoviridae: The viruses and their replication. In Fields Virology, Knipe, D.M. and Howley, P.M., eds (Philadelphia, Lippincott Williams & Wilkins), pp. 1648–1689.

Palese, P., Tobita, K., Ueda, M., and Compans, R.W. (1974b). Characterization of temperature sensitive influenza virus mutants defective in neuraminidase. Virology *61*, 397–410.

Pappas, C., Aguilar, P.V., Basler, C.F., Solorzano, A., Zeng, H., Perrone, L.A., Palese, P., Garcia-Sastre, A., Katz, J.M., and Tumpey, T.M. (2008). Single gene reassortants identify a critical role for PB*1*, HA, and NA in the high virulence of the 1918 pandemic influenza virus. Proc. Natl. Acad. Sci. U.S.A. *105*, 3064–3069.

Perrone, L.A., Plowden, J.K., Garcia-Sastre, A., Katz, J.M., and Tumpey, T.M. (2008). H5N1 and 1918 pandemic influenza virus infection results in early and excessive infiltration of macrophages and neutrophils in the lungs of mice. PLoS Pathog. *4*, e1000115.

Pichlmair, A., Schulz, O., Tan, C.P., Naslund, T.I., Liljestrom, P., Weber, F., and Reis e Sousa, C. (2006). RIG-I-mediated antiviral responses to single-stranded RNA bearing 5′-phosphates. Science *314*, 997–1001.

Reid, A.H., Fanning, T.G., Hultin, J.V., and Taubenberger, J.K. (1999). Origin and evolution of the 1918 'Spanish' influenza virus hemagglutinin gene. Proc. Natl. Acad. Sci. U.S.A. *96*, 1651–1656.

Reid, A.H., Fanning, T.G., Janczewski, T.A., Lourens, R.M., and Taubenberger, J.K. (2004). Novel origin of the 1918 pandemic influenza virus nucleoprotein gene. J. Virol. *78*, 12462–12470.

Reid, A.H., Fanning, T.G., Janczewski, T.A., McCall, S., and Taubenberger, J.K. (2002). Characterization of the 1918 'Spanish' influenza virus matrix gene segment. J. Virol. *76*, 10717–10723.

Reid, A.H., Fanning, T.G., Janczewski, T.A., and Taubenberger, J.K. (2000). Characterization of the 1918 'Spanish' influenza virus neuraminidase gene. Proc. Natl. Acad. Sci. U.S.A. *97*, 6785–6790.

Sadler, A.J., and Williams, B.R. (2008). Interferon-inducible antiviral effectors. Nat. Rev. Immunol. *8*, 559–568.

Satterly, N., Tsai, P.L., van Deursen, J., Nussenzveig, D.R., Wang, Y., Faria, P.A., Levay, A., Levy, D.E., and Fontoura, B.M. (2007). Influenza virus targets the mRNA export machinery and the nuclear pore complex. Proc. Natl. Acad. Sci. U.S.A. *104*, 1853–1858.

Shinya, K., Ebina, M., Yamada, S., Ono, M., Kasai, N., and Kawaoka, Y. (2006). Avian flu: influenza virus receptors in the human airway. Nature *440*, 435–436.

Smith, M.W., Schmidt, J.E., Rehg, J.E., Orihuela, C.J., and McCullers, J.A. (2007). Induction of pro- and anti-inflammatory molecules in a mouse model of pneumococcal pneumonia after influenza. Comp. Med. *57*, 82–89.

Srinivasan, A., Viswanathan, K., Raman, R., Chandrasekaran, A., Raguram, S., Tumpey, T.M., Sasisekharan, V., and Sasisekharan, R. (2008). Quantitative biochemical rationale for differences in transmissibility of 1918 pandemic influenza A viruses. Proc. Natl. Acad. Sci. U.S.A. *105*, 2800–2805.

Staeheli, P., Grob, R., Meier, E., Sutcliffe, J.G., and Haller, O. (1988). Influenza virus-susceptible mice carry Mx genes with a large deletion or a nonsense mutation. Mol. Cell. Biol. *8*, 4518–4523.

Steinhauer, D.A. (1999). Role of hemagglutinin cleavage for the pathogenicity of influenza virus. Virology *258*, 1–20.

Stevens, J., Blixt, O., Glaser, L., Taubenberger, J.K., Palese, P., Paulson, J.C., and Wilson, I.A. (2006). Glycan microarray analysis of the hemagglutinins from modern and pandemic influenza viruses reveals different receptor specificities. J. Mol. Biol. 355, 1143–1155.

Stieneke-Grober, A., Vey, M., Angliker, H., Shaw, E., Thomas, G., Roberts, C., Klenk, H.D., and Garten, W. (1992). Influenza virus hemagglutinin with multibasic cleavage site is activated by furin, a subtilisin-like endoprotease. EMBO J. *11*, 2407–2414.

Taubenberger, J.K., and Morens, D.M. (2008). The pathology of influenza virus infections. Annu. Rev. Pathol. 3, 499–522.

Taubenberger, J.K., Reid, A.H., and Fanning, T.G. (2000). The 1918 influenza virus: A killer comes into view. Virology *274*, 241–245.

Taubenberger, J.K., Reid, A.H., Krafft, A.E., Bijwaard, K.E., and Fanning, T.G. (1997). Initial genetic characterization of the 1918 'Spanish' influenza virus. Science *275*, 1793–1796.

Taubenberger, J.K., Reid, A.H., Lourens, R.M., Wang, R., Jin, G., and Fanning, T.G. (2005). Characterization of the 1918 influenza virus polymerase genes. Nature *437*, 889–893.

Tumpey, T.M., Basler, C.F., Aguilar, P.V., Zeng, H., Solorzano, A., Swayne, D.E., Cox, N.J., Katz, J.M., Taubenberger, J.K., Palese, P., *et al.* (2005). Characterization of the reconstructed 1918 Spanish influenza pandemic virus. Science *310*, 77–80.

Tumpey, T.M., Garcia-Sastre, A., Taubenberger, J.K., Palese, P., Swayne, D.E., and Basler, C.F. (2004). Pathogenicity and immunogenicity of influenza viruses with genes from the 1918 pandemic virus. Proc. Natl. Acad. Sci. U.S.A. *101*, 3166–3171.

Tumpey, T.M., Maines, T.R., Van Hoeven, N., Glaser, L., Solorzano, A., Pappas, C., Cox, N.J., Swayne, D.E., Palese, P., Katz, J.M., *et al.* (2007a). A two-amino acid change in the hemagglutinin of the 1918 influenza virus abolishes transmission. Science *315*, 655–659.

Tumpey, T.M., Szretter, K.J., Van Hoeven, N., Katz, J.M., Kochs, G., Haller, O., Garcia-Sastre, A., and Staeheli, P. (2007b). The Mx1 gene protects mice against the pandemic 1918 and highly lethal human H5N1 influenza viruses. J. Virol. *81*, 10818–10821.

Van Hoeven, N., Besler, J.A., Szretter, K.J., Zeng, H., Staeheli, P., Swayne, D., Katz, J.M., and Tumpey, T. (2009). Pathogenesis of the 1918 pandemic and H5N1 influenza virus infection in a guinea pig model: The antiviral potential of exogenous alpha-interferon to reduce virus shedding. J. Virol. *83*, 2851–2861.

Wang, X., Basler, C.F., Williams, B.R., Silverman, R.H., Palese, P., and Garcia-Sastre, A. (2002). Functional replacement of the carboxy-terminal two-thirds of the influenza A virus NS1 protein with short heterologous dimerization domains. J. Virol. *76*, 12951–12962.

Watanabe, T., Watanabe, S., Shinya, K., Hyun Kim, J., Hatta, M., and Kawaoka, Y. (2008). Viral RNA polymerase complex promotes optimal growth of 1918 virus in the lower respiratory tract of ferrets. Proc. Natl. Acad. Sci. U.S.A. *106*, 588–592.

Winternitz, M., Wason, I., and McNamara, F. (1920). The Pathology of Influenza (New Haven, Yale University Press).

Wright, P., Neumann, G., and Kawaoka, Y. (2006). Orthomyxoviruses. In Fields Virology, Knipe, D.M. and Howley, P.M., eds (Philadelphia, Lippincott Williams & Wilkins), pp. 1691–1740.

Yamada, H., Chounan, R., Higashi, Y., Kurihara, N., and Kido, H. (2004). Mitochondrial targeting sequence of the influenza A virus PB1-F2 protein and its function in mitochondria. FEBS Lett *578*, 331–336.

Yoneyama, M., Kikuchi, M., Natsukawa, T., Shinobu, N., Imaizumi, T., Miyagishi, M., Taira, K., Akira, S., and Fujita, T. (2004). The RNA helicase RIG-I has an essential function in double-stranded RNA-induced innate antiviral responses. Nat. Immunol. 5, 730–737.

Yu, X., Tsibane, T., McGraw, P.A., House, F.S., Keefer, C.J., Hicar, M.D., Tumpey, T.M., Pappas, C., Perrone, L.A., Martinez, O., *et al.* (2008). Neutralizing antibodies derived from the B cells of 1918 influenza pandemic survivors. Nature *455*, 532–536.

Zamarin, D., Garcia-Sastre, A., Xiao, X., Wang, R., and Palese, P. (2005). Influenza virus PB1-F2 protein induces cell death through mitochondrial ANT3 and VDAC1. PLoS Pathog. *1*, e4.

Zamarin, D., Ortigoza, M.B., and Palese, P. (2006). Influenza A virus PB1-F2 protein contributes to viral pathogenesis in mice. J. Virol. *80*, 7976–7983.

Zamzami, N., and Kroemer, G. (2001). The mitochondrion in apoptosis: how Pandora's box opens. Nat Rev Mol. Cell Biol. 2, 67–71.

Zell, R., Krumbholz, A., Eitner, A., Krieg, R., Halbhuber, K.J., and Wutzler, P. (2007). Prevalence of PB1-F2 of influenza A viruses. J. Gen. Virol. *88*, 536–546.

Zhou, A., Paranjape, J., Brown, T.L., Nie, H., Naik, S., Dong, B., Chang, A., Trapp, B., Fairchild, R., Colmenares, C., *et al.* (1997). Interferon action and apoptosis are defective in mice devoid of 2′,5′-oligoadenylate-dependent RNase L. EMBO J. *16*, 6355–6363.

Polymerase Structure and Function

8

Mark Bartlam, Zhiyong Lou, Yingfang Liu and Zihe Rao

Abstract

The influenza virus RNA-dependent RNA polymerase is a heterotrimeric complex (PA, PB1 and PB2) with multiple enzymatic activities for catalysing viral RNA transcription and replication. Its critical roles in the influenza virus life cycle and high sequence conservation suggest it should be a major target for therapeutic intervention. However, until very recently, functional studies and drug discovery targeting the influenza polymerase have been hampered by the lack of three-dimensional structural information. In this chapter, we will review the recent progress in the structure and function of the influenza polymerase, and discuss prospects for the development of anti-influenza therapeutics.

Introduction

The emergence of avian influenza H5N1 subtype viruses with high pathogenicity poses a global threat to human health (Gambotto *et al.*, 2008), with 256 of the 412 reported human cases since 2003 proving fatal (WHO, March 2009). Currently approved influenza drugs target regions of the virus with substantial sequence variation (De Clercq, 2006), and resistant strains of H5N1 subtype influenza viruses are already emerging. There is therefore an urgent need for new and more effective therapeutics to combat influenza virus infection. Elucidating the underlying mechanisms of the virus life cycle is crucial to identifying new targets and developing new approaches for therapeutic intervention against the influenza virus.

The influenza virus RNA-dependent RNA polymerase is a heterotrimeric complex with a molecular mass of around 250 kDa and consisting of three subunits: PA, PB1 and PB2. The complex harbours multiple enzymatic activities for catalysing both viral RNA transcription (vRNA → mRNA) and replication (vRNA → cRNA → vRNA). However, the mechanisms by which these two different RNA synthesis functions are regulated within the large polymerase complex remain unclear.

There is still some controversy about the functions of the various subunits, but those of PB1 and PB2 have been more clearly defined (Li *et al.*, 2001). PB1 contains conserved, well-characterized RNA-dependent RNA polymerase motifs (Biswas and Nayak, 1994) and is the central protein subunit for the RNA polymerase activity. It binds to the viral promoter and is responsible for viral RNA elongation and cap RNA cleavage activities (Biswas and Nayak, 1994). The PB2 subunit is required for transcription of vRNA (Li *et al.*, 2001) and can bind to the methylated cap-1 structure of host RNAs for cleavage by the PB1 subunit (Fechter *et al.*, 2003). In contrast to PB1 and PB2, however, the role of PA in the polymerase heterotrimer has only been partly outlined. It has been reported to be required for replication and for transcription of vRNA as well as endonuclease cleavage of the cap RNA primer (Fodor *et al.*, 2002; Fodor *et al.*, 2003; Huarte *et al.*, 2003; Jung and Brownlee, 2006; Kawaguchi *et al.*, 2005). It reportedly induces proteolysis of viral and host proteins (Hara *et al.*, 2001; Rodriguez *et al.*, 2007; Zurcher *et al.*, 1996) and may also be involved in virus assembly (Regan *et al.*, 2006).

The polymerase heterotrimer has been identified in two forms in the nucleus of infected cells: as part of a virion ribonucleoprotein (vRNP) complex in which the eight RNA segments of the influenza genome are packaged together with the nucleoprotein (NP) (Klumpp *et al.*, 1997); and also as a non-nucleocapsid polymerase complex that is not associated with vRNP (Detjen *et al.*, 1987). Each vRNP complex, which serves as an independent template for transcription and replication, has a supercoiled ribbon-like structure (Jennings *et al.*, 1983; Pons *et al.*, 1969) in which ~24 nucleotides of RNA are bound to each NP monomer (Ortega *et al.*, 2000). Initiation of transcription occurs by cap-snatching primers from cellular pre-mRNAs, and transcription is terminated by polyadenylation at an oligo-U signal before the end of the template. The replication process is initiated de novo and newly synthesized cRNA associates with NP and the polymerase in complementary RNPs (cRNP), which then act as replication intermediates for generation of progeny vRNP. The polymerase locates at one end of the vRNP supercoil and helps to maintain the termini of the RNA in proximity (Murti *et al.*, 1988). RNA recognition by the polymerase is essential for RNA synthesis, and the polymerase complex is able to recognize and bind both viral RNA and cRNA panhandle RNA structures (Gonzalez and Ortin, 1999). Viral RNAs share the same 5′ and 3′ sequences, and the two ends form a panhandle structure which is essential to initiate transcription (Fodor *et al.*, 1994). The soluble non-nucleocapsid polymerase heterotrimers in the nuclei of infected cells are proposed to play a role in recognizing newly synthesized replication intermediates and progeny RNAs, as they bind specifically to the 5′-terminal conserved sequences and the 5′–3′ panhandle structure (Gonzalez and Ortin, 1999; Lee *et al.*, 2003; Tiley *et al.*, 1994).

The architecture of the polymerase complex has been intensively studied in order to understand the minimum components required for viral transcription or replication. Nakagawa and colleagues (Nakagawa *et al.*, 1995; Nakagawa *et al.*, 1996) reported that PB2 is not required for replication and PB1 alone can synthesize RNA. A PB1-PA dimeric protein complex was shown to synthesize a 53-nt-long transcript directed by a vRNA promoter de novo, but could not efficiently synthesize RNA under a cRNA promoter (Honda *et al.*, 2002). Deng and colleagues reported that, using a recombinant protein complex purified from 293T cells, only the heterotrimer could synthesize pppApG dinucleotides, not the PB1-PB2 dimer or the PB1-PA dimeric RNA polymerase (Deng *et al.*, 2006b), which is consistent with other previous reports (Deng *et al.*, 2005; Lee *et al.*, 2002) and also with a previous study in which PB2 was also found to be required for replication (Perales and Ortin, 1997). A point mutation at position 510 in PA resulting in impaired viral mRNA synthesis indicates that PA is also required for transcription (Fodor *et al.*, 2003). It would appear that all three subunits are thus required for effective transcription and replication activity.

Certain strains of influenza, including the H1N1 virus responsible for the 1918 ‘Spanish flu’ epidemic and the avian H5N1 viruses, have increased virulence, although the molecular determinants of their virulence remain unclear. The polymerase heterotrimer has been implicated in the high pathogenicity of the influenza virus. Most notably, K627 of the PB2 protein is associated with increased virulence of H5N1 viruses (Subbarao *et al.*, 1993) and is found in most pathogenic human influenza viruses. Avian viruses with glutamic acid in position 627 exhibited limited replication in mammalian cells, but their viability was restored by substitution to lysine. Non-virulent H5N1 viruses in mice contain E627, but those with K627 proved lethal (Hatta *et al.*, 2001). The PA and PB1 subunits have also been implicated in highly pathogenic viruses, particularly the amino acids T515 of PA and Y436 of PB1 in mallard ducks (Hulse-Post *et al.*, 2007). While PA T515A and PB1 Y436A mutant genes abolished pathogenicity of the virus in wild fowl, other properties including replication, transmission and lethality via inoculation remained unaffected, suggesting the mutations may have affected an early step of the infection process. The single T515A mutation in PA does not affect pathogenicity of the influenza virus in ferrets or mice, however. RNA synthesis by the influenza virus has also been postulated to involve host factors that differ between species, with accumulating evidence to support the participation of host factors in influenza virus RNA synthesis.

Polymerase structure and organization

The polymerase genes are the most widely conserved in the influenza virus. Unlike other parts of the influenza virus, particularly the well-characterized haemagglutinin and neuraminidase, no detailed three-dimensional structures of the influenza polymerase heterotrimer or its individual subunits were reported until very recently. Nevertheless, knowledge of the influenza polymerase organization and function has increased over the years. The regions of the three subunits – PA, PB1 and PB2 – involved in interactions that form the polymerase complex have been mapped and defined more clearly through structural and functional analysis (Fig. 8.1A).

The earliest published structural details of the influenza polymerase complex resulted from low-resolution (26 Å and 23 Å) electron microscopy studies of the isolated polymerase complex (Torreira *et al.*, 2007) and the ribonucleoprotein (RNP) complex (Area *et al.*, 2004), in which the RNA is bound to the NP and associated with the polymerase complex. The polymerase

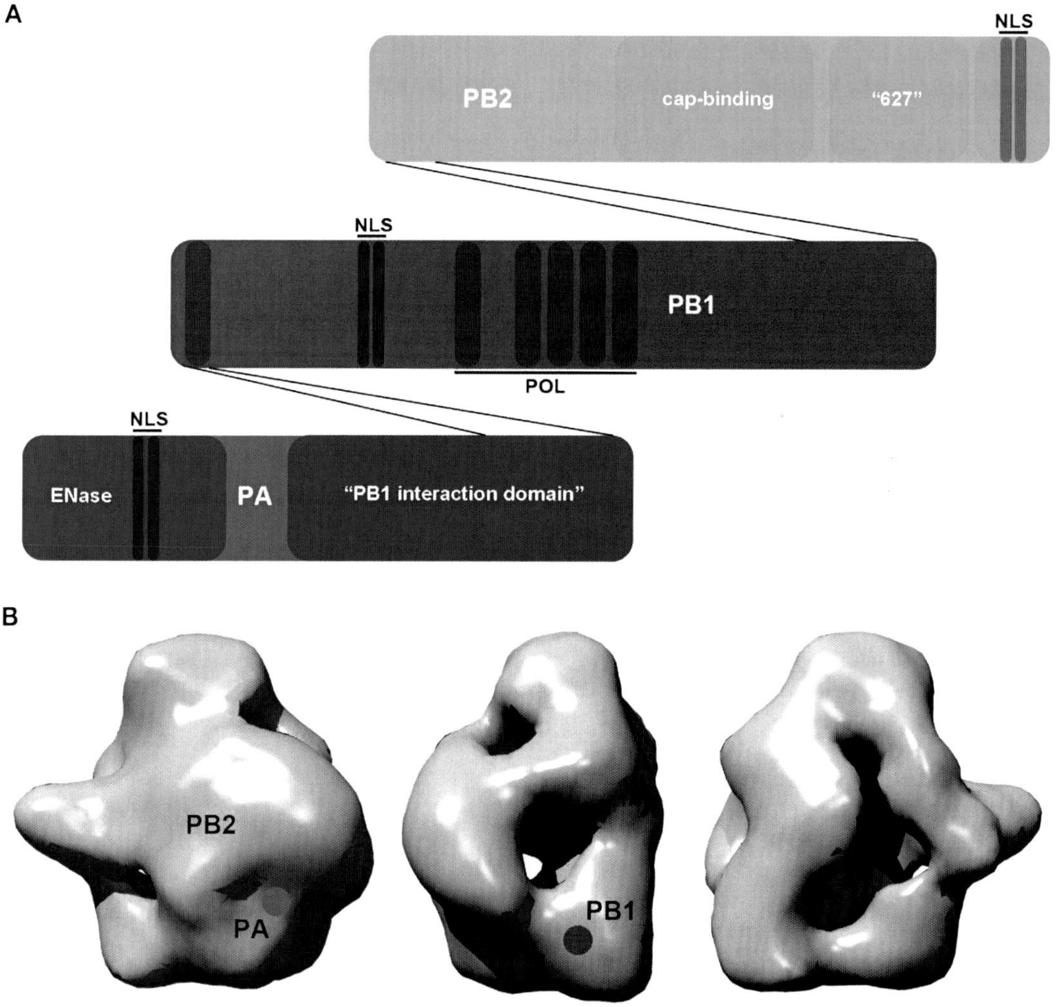

Figure 8.1 Organization of the influenza polymerase heterotrimer. (A) The PA, PB1 and PB2 subunits are shown and putative functional or interaction domains are indicated. Key: NLS: nuclear localization signal; POL, polymerase motifs; ENase, endonuclease. NLS are shown by two bands to represent bipartite signals. (B) Electron microscopy model of the isolated polymerase complex, shown from the front (left), side (centre) and reverse (right). The approximate locations of PA, PB1 and PB2 from antibody labelling are indicated. The electron microscopy data were generously provided by Juan Ortin and Oscar Llorca.

heterotrimer is tightly packed with a cage-like hollow core and no obvious boundaries between domains and subunits (Fig. 8.1B). In both models, structural organization of the isolated polymerase heterotrimer indicates that PB1 interacts directly with PA and PB2, though it would appear no direct interaction exists between PA and PB2, which is consistent with previous studies (Toyoda *et al.*, 1996). PA–PB1 and PB1–PB2 are known to form stable complexes with limited functionality. A recent study suggests that PA and PB2 do interact directly in living cells, albeit with an affinity 2- to 3-fold lower than for PA–PB1 and PB1–PB2, and that this interaction might be transient resulting from conformational changes in the polymerase complex (Hamerka *et al.*, 2009). Furthermore, the PA–PB1 complex binds to the vRNA promoter, but not the individual subunits or the PB1–PB2 complex (Deng *et al.*, 2005; Lee *et al.*, 2002).

Structure and function of PA

As described above, PA is a key protein in the influenza virus polymerase complex, although its role was only partly outlined until recently and remains less clear than PB1 and PB2. Functions associated with PA include replication and transcription of vRNA as well as endonuclease cleavage of the cap RNA primer (Fodor *et al.*, 2002, 2003; Huarte *et al.*, 2003; Jung and Brownlee, 2006; Kawaguchi *et al.*, 2005). It reportedly induces proteolysis of viral and host proteins (Hara *et al.*, 2001; Rodriguez *et al.*, 2007; Zurcher *et al.*, 1996) and may also be involved in virus assembly (Regan *et al.*, 2006).

The PA protein can be cleaved by limited tryptic digestion into two domains: a smaller N-terminal domain of ~25 kDa and a larger C-terminal of ~55 kDa (Guu *et al.*, 2008; Hara *et al.*, 2006). Structures for both the N- and C-terminal domains of PA have been reported and will be described in greater detail below, but they indicate that the two domains are separated by a long linker peptide. Guu and colleagues highlighted the importance of this linker in the PA–PB1 interface (Guu *et al.*, 2008) by showing that neither the N-terminal nor C-terminal domains of PA could ensure a stable interaction with PB1 without the presence of the linker. Furthermore, protease treatment of the PA–PB1 complex showed that its PA protein is markedly more stable than free PA, implying that the linker is protected from digestion by PB1 and forms an essential part of the subunit interface. This interface spans much of the PA sequence, with the exception of the N-terminal 154 amino acids, which is consistent with previous reports mapping the interaction domain of PA and PB1 to the C-terminal two thirds of PA (Toyoda *et al.*, 1996; Zurcher *et al.*, 1996). The presence of the linker connecting PAN and PAC should afford PA with a degree of conformational flexibility, and may provide an advantage for roles in regulating polymerase functions through conformational changes of polymerase complexes (Kawaguchi *et al.*, 2005).

The N-terminal domain of PA is a cap-dependent endonuclease

The crystal structure of the N-terminal domain of PA (PAN hereafter) which has been implicated in a range of functions (Fodor *et al.*, 2002; Fodor *et al.*, 2003; Hara *et al.*, 2001; Huarte *et al.*, 2003; Jung and Brownlee, 2006; Kawaguchi *et al.*, 2005; Rodriguez *et al.*, 2007; Zurcher *et al.*, 1996), was reported in early 2009 by two groups using different constructs and virus strains. Yuan and colleagues used residues 1–256 cloned from an avian type A virus isolate (A/goose/Guangdong/1/96 (H5N1)) (Yuan *et al.*, 2009), while Dias and colleagues used residues 1–209 from a human type A isolate (A/Victoria/3/1975 (H3N2)) (Dias *et al.*, 2009). Both proteolytically stable structures were determined to 2.2 Å and 2.0 Å respectively. The structure of PAN from Yuan *et al.* was visible from residues 1–197, while the equivalent structure from Dias *et al.* was visible from residues 1–196; no residues were visible beyond position 197.

The PAN structure has an α/β architecture in which five mixed β-strands (β1–5) form a twisted plane surrounded by seven α-helices (α1–7) (Fig. 8.2A). A strongly negatively charged cavity, formed by a concentration of acidic residues, is surrounded by helices α2-α5 and strand β3 and houses a metal binding site. Structural comparison of PAN revealed a close match with other endonucleases, pointing towards PAN as a new member of the (P)DXN(D/E)XK endonuclease superfamily. A number of catalytically important residues were identified, including P107, D108

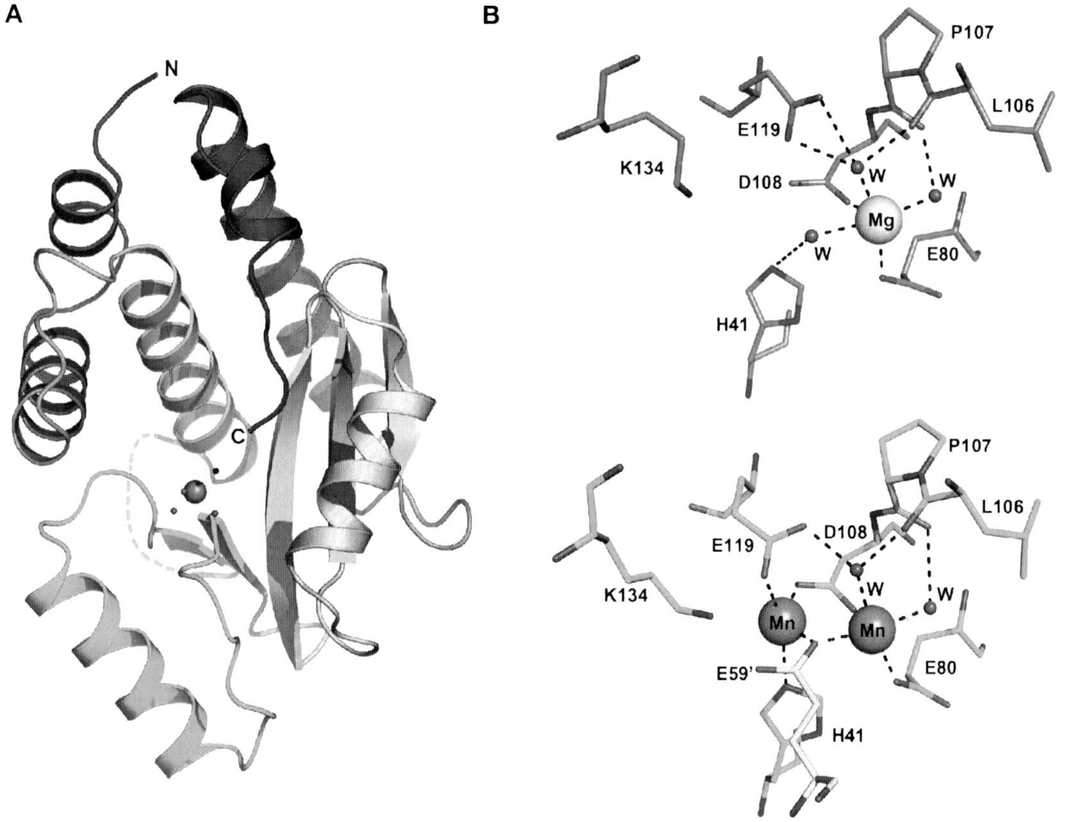

Figure 8.2 Structure of the PA cap-dependent endonuclease. (A) The N-terminal cap-dependent or 'cap-snatching' endonuclease. The structure is shown in ribbon representation and coloured from blue at the N-terminal to red at the C-terminal. B. Comparison of the Mg2+ and Mn2+ bound endonuclease structures from Yuan *et al.* (top) and Dias *et al.* (bottom), respectively. In the top panel, residues coordinating the metal are shown as blue sticks. In the lower panel, residues coordinating the metal are shown as green sticks; residue E59′ from a neighbouring monomer is shown as a yellow stick. Metal ions are shown as grey spheres; bound waters are shown as red spheres; and interactions are shown by dashed black lines.

and E119 in the (P)DXN(D/E)XK motif. From structural analysis, the putative lysine is K134 on the basis of its close proximity to the metal binding site.

Both PAN structures share a similar overall structure, but with one key difference in the metal binding site: the structure by Yuan *et al.* included one magnesium (Mg^{2+}) ion, whereas the structure by Dias *et al.* featured two manganese (Mn^{2+}) ions (Fig. 8.2B). In each case, the metal ions are coordinated by the same residues, including H41, D108, E119 and K134. In the Mg^{2+} bound structure determined by Yuan and colleagues (Yuan *et al.*, 2009), the metal is directly coordinated by five ligands: the acidic residues E80 and D108, and three water molecules stabilized by residue H41, E119 and the carbonyl oxygens of L106 and P107. All six amino acids involved in coordinating Mg^{2+} are conserved among influenza A, B, and C viruses, with the exception of P107 that is replaced with alanine and cysteine in influenza B and C viruses, respectively. With two bound Mn^{2+} ions, as in the structure by Dias and colleagues (Dias *et al.*, 2009), residue D108 serves as a bridge between the two metals. Furthermore, residue E59 from a neighbouring monomer is involved in coordination of the second Mn^{2+} ion, occupying an equivalent position to the water molecule coordinated by H41 in the Yuan structure.

Doan and colleagues previously reported that cleavage of RNA by the endonuclease is activated by the binding of divalent cations and should involve a two-metal-dependent mechanism in common with other nucleases. Optimal endonuclease activity was observed for Mn^{2+} and was reduced to around 50% activity for Mg^{2+} (Doan *et al.*, 1999). However, given that the concentration of Mg^{2+} ions is higher than Mn^{2+} ions under physiological conditions, it may be that Mg^{2+} is the more likely cofactor in the cell. Further work is therefore needed to elucidate the mechanism of endonuclease cleavage. The influenza virus endonuclease has also been shown to cleave single-stranded DNA (ssDNA) with slightly reduced activity than for RNA, which is consistent with the observation that capped ssDNA endonuclease products can function as primers for transcription initiation by the influenza virus polymerase (Klumpp *et al.*, 2000). Our preliminary results suggest that W88 and R84 could be directly involved in nucleotide binding for transcription termination. W88A and R84A mutants result in an almost complete loss of nuclease activity, similar to the E80A and E119A mutants (data not shown).

The endonuclease activity of the influenza virus polymerase is critical for snatching capped primers from host mRNA to initiate mRNA transcription. PA was reported to be involved in endonuclease activity, either via its N- or C-terminal domain (Fodor *et al.*, 2002; Hara *et al.*, 2006); in particular, a number of mutations in the N-terminal of PA inhibited this activity (Hara *et al.*, 2006). Li and colleagues, however, reported that endonuclease activity of the influenza polymerase is located in PB1, and the acidic residues E508, E519 and D522 are essential for the activity (Li *et al.*, 2001), while earlier studies suggested that the PB2 subunit might contain the endonuclease active site (Blok *et al.*, 1996; Honda *et al.*, 2002; Shi *et al.*, 1995). In their study, Yuan and colleagues used mutant RNP complexes to examine the putative endonuclease site in PB1, showing using primer extension assays that E508A, E519A and D522A mutations resulted in significant levels of mRNA, cRNA and vRNA synthesis activity. The availability of PAN structures, therefore, clearly delineate the endonuclease active site of the influenza polymerase and help to resolve this particular controversy. Furthermore, the identification of an enzymatic active site in the polymerase complex provides another target for the design of anti-viral inhibitors. A number of compounds have been reported to inhibit the endonuclease activity of the influenza virus polymerase (Cianci *et al.*, 1996; Nakazawa *et al.*, 2008; Tomassini *et al.*, 1994; Tomassini, 1996), but their further development and optimization was presumably limited by the lack of an available structural target.

An essential subunit interaction between PA and PB1

As described above, PA and PB1 directly interact to form a stable complex which binds to the vRNA promoter. This interaction has been extensively characterized and is known to occur between the C-terminal region of PA and the N-terminal region of PB1. More specifically, the N-terminal 25-residues of PB1 bind to the C-terminal region of PA. This PB1 N-terminal peptide (PB1N hereafter) is also known to act as an inhibitor and abolishes viral replication by blocking the polymerase heterotrimer.

Using this information as a basis, two groups determined crystal structures showing the essential subunit interaction between PA and PB1. In the first study, He and colleagues used an avian type A influenza virus subtype (A/goose/Guangdong/1/96 (H5N1)) to clone PA from residues 257 to 716 and PB1 from residues 2 to25 (He *et al.*, 2008). In the second study, Obayashi and colleagues used a human avian type A influenza virus subtype (A/Puerto Rico/8/1934 (H1N1)) to clone PA from residues 239–716 and PB1 from 1–81 (Obayashi *et al.*, 2008). The structures were determined to 2.9 Å and 3.2 Å resolution, respectively.

The C-terminal domain of PA, hereafter referred to as PAC, consists of 13 α-helices, one short 310 helix, nine β-strands and several loops/turns. The overall structure of PAC resembles the head of a dragon and can be subdivided into two parts: domain I, the 'brain', and domain II, the 'mouth' (Fig. 8.3A). The seven β-strands of the 'brain' form a twisted plane surrounded by five α-helices and a short 310 helix. The 'mouth' consists of strands two β-strands and eight α-helices. Structural comparison indicated that PAC bears no similarity to other structures in the Protein Data Bank (PDB) and thus should have a new fold.

The N-terminal PB1 peptide, PB1N, binds obliquely between the jaws of PAC with its N-terminus pointing towards the back of the

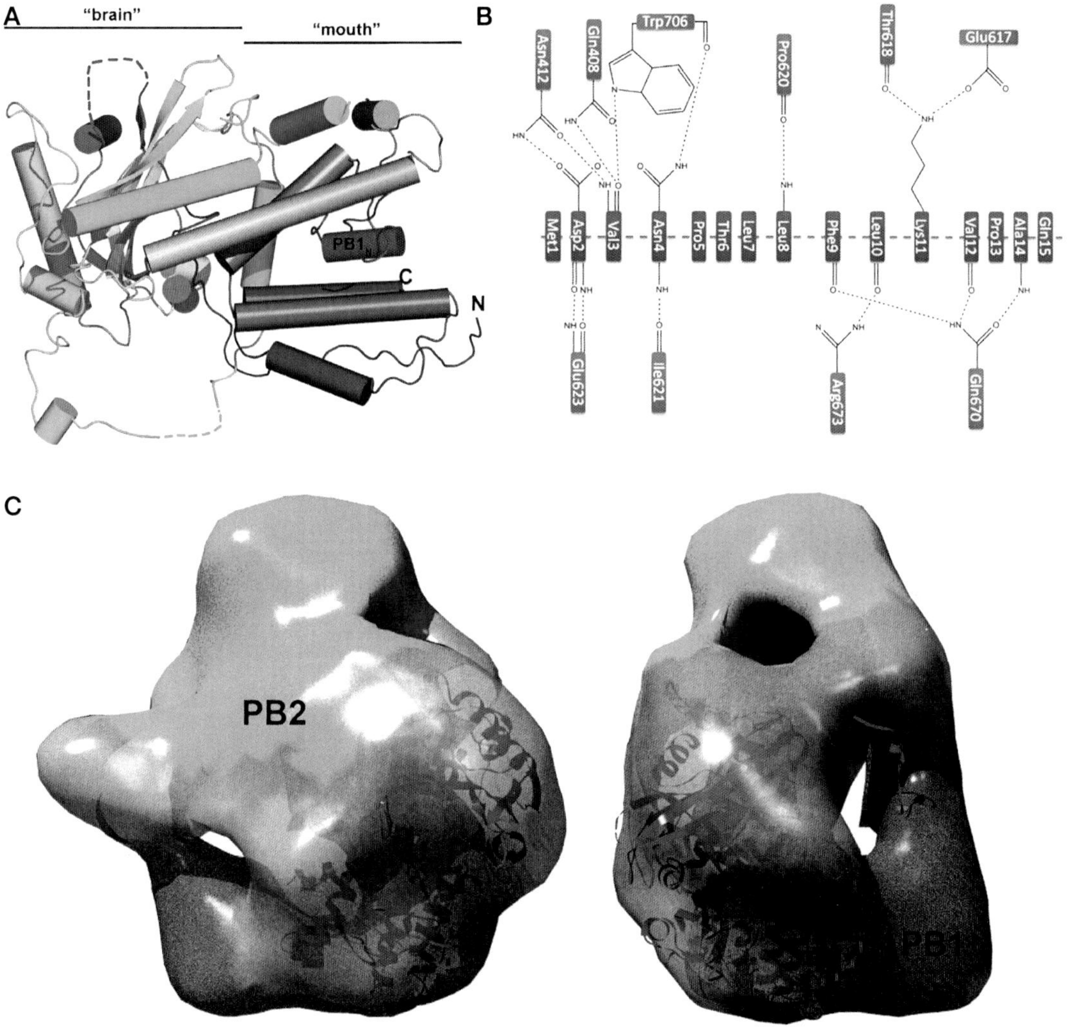

Figure 8.3 Structure of the PB1 interaction domain of PA. (A) The C-terminal PB1 interaction domain of PA. The PA structure is shown in ribbon representation and coloured from blue at the N-terminal to red at the C-terminal. The N-terminal PB1 peptide is coloured magenta. (B) Schematic showing the interaction between PAC and the PB1N peptide. Residues in the PB1N peptide are shown in magenta; residues in PA are shown in blue. C. Front and side views of the PAC domain docked into the electron microscopy model of the isolated polymerase heterotrimer. The PAC structure is shown in blue ribbon representation; the PB1N peptide is shown in red ribbon representation; approximate locations of the PB1 and PB2 subunits are shown in red and green, respectively.

'mouth' and its C-terminus extending outwards. Residues 5 through 11 of PB1N fold into a short ordered helix. The interaction with PB1N is mediated by four helices (α8, α10, α11 and α13), which form a hydrophobic core at the tip of the 'mouth' and tightly interact with PB1N via hydrophobic interactions, hydrogen bonds and van der Waals forces (Fig. 8.3B). Other minor interactions are provided by α4 and the α9-α10 loop. A short –LLFL- motif from residues 7–10 of PB1N, known to be important for the interaction with PA (Perez and Donis, 2001), interacts with the PAC hydrophobic core formed by F411, M595, L666, W706, F710, V636 and L640. W706 also interacts with residues V3 and N4; Q408 and N412 interact with V3 and D2; and Q670 interacts with PB1N residues F9, V12, P13 and A14. He and colleagues showed that the W706A/Q670A double mutation disrupts the binding of PB1N to PAC, as do the L666G/F710E, L666G/F710G and W706A/F710Q double mutations (He *et al.*, 2008). V636S, L640D, L666D and W706A mutations were also found to reduce the yield levels of vRNA, mRNA and cRNA in reporter assays by at least 40% (Obayashi *et al.*, 2008).

PB1N inhibits influenza A viral replication by interfering with polymerase activity, presumably by blocking assembly of the polymerase heterotrimer (Fig. 8.3C) (Ghanem *et al.*, 2007). The available crystal structures of PAC therefore identify a critical PB1 binding region of PA, and can thus be used as a basis to design novel anti-influenza compounds that inhibit polymerase assembly and function. Protein–protein interfaces often involve a large surface area, which can pose problems for drug discovery. However, relatively few residues drive the binding of PB1N to PAC, implying that designing small molecule inhibitors of this interaction is feasible. Mutation of specific residues in PB1N, including V3, N4, P5, L7, L8, F9 and L10, has been shown to result in loss of more than two-thirds of the binding affinity, with significant reductions in polymerase activity and virus production (Perez and Donis, 2001). D2V and A14D mutations in PB1N do not significantly influence the binding affinity to PA, but they do abolish polymerase activity and virus production, suggesting that D2 and A14 have other crucial functions (Perez and Donis, 2001). An L13P mutation in PB1 from a H7N7 virus is associated with increased virulence (Gabriel *et al.*, 2007). In both crystal structures, a L13P substitution in PB1N would break the ordered helix and the associated structural changes may result in increased RNA synthesis by PB1. PB1N residues interacting with PA are conserved across type A, B and C influenza viruses, and PA residues shown to interact with PB1N are similarly conserved. It is expected that inhibitory peptides or compounds designed on the basis of PB1N, therefore, would be effective against the majority of influenza A strains. The high conservation of the PB1N binding site on PA suggests that antivirals targeting this site may be less susceptible to problems of resistance associated with drugs targeting the neuraminidase.

In addition to housing a key interaction site for PB1, the C-terminal domain of PA has been associated with other functions, including polymerase activity, viral pathogenicity and virus assembly. It reportedly interacts with the host protein hCLE via two regions, residues 493–512 and 557–574 (Huarte *et al.*, 2001), and with the microchromosomal maintenance complex (MCM) (Kawaguchi and Nagata, 2007). The precise role of hCLE in viral replication remains unclear but it shares homology with a family of transcriptional activators and has been shown to function as a modulator of mRNA transcription (Perez-Gonzalez *et al.*, 2006). The MCM complex, thought to be a DNA replicative helicase, is another of the host factors believed to be hijacked by the influenza virus for genome replication and may serve as a scaffold between the polymerase and nascent RNA chains. A H510A mutation in PAC impairs nuclease activity (Fodor *et al.*, 2002) and was initially thought to be a possible centre for the endonuclease activity. However, given the identification of the endonuclease site in PAN, the role of H510 may be linked with a putative nucleotide binding motif from residues 501–509 (de la Luna *et al.*, 1989). In the same region, positions 507 and 508 are associated with virus assembly defects (Regan *et al.*, 2006) and position 515 is linked with highly pathogenic H5N1 viruses (Hulse-Post *et al.*, 2007).

Structure and function of PB2

The PB2 subunit has been associated with transcription of vRNA (Li *et al.*, 2001) and recognizing the methylated cap-1 structure of host RNAs for cleavage by the PB1 subunit (Fechter and Brownlee, 2005). Amino acid 627 of PB2 has also been implicated as one of the key determinants of increased virulence of H5N1 influenza viruses. No structure exists for the full-length subunit, but structures have been reported for three functional domains of PB2 and will be described in more detail below.

Structural basis for 'cap-snatching' by the PB2 subunit

The influenza polymerase utilizes a 'cap-snatching' mechanism for short, capped primers in the synthesis of influenza virus mRNAs. The PB2 subunit is known to bind the 5′ cap of pre-mRNAs, which are subsequently cleaved after 10–13 nucleotides by the PB1 subunit. These capped oligoribonucleotides then prime synthesis of viral mRNAs, which become polyadenylated at the 3′

ends by stuttering of the polymerase at an oligo-U motif at the 5′ end of the template to render them competent for translation. The structure of the PB2 subunit cap-binding domain was reported by Guilligay and colleagues from a human influenza A virus (A/Victoria/3/1975 (H3N2)) in 2008 (Guilligay *et al.*, 2008). Prior to this, the location of the PB2 cap-binding domain was controversial and three potential sites were identified from cross-linking and mutagenesis: an N-terminal site from 242–252, a C-terminal site from 533–577, and a central site involving amino acids 363 and 404. Guilligay *et al.* showed that a fragment of PB2 from amino acids 318–483 forms an independently folded domain and has cap-binding activity for an m7GTP analogue *in vitro*, thus confirming that the central site of PB2 is responsible for cap-binding. The authors further showed, using binding and functional studies with point mutants, that this central cap-binding site is also essential for cap-dependent transcription *in vivo* by the influenza polymerase heterotrimer.

The structure of the cap-binding domain, determined to 2.3 Å resolution, presents a compact and well-ordered α-β fold unrelated to other structures in the PDB. An anti-parallel five-strand β-sheet is packed against a four-helix bundle and flanked by a small four-strand C-terminal β-sheet. A bound m7GTP analogue identified a binding site at one end of the molecule between the β-sheet and helical subdomains, with a long β-hairpin completing the binding site (Fig. 8.4A).

The mode of binding by the m7GTP analogue to the PB2 cap-binding domain resembles that employed by other cap-binding proteins. The m7G moiety is sandwiched between two aromatic residues, H357 and F404, resulting in a strong π-interaction with a 100-fold discrimination against a non-methylated base. Histidine in this position is unique and has not been observed in an equivalent role in other cap-binding proteins. F323 stacks against the ribose and, together with F404, forms part of a hydrophobic cluster of five phenylalanines. E361 is the key acidic residue and forms hydrogen bonds with the N1 and N2 positions of the guanine, while K376 is also involved in base recognition via the O6 position. The methyl group of m7GTP forms van der Waals contacts with Q406, and F404 is in contact with M431. However, specificity for the methyl group should be provided largely by the positive charge induced by methylation, rather than by the weak interactions of the methyl group. The triphosphate interacts with H432 and N429 (α-phosphate), and H357, K339 and R355 (γ-phosphate), although it may adopt a different conformation in the context of a bound capped oligoribonucleotide. Compounds designed to block the cap-binding by PB2 may prove to be effective inhibitors of influenza virus replication.

The '627' domain of PB2

As described above, position 627 of the PB2 subunit is an important determinant of host specificity in influenza viruses. Efficient viral replication at lower temperature in the human respiratory tract requires lysine in position 627, rather than glutamic acid systematically found in avian influenza viruses. In order to examine how the 627 site in PB2 determines host range, Tarandeau and colleagues determined the crystal structure of an independently folded domain from residues 538–676, including K627, hereafter known as the '627' domain (Tarendeau *et al.*, 2008). The resulting structure exhibits a novel fold and can be divided into two halves: an N-terminal half comprising a six-helical cluster with a hydrophobic core rich in aromatic residues; and a C-terminal half consisting of five short β-strands wrapped around one side of the N-terminal helical bundle (Fig. 8.4B). An extended peptide, containing the host-specific residue 627, wraps around part of the helical cluster such that K627 is fully solvent exposed. A K627E mutant structure is essentially identical and E627 points into solvent. Substitution at position 627 does not result in local or global conformational changes, but the charge reversal does significantly alter the electrostatic surface potential of the domain. The structure alone provides few clues for the further role of position 627 in host specificity, but its surface exposed location may provide a binding surface for interaction with other viral or host proteins.

The nuclear import domain of PB2

The polymerase is active in the nucleus and nuclear localization signals have been identified in PA (Nieto *et al.*, 1992), PB1 (Nath and Nayak, 1990)

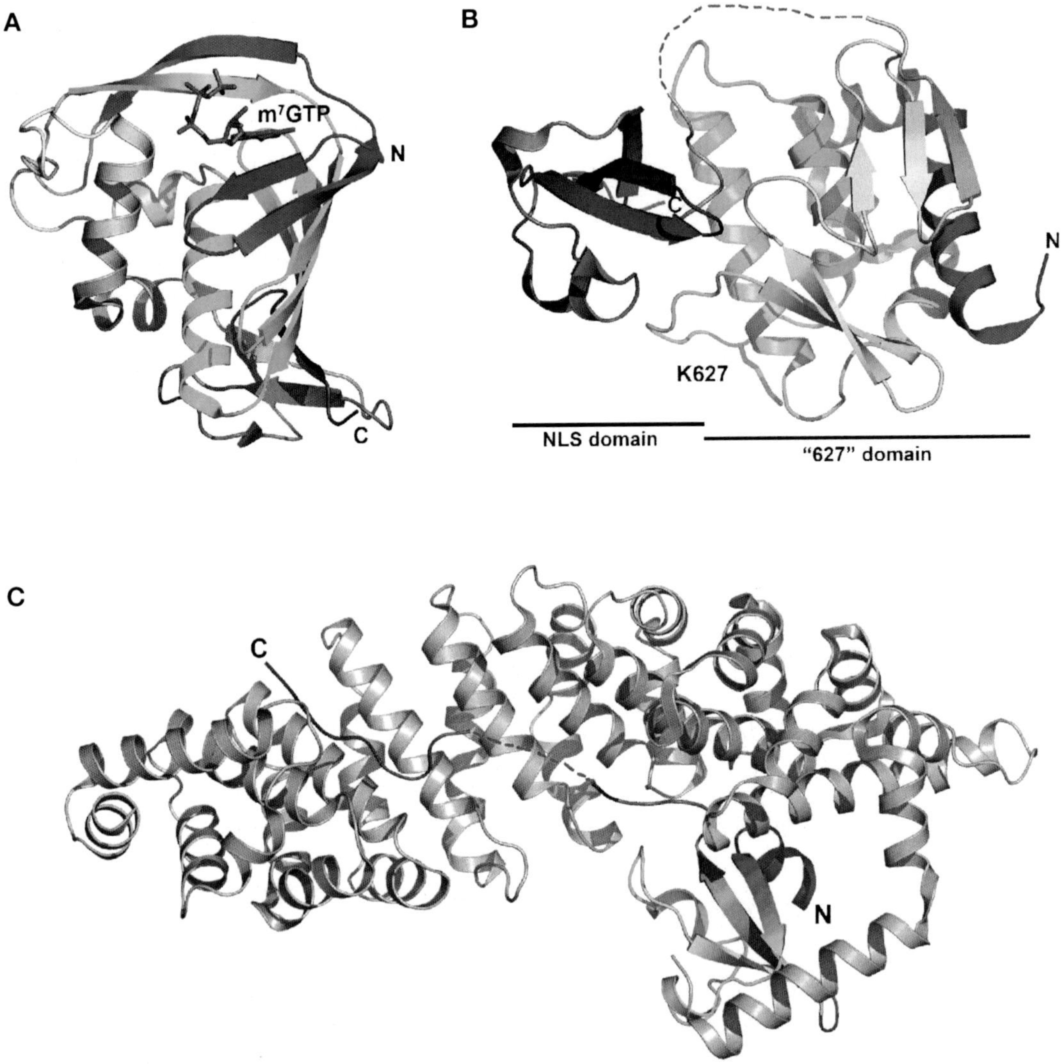

Figure 8.4 Structures of the PB2 subunit. (A) The cap-snatching domain of PB2. The structure is shown in ribbon representation and coloured from blue at the N-terminal to red at the C-terminal. The m7GTP is shown in magenta. (B) The structure of the '627' and nuclear import (NLS) domains. The structure is shown in ribbon representation and coloured from blue at the N-terminal to red at the C-terminal. Residue K627 is shown in stick representation and labelled. (C) The nuclear import domain of PB2 in complex with human importin α5. The structure of the PB2 nuclear import domain is coloured from blue at the N-terminal to red at the C-terminal. Importin α5 is shown in ribbon representation and coloured grey.

and PB2 (Mukaigawa and Nayak, 1991). It is believed that PB1 requires PA for efficient nuclear accumulation and the two subunits should be imported together as a subcomplex (Fodor and Smith, 2004), possibly by the RanBP5 importin (Deng *et al.*, 2006a). PB2, on the other hand, is understood to be imported alone and then assembles with the PA-PB1 subcomplex within the nucleus. The nuclear localization signal of PB2 has been mapped to its C-terminal, and the structure of the isolated nuclear import domain from amino acids 678 to 759 has been determined alone by NMR and as a crystal structure in complex with the nuclear import factor importin α5 (Tarendeau *et al.*, 2007). The solution structure of the isolated nuclear import domain has a compact α-β fold in which a three-strand anti-parallel β-sheet is packed with a short amphipathic helix. The PB2 nuclear import domain is unrelated to other structures in the PDB and most likely represents a novel structural motif.

PB2 was originally predicted to contain a monopartite -KRKR- nuclear import motif from amino acids 736 to 739 that is highly conserved among influenza viruses. The PB2 nuclear import fragment was shown to form a stable complex with human importin α5 *in vitro*. However, the monopartite motif alone is insufficient for nuclear import, and a bipartite -KRX12KRIR- motif from amino acids 738 to 755 was found to function as an efficient nuclear import signal. From the complex structure determined by Tarendeau and colleagues (Tarendeau *et al.*, 2007), it is evident that the C-terminal region of PB2 from residue 736 onwards mediates the interaction with importin α5, and that the bipartite NLS motifs can interact with two distinct regions in the superhelical groove of importin α5 as a result of the extended conformation adopted by the nuclear import domain (Fig. 8.4C).

A double structure containing the '627' and nuclear import domains was also determined by Tarendeau and colleagues from residues 539–753 (Tarendeau *et al.*, 2008), and by Kuzuhara and colleagues from residues 535 to 759 (Kuzuhara *et al.*, 2009). In both structures, the '627' and nuclear import domains pack side by side to form a single module (Fig. 8.4B). The ordered parts of each domain exhibit only minor differences from their individual counterparts. The double-domain complex has been shown to interact strongly with human importin α1, which is consistent with observations of a functional PB2-importin α1 interaction from cellular studies.

Interestingly, Kuzuhara and colleagues showed that a K627E mutant has significantly lower RNA binding activity, which might suggest a strong correlation between RNA-binding activity at position 627 and host specificity. Kuzuhara and colleagues further suggested that this domain has preferentially higher binding affinity for the vRNA promoter, consistent with previous reports that PB2 binds to the vRNA promoter via an unidentified region (Jung and Brownlee, 2006). Several aromatic residues in the '627' domain have been implicated in capped RNA binding, while Kuzuhara and colleagues suggested similarity between the helical bundles of PB2 with the activator 1 subunit of the DNA replication clamp loader (replication factor C), which also functions as an RNA-binding protein. Further work is needed, therefore, to explore the structure–function relationship between RNA-binding mechanisms and influenza virus transcription and cellular DNA replication. It is also important to confirm the effect of position 627 in determining host specificity and pathogenicity, either via interaction with viral or host proteins, or by altering the RNA binding activity.

The influenza polymerase as a target for drug discovery

The majority of approved drugs for influenza work by targeting the neuraminidase (NA), although other targets such as the M2 proton channel also exist in the influenza virus (see De Clercq, 2006, for a review). Sialic acid receptors on the surface of the host cell are recognized by the influenza virus haemagglutinin (HA). The sialic acid is then cleaved off from the cell-surface glycoprotein by the NA, after which newly-formed virus particles are released for spread to other host cells. Neuraminidase inhibitors work by effectively blocking the release of progeny virus particles. However, the currently licensed influenza drugs oseltamivir (Tamiflu®) and zanamivir (Relenza®) are beset by problems of resistance due to high mutation rates or selection pressures on the influenza virus, limiting their clinical effectiveness. Their efficacy also becomes significantly diminished if they are not taken within 48 hours of the onset of symptoms. Furthermore, the currently approved drugs have limited effectiveness against the H5N1 influenza virus subtypes.

In contrast to drugs targeting the NA, very few compounds have been reported to operate against the influenza polymerase. Given that the polymerase complex harbours multiple enzymatic activities, together with its critical role in the influenza virus life cycle and the high conservation of its genes, it should constitute a major target or targets for anti-influenza drug discovery. Progress in the development of polymerase inhibitors to date has presumably been hindered by the lack of detailed three-dimensional structural information. The few compounds that have been reported can be divided into nucleosides and non-nucleosides, as in the case of inhibitors targeting the HIV reverse transcriptase or the HCV RNA replicase.

One specific example of a nucleoside inhibitor of the influenza polymerase is a substituted pyrazine compound called T-705, which has potent *in vitro* and *in vivo* anti-influenza activity (Furuta *et al.*, 2002, 2005; Takahashi *et al.*, 2003). T-705 is believed to be converted intracellularly by a phosphoribosyl transferase to a ribonucleotide T-705-4-ribofuranosyl-5′-monophosphate (T-705 RMB) form. After further phosphorylation of T-705 RMP to 5′-triphosphate, the T-705 RTP form should then inhibit the influenza polymerase in a GTP-competitive manner. Its anti-viral activity has been attributed largely to the influenza polymerase; it does not inhibit the cellular enzyme inosine 5′-monophosphate (IMP) dehydrogenase, which is key for biosynthesis of GTP and viral RNA synthesis.

Of the polymerase structures reported to date, PA contains two significant new targets for inhibitor development. The N-terminal domain of PA is a cap-dependent or 'cap-snatching' endonuclease with a highly conserved active site (Dias *et al.*, 2009; Yuan *et al.*, 2009). Specific inhibitors of the polymerase cap-dependent endonuclease activity have been described previously and can be divided into several classes of compounds: 4-substituted 2,4-dioxobutanoic acid derivatives (Tomassini *et al.*, 1994); N-hydroxamic acid and N-hydroxy-imide derivatives (Cianci *et al.*, 1996); and a 2,6-diketopiperazine natural product of the fungus *Delitschia confertaspora* called flutimide (Tomassini, 1996). Flutimide in particular has also been shown to inhibit the replication of influenza A virus in cell culture. One 2,4-diketobutanoic acid analogue targeting PA has shown potent polymerase inhibition without any cytotoxicity (Nakazawa *et al.*, 2008). The availability of high-resolution three-dimensional structures for the cap-dependent endonuclease in PA should accelerate development of anti-influenza compounds targeting this region of the polymerase. The C-terminal domain of PA mediates the interaction with the PB1 subunit, and two crystal structures have revealed the molecular basis for the interaction of PA with the N-terminal peptide of PB1 via a highly conserved region. A peptide derived from the N-terminal of PB1 has been shown to bind to PA and inhibit the polymerase function, which should be achieved by interfering with the assembly of the polymerase complex. The structures showing the PA–PB1 interaction should provide a starting point for the design of inhibitors aimed at blocking assembly of the polymerase complex.

In PB2, the minimal structure of the cap-binding domain presents another target that can be exploited for the development of novel anti-influenza compounds. Mutation of residues involved in cap-binding resulted in severely impaired transcription activity using β-globin mRNA as a cap donor (Guilligay *et al.*, 2008), which would suggest that blocking the cap-binding activity of PB2 may be another effective means of inhibiting viral transcription initiation. The PB2 cap binding domain has a particular advantage as a therapeutic target since it is structurally distinct from known host cap binding factors.

Future prospects

The recent glut of structures from the influenza virus polymerase help to paint a more detailed picture of the structure, function and organization of the complex, yet many more questions remains to be answered. Low-resolution electron microscopy models of the isolated polymerase heterotrimer and the vRNP complex have been reported, but no more detailed study of the polymerase heterotrimer exist at present. There will be considerable value in observing the conformations of the polymerase heterotrimer in its soluble and vRNP-associated forms at higher detail.

At the time of writing, no structural information is available for the PB1 subunit, which is perhaps the most critical subunit in the polymerase heterotrimer and should be another major target for therapeutic intervention. PB1 is known to contain conserved, well-characterized RNA-dependent RNA polymerase and is central to the polymerase activity of the influenza virus. Understanding the three-dimensional structure and organization of the PB1 subunit should help to elucidate the mechanisms of influenza virus replication and transcription, as well as open further avenues for therapeutic intervention. Analogous to the PA-PB1 subunit interaction described above, another interaction exists between the PB1 and PB2 subunits. Understanding and exploring the structural basis for this subunit interaction in greater detail should help to complete the overall picture of the polymerase structure and function, and possibly reveal another means to block assembly of the heterotrimer complex.

Finally, relatively little is understood about how the polymerase complex interacts with or hijacks host factors in the infected cell for viral replication and transcription. For instance, as described above, the PA subunit has been reported to interact with several host factors, including the microchromosomal maintenance complex (MCM) (Kawaguchi and Nagata, 2007) and hCLE (Huarte *et al.*, 2001; Perez-Gonzalez *et al.*, 2006), although the precise nature and role of these interactions remains unclear. The influenza polymerase was recently reported to form part of a complex between the influenza NS1A protein and the cellular CPSF30 protein (Kuo and Krug, 2009). The binding of NS1A to CPSF30 is an important mechanism employed by the virus in suppression of the host anti-viral response, in this case the production of beta-interferon (IFN-β) mRNA. A recent crystal structure of the NS1A–CPSF30 complex indicates a common binding site that may be employed by all influenza A viruses to inhibit production of anti-viral mRNA (Das *et al.*, 2008). Kuo and Krug showed that the polymerase subunits and the NP are an integral part of the complex in infected cells and help to stabilize the NS1A–CPSF30 complex, and furthermore showed that NS1A interacts primarily with the PA subunit and the NP.

Acknowledgements

The authors are grateful for support from the National Natural Science Foundation of China (grant numbers 30599432 and 30221003), the Ministry of Science and Technology International Cooperation Project (grant number 2006DFB32420), the Ministry of Science and Technology 863 Project (grant numbers 2006AA02A314 and 2006AA02A322), and the Ministry of Science and Technology 973 Project (grant numbers 2006CB504300 and 2007CB914300).

References

Area, E., Martin-Benito, J., Gastaminza, P., Torreira, E., Valpuesta, J.M., Carrascosa, J.L., and Ortin, J. (2004). 3D structure of the influenza virus polymerase complex: localization of subunit domains. Proc. Natl. Acad. Sci. U.S.A. *101*, 308–313.

Biswas, S.K., and Nayak, D.P. (1994). Mutational analysis of the conserved motifs of influenza A virus polymerase basic protein 1. J. Virol. *68*, 1819–1826.

Blok, V., Cianci, C., Tibbles, K.W., Inglis, S.C., Krystal, M., and Digard, P. (1996). Inhibition of the influenza virus RNA-dependent RNA polymerase by antisera directed against the carboxy-terminal region of the PB2 subunit. J. Gen. Virol. 77 (Pt 5), 1025–1033.

Cianci, C., Chung, T.D.Y., Meanwell, N., Putz, H., Hagen, M., Colonno, R.J., and Krystal, M. (1996). Identification of N-hydroxamic acid and N-hydroxyimide compounds that inhibit the influenza virus polymerase. Antiviral Chem. Chemother. 7, 353–360.

Das, K., Ma, L.C., Xiao, R., Radvansky, B., Aramini, J., Zhao, L., Marklund, J., Kuo, R.L., Twu, K.Y., Arnold, E., *et al.* (2008). Structural basis for suppression of a host antiviral response by influenza A virus. Proc. Natl. Acad. Sci. U.S.A. *105*, 13093–13098.

De Clercq, E. (2006). Antiviral agents active against influenza A viruses. Nat. Rev. Drug. Discov. *5*, 1015–1025.

de la Luna, S., Martinez, C., and Ortin, J. (1989). Molecular cloning and sequencing of influenza virus A/Victoria/3/75 polymerase genes: sequence evolution and prediction of possible functional domains. Virus. Res. *13*, 143–155.

Deng, T., Engelhardt, O.G., Thomas, B., Akoulitchev, A.V., Brownlee, G.G., and Fodor, E. (2006a). Role of ran binding protein 5 in nuclear import and assembly of the influenza virus RNA polymerase complex. J. Virol. *80*, 11911–11919.

Deng, T., Sharps, J., Fodor, E., and Brownlee, G.G. (2005). *In vitro* assembly of PB2 with a PB1-PA dimer supports a new model of assembly of influenza A virus polymerase subunits into a functional trimeric complex. J. Virol. *79*, 8669–8674.

Deng, T., Sharps, J.L., and Brownlee, G.G. (2006b). Role of the influenza virus heterotrimeric RNA polymerase complex in the initiation of replication. J. Gen. Virol. *87*, 3373–3377.

Detjen, B.M., St Angelo, C., Katze, M.G., and Krug, R.M. (1987). The three influenza virus polymerase (P) proteins not associated with viral nucleocapsids in the infected cell are in the form of a complex. J. Virol. *61*, 16–22.

Dias, A., Bouvier, D., Crepin, T., McCarthy, A.A., Hart, D.J., Baudin, F., Cusack, S., and Ruigrok, R.W. (2009). The cap-snatching endonuclease of influenza virus polymerase resides in the PA subunit. Nature *458*, 914–918.

Doan, L., Handa, B., Roberts, N.A., and Klumpp, K. (1999). Metal ion catalysis of RNA cleavage by the influenza virus endonuclease. Biochemistry *38*, 5612–5619.

Fechter, P., and Brownlee, G.G. (2005). Recognition of mRNA cap structures by viral and cellular proteins. J. Gen. Virol. *86*, 1239–1249.

Fechter, P., Mingay, L., Sharps, J., Chambers, A., Fodor, E., and Brownlee, G.G. (2003). Two aromatic residues in the PB2 subunit of influenza A RNA polymerase are crucial for cap binding. J. Biol. Chem. *278*, 20381–20388.

Fodor, E., Crow, M., Mingay, L.J., Deng, T., Sharps, J., Fechter, P., and Brownlee, G.G. (2002). A single amino acid mutation in the PA subunit of the influenza virus RNA polymerase inhibits endonucleolytic cleavage of capped RNAs. J. Virol. *76*, 8989–9001.

Fodor, E., Mingay, L.J., Crow, M., Deng, T., and Brownlee, G.G. (2003). A single amino acid mutation in the PA subunit of the influenza virus RNA polymerase promotes the generation of defective interfering RNAs. J. Virol. *77*, 5017–5020.

Fodor, E., Pritlove, D.C., and Brownlee, G.G. (1994). The influenza virus panhandle is involved in the initiation of transcription. J. Virol. *68*, 4092–4096.

Fodor, E., and Smith, M. (2004). The PA subunit is required for efficient nuclear accumulation of the PB1 subunit of the influenza A virus RNA polymerase complex. J. Virol. *78*, 9144–9153.

Furuta, Y., Takahashi, K., Fukuda, Y., Kuno, M., Kamiyama, T., Kozaki, K., Nomura, N., Egawa, H., Minami, S., Watanabe, Y., *et al.* (2002). *In vitro* and *in vivo* activities of anti-influenza virus compound T-705. Antimicrob. Agents Chemother. *46*, 977–981.

Furuta, Y., Takahashi, K., Kuno-Maekawa, M., Sangawa, H., Uehara, S., Kozaki, K., Nomura, N., Egawa, H., and Shiraki, K. (2005). Mechanism of action of T-705 against influenza virus. Antimicrob. Agents Chemother. *49*, 981–986.

Gabriel, G., Abram, M., Keiner, B., Wagner, R., Klenk, H.D., and Stech, J. (2007). Differential polymerase activity in avian and mammalian cells determines host range of influenza virus. J. Virol. *81*, 9601–9604.

Gambotto, A., Barratt-Boyes, S.M., de Jong, M.D., Neumann, G., and Kawaoka, Y. (2008). Human infection with highly pathogenic H5N1 influenza virus. Lancet *371*, 1464–1475.

Ghanem, A., Mayer, D., Chase, G., Tegge, W., Frank, R., Kochs, G., Garcia-Sastre, A., and Schwemmle, M. (2007). Peptide-mediated interference with influenza A virus polymerase. J. Virol. *81*, 7801–7804.

Gonzalez, S., and Ortin, J. (1999). Distinct regions of influenza virus PB1 polymerase subunit recognize vRNA and cRNA templates. EMBO J. *18*, 3767–3775.

Guilligay, D., Tarendeau, F., Resa-Infante, P., Coloma, R., Crepin, T., Sehr, P., Lewis, J., Ruigrok, R.W., Ortin, J., Hart, D.J., and Cusack, S. (2008). The structural basis for cap binding by influenza virus polymerase subunit PB2. Nat. Struct. Mol. Biol. *15*, 500–506.

Guu, T.S., Dong, L., Wittung-Stafshede, P., and Tao, Y.J. (2008). Mapping the domain structure of the influenza A virus polymerase acidic protein (PA) and its interaction with the basic protein 1 (PB1) subunit. Virology 379, 135–142.

Hara, K., Schmidt, F.I., Crow, M., and Brownlee, G.G. (2006). Amino acid residues in the N-terminal region of the PA subunit of influenza A virus RNA polymerase play a critical role in protein stability, endonuclease activity, cap binding, and virion RNA promoter binding. J. Virol. *80*, 7789–7798.

Hara, K., Shiota, M., Kido, H., Ohtsu, Y., Kashiwagi, T., Iwahashi, J., Hamada, N., Mizoue, K., Tsumura, N., Kato, H., and Toyoda, T. (2001). Influenza virus RNA polymerase PA subunit is a novel serine protease with Ser624 at the active site. Genes Cells *6*, 87–97.

Hatta, M., Gao, P., Halfmann, P., and Kawaoka, Y. (2001). Molecular basis for high virulence of Hong Kong H5N1 influenza A viruses. Science *293*, 1840–1842.

He, X., Zhou, J., Bartlam, M., Zhang, R., Ma, J., Lou, Z., Li, X., Li, J., Joachimiak, A., Zeng, Z., *et al.* (2008). Crystal structure of the polymerase PA(C)–PB1(N) complex from an avian influenza H5N1 virus. Nature *454*, 1123–1126.

Honda, A., Mizumoto, K., and Ishihama, A. (2002). Minimum molecular architectures for transcription and replication of the influenza virus. Proc. Natl. Acad. Sci. U.S.A. *99*, 13166–13171.

Huarte, M., Falcon, A., Nakaya, Y., Ortin, J., Garcia-Sastre, A., and Nieto, A. (2003). Threonine 157 of influenza virus PA polymerase subunit modulates RNA replication in infectious viruses. J. Virol. *77*, 6007–6013.

Huarte, M., Sanz-Ezquerro, J.J., Roncal, F., Ortin, J., and Nieto, A. (2001). PA subunit from influenza virus polymerase complex interacts with a cellular protein with homology to a family of transcriptional activators. J. Virol. *75*, 8597–8604.

Hulse-Post, D.J., Franks, J., Boyd, K., Salomon, R., Hoffmann, E., Yen, H.L., Webby, R.J., Walker, D., Nguyen, T.D., and Webster, R.G. (2007). Molecular changes in the polymerase genes (PA and PB1) associated with high pathogenicity of H5N1 influenza virus in mallard ducks. J. Virol. *81*, 8515–8524.

Jennings, P.A., Finch, J.T., Winter, G., and Robertson, J.S. (1983). Does the higher order structure of the influenza virus ribonucleoprotein guide sequence rearrangements in influenza viral RNA? Cell *34*, 619–627.

Jung, T.E., and Brownlee, G.G. (2006). A new promoter-binding site in the PB1 subunit of the influenza A virus polymerase. J. Gen. Virol. *87*, 679–688.

Kawaguchi, A., and Nagata, K. (2007). De novo replication of the influenza virus RNA genome is regulated by DNA replicative helicase, MCM. EMBO J. *26*, 4566–4575.

Kawaguchi, A., Naito, T., and Nagata, K. (2005). Involvement of influenza virus PA subunit in assembly of functional RNA polymerase complexes. J. Virol. *79*, 732–744.

Klumpp, K., Doan, L., Roberts, N.A., and Handa, B. (2000). RNA and DNA hydrolysis are catalyzed by the influenza virus endonuclease. J. Biol. Chem. *275*, 6181–6188.

Klumpp, K., Ruigrok, R.W., and Baudin, F. (1997). Roles of the influenza virus polymerase and nucleoprotein in forming a functional RNP structure. EMBO J. *16*, 1248–1257.

Kuo, R.L., and Krug, R.M. (2009). Influenza a virus polymerase is an integral component of the CPSF30-

NS1A protein complex in infected cells. J. Virol. *83*, 1611–1616.

Kuzuhara, T., Kise, D., Yoshida, H., Horita, T., Murazaki, Y., Nishimura, A., Echigo, N., Utsunomiya, H., and Tsuge, H. (2009). Structural Basis of the Influenza A Virus RNA Polymerase PB2 RNA-binding Domain Containing the Pathogenicity-determinant Lysine 627 Residue. J. Biol. Chem. *284*, 6855–6860.

Lee, M.T., Bishop, K., Medcalf, L., Elton, D., Digard, P., and Tiley, L. (2002). Definition of the minimal viral components required for the initiation of unprimed RNA synthesis by influenza virus RNA polymerase. Nucleic Acids Res. *30*, 429–438.

Lee, M.T., Klumpp, K., Digard, P., and Tiley, L. (2003). Activation of influenza virus RNA polymerase by the 5′ and 3′ terminal duplex of genomic RNA. Nucleic Acids Res. *31*, 1624–1632.

Li, M.L., Rao, P., and Krug, R.M. (2001). The active sites of the influenza cap-dependent endonuclease are on different polymerase subunits. EMBO J. *20*, 2078–2086.

Mukaigawa, J., and Nayak, D.P. (1991). Two signals mediate nuclear localization of influenza virus (A/WSN/33) polymerase basic protein 2. J. Virol. *65*, 245–253.

Murti, K.G., Webster, R.G., and Jones, I.M. (1988). Localization of RNA polymerases on influenza viral ribonucleoproteins by immunogold labeling. Virology *164*, 562–566.

Nakagawa, Y., Kimura, N., Toyoda, T., Mizumoto, K., Ishihama, A., Oda, K., and Nakada, S. (1995). The RNA polymerase PB2 subunit is not required for replication of the influenza virus genome but is involved in capped mRNA synthesis. J. Virol. *69*, 728–733.

Nakagawa, Y., Oda, K., and Nakada, S. (1996). The PB1 subunit alone can catalyze cRNA synthesis, and the PA subunit in addition to the PB1 subunit is required for viral RNA synthesis in replication of the influenza virus genome. J. Virol. *70*, 6390–6394.

Nakazawa, M., Kadowaki, S.E., Watanabe, I., Kadowaki, Y., Takei, M., and Fukuda, H. (2008). PA subunit of RNA polymerase as a promising target for anti-influenza virus agents. Antiviral Res. *78*, 194–201.

Nath, S.T., and Nayak, D.P. (1990). Function of two discrete regions is required for nuclear localization of polymerase basic protein 1 of A/WSN/33 influenza virus (H1 N1). Mol. Cell. Biol. *10*, 4139–4145.

Nieto, A., de la Luna, S., Barcena, J., Portela, A., Valcarcel, J., Melero, J.A., and Ortin, J. (1992). Nuclear transport of influenza virus polymerase PA protein. Virus Res. *24*, 65–75.

Obayashi, E., Yoshida, H., Kawai, F., Shibayama, N., Kawaguchi, A., Nagata, K., Tame, J.R., and Park, S.Y. (2008). The structural basis for an essential subunit interaction in influenza virus RNA polymerase. Nature *454*, 1127–1131.

Ortega, J., Martin-Benito, J., Zurcher, T., Valpuesta, J.M., Carrascosa, J.L., and Ortin, J. (2000). Ultrastructural and functional analyses of recombinant influenza virus ribonucleoproteins suggest dimerization of nucleoprotein during virus amplification. J. Virol. *74*, 156–163.

Perales, B., and Ortin, J. (1997). The influenza A virus PB2 polymerase subunit is required for the replication of viral RNA. J. Virol. *71*, 1381–1385.

Perez-Gonzalez, A., Rodriguez, A., Huarte, M., Salanueva, I.J., and Nieto, A. (2006). hCLE/CGI-99, a human protein that interacts with the influenza virus polymerase, is a mRNA transcription modulator. J. Mol. Biol. *362*, 887–900.

Perez, D.R., and Donis, R.O. (2001). Functional analysis of PA binding by influenza a virus PB1: effects on polymerase activity and viral infectivity. J. Virol. *75*, 8127–8136.

Pons, M.W., Schulze, I.T., Hirst, G.K., and Hauser, R. (1969). Isolation and characterization of the ribonucleoprotein of influenza virus. Virology *39*, 250–259.

Regan, J.F., Liang, Y., and Parslow, T.G. (2006). Defective assembly of influenza A virus due to a mutation in the polymerase subunit PA. J. Virol. *80*, 252–261.

Rodriguez, A., Perez-Gonzalez, A., and Nieto, A. (2007). Influenza virus infection causes specific degradation of the largest subunit of cellular RNA polymerase II. J. Virol. *81*, 5315–5324.

Shi, L., Summers, D.F., Peng, Q., and Galarz, J.M. (1995). Influenza A virus RNA polymerase subunit PB2 is the endonuclease which cleaves host cell mRNA and functions only as the trimeric enzyme. Virology *208*, 38–47.

Subbarao, E.K., London, W., and Murphy, B.R. (1993). A single amino acid in the PB2 gene of influenza A virus is a determinant of host range. J. Virol. *67*, 1761–1764.

Takahashi, K., Furuta, Y., Fukuda, Y., Kuno, M., Kamiyama, T., Kozaki, K., Nomura, N., Egawa, H., Minami, S., and Shiraki, K. (2003). *In vitro* and *in vivo* activities of T-705 and oseltamivir against influenza virus. Antivir. Chem. Chemother. *14*, 235–241.

Tarendeau, F., Boudet, J., Guilligay, D., Mas, P.J., Bougault, C.M., Boulo, S., Baudin, F., Ruigrok, R.W., Daigle, N., Ellenberg, J., *et al.* (2007). Structure and nuclear import function of the C-terminal domain of influenza virus polymerase PB2 subunit. Nat. Struct. Mol. Biol. *14*, 229–233.

Tarendeau, F., Crepin, T., Guilligay, D., Ruigrok, R.W., Cusack, S., and Hart, D.J. (2008). Host determinant residue lysine 627 lies on the surface of a discrete, folded domain of influenza virus polymerase PB2 subunit. PLoS Pathog. *4*, e1000136.

Tiley, L.S., Hagen, M., Matthews, J.T., and Krystal, M. (1994). Sequence-specific binding of the influenza virus RNA polymerase to sequences located at the 5′ ends of the viral RNAs. J. Virol. *68*, 5108–5116.

Tomassini, J., Selnick, H., Davies, M.E., Armstrong, M.E., Baldwin, J., Bourgeois, M., Hastings, J., Hazuda, D., Lewis, J., McClements, W., and *et al.* (1994). Inhibition of cap (m7GpppXm)-dependent endonuclease of influenza virus by 4-substituted 2,4-dioxobutanoic acid compounds. Antimicrob. Agents Chemother. *38*, 2827–2837.

Tomassini, J.E. (1996). Expression, purification, and characterization of orthomyxovirus: influenza transcriptase. Methods Enzymol. *275*, 90–99.

Torreira, E., Schoehn, G., Fernandez, Y., Jorba, N., Ruigrok, R.W., Cusack, S., Ortin, J., and Llorca, O.

(2007). Three-dimensional model for the isolated recombinant influenza virus polymerase heterotrimer. Nucleic Acids Res. 35, 3774–3783.

Toyoda, T., Adyshev, D.M., Kobayashi, M., Iwata, A., and Ishihama, A. (1996). Molecular assembly of the influenza virus RNA polymerase: determination of the subunit–subunit contact sites. J. Gen. Virol. 77, 2149–2157.

Yuan, P., Bartlam, M., Lou, Z., Chen, S., Zhou, J., He, X., Lv, Z., Ge, R., Li, X., Deng, T., *et al.* (2009). Crystal structure of an avian influenza polymerase PA(N) reveals an endonuclease active site. Nature 458, 909–913.

Zurcher, T., de la Luna, S., Sanz-Ezquerro, J.J., Nieto, A., and Ortin, J. (1996). Mutational analysis of the influenza virus A/Victoria/3/75 PA protein: studies of interaction with PB1 protein and identification of a dominant negative mutant. J. Gen. Virol. 77 (Pt 8), 1745–1749.

MChip: a Single-gene Diagnostic Microarray for Influenza A

9

Erica D. Dawson and Kathy L. Rowlen

Abstract

Rapid and accurate diagnostic methods for typing and subtyping influenza viruses are needed for improved worldwide surveillance. Although molecular-based diagnostic methods are becoming more widespread in influenza diagnosis, they generally involve amplification of the haemagglutinin (HA) and/or neuraminidase (NA) gene segments for subtyping. We have developed a low-density microarray (MChip) and accompanying pattern recognition algorithm combined with a reverse transcription PCR-based assay that allows for the identification and subtyping of influenza A viruses in approximately seven hours. MChip is unique in that it is based solely on the matrix (M) gene segment which has enough genetic diversity for subtype analysis but sufficient genetic stability to circumvent the need for continual redesign of primers and microarray probes. An overview is presented here of several MChip studies undertaken over the last several years, highlighting the ability of MChip to identify and subtype influenza A/H1N1, H3N2, and H5N1 viruses. In addition, we present data indicating that the MChip performs well relative to gold standard methods for identification of influenza A. It is our hope that improved diagnostic assays coupled with improvements in microarray detection methodology will enable better worldwide surveillance and diagnosis of influenza viral infections.

Introduction

Subtype determination ('subtyping') of influenza viruses is important for the epidemiological surveillance of currently-circulating and emerging strains, and for the design of appropriate influenza vaccines (Lu, 2006). The current reference standards for complete determination of HA and NA subtypes of influenza A involve virus replication in egg or cell culture followed by either immunofluorescence assay with strain-specific monoclonal antibodies or haemagglutinin inhibition (HI) assay using a panel of reference antisera (WHO, 2005). Traditional viral culture is tedious and requires several days (Bourinbaiar *et al.*, 2006), and analysis can often extend to several weeks for antigenically novel viruses (Taubenberger, 2001). Newer shell-vial culture techniques are an improvement, but still require 2–3 days (Chomel *et al.*, 1991; Matthey *et al.*, 1992). Rapid immunoassays for influenza typing are available from several manufacturers and are very useful for routine point of care diagnosis, but their lack of subtyping capability make them less useful from a surveillance standpoint. In addition, rapid immunoassays have been reported to have lower sensitivity than other diagnostic tests (Storch, 2003; Templeton *et al.*, 2004). Several rapid immunoassays for influenza have been compared in the literature to viral culture or RT-PCR (Dominguez *et al.*, 1993; Herrmann *et al.*, 2001; Leonardi *et al.*, 1994; Liolios *et al.*, 2001; Reina *et al.*, 1996; Ruest *et al.*, 2003; Schultze *et al.*, 2001; Storch, 2003), with reported sensitivities of 75–94% and specificities of 86–100%.

RT-PCR-based diagnostic assays

Molecular-based methods based on RT-PCR (reverse transcription polymerase chain reaction) and real-time RT-PCR are becoming increasingly popular as diagnostic assays for influenza virus, and are generally available for influenza typing (A or B) or as presumptive tests for a specific subtype in which one gene segment is detected (Daum *et al.*, 2007; Wu *et al.*, 2008; Thontiravong *et al.*, 2007; Wei *et al.*, 2006; DiTrani *et al.*, 2006; Fouchier *et al.*, 2000; Ng *et al.*, 2006; Payungporn, *et al.*, 2004, 2006; Poddar, 2002; Spackman *et al.*, 2002; Whiley *et al.*, 2005; Xie *et al.*, 2006; Beladi *et al.*, 2005). RT-PCR-based typing of influenza A generally utilizes a single primer pair that targets a short, well-conserved portion of the matrix (M), nucleoprotein (NP), or non-structural (NS) gene segments (Suarez *et al.*, 2007; Stone *et al.*, 2004; DiTrani *et al.*, 2006; Fouchier *et al.*, 2000; Spackman *et al.*, 2002; Whiley *et al.*, 2005; Smith *et al.*, 2003; Beladi *et al.*, 2005), whereas PCR-based presumptive tests for a specific subtype utilize a single primer pair that targets a portion of the HA or NA gene that is well-conserved only for a particular subtype (Wei *et al.*, 2006; Thontiravong *et al.*, 2007; Poddar, 2002; Payungporn, *et al.*, 2006; Daum *et al.*, 2007; Ng *et al.*, 2006; Spackman *et al.*, 2002; Xie *et al.*, 2006). These typing and subtyping RT-PCR-based assays are frequently multiplexed in an effort to maximize the information content obtained while minimizing time to result. The reader is also directed to a recent review by Bourinbaiar *et al.* (2006) for an introduction to the numerous RT-PCR-based diagnostic assays, particularly for avian influenza, that are either commercially-available or in development for commercialization.

A limitation of these RT-PCR-based tests for influenza subtyping, whether singleplexed or multiplexed, is the reliance on the HA and NA gene segment sequences that are under immune selective pressure and which have relatively high rates of mutation of 6.7×10^{-3} and 3.2×10^{-3} nucleotide substitutions per nucleotide per year, respectively (Steinhauer *et al.*, 2002). Mutations in the priming regions may cause the assay to have reduced analytical sensitivity or to fail entirely. In an external quality assessment study examining the ability of laboratories worldwide to detect and subtype influenza using molecular-based methods, MacKay *et al.* (2008) highlighted the need for improved molecular-based subtyping for influenza A. The participating laboratories returned type data (A or B), as well as HA subtype data (obtained using HA-specific RT-PCR assays, although the particular methods utilized by each laboratory were not disclosed) for a panel of 14 samples provided to them in a blinded fashion. Influenza-positive specimens were correctly typed by 87.6% (mean) of the laboratories, but correct HA subtype was only obtained by 55% (mean). The blinded panel of viruses included several of the same HA subtype with a range of viral loads, and a strong correlation existed between decreased ability to subtype and decreased viral load. The authors of the study note that a likely explanation for this is that an incomplete match of the HA primers and/or probe resulted in poor analytical sensitivity. Another inherent shortcoming in these HA/NA-based presumptive assays is that each assay is limited to identification of a specific subtype due to the significant diversity in HA and NA between subtypes (Steinhauer *et al.*, 2002). Several assays targeting different gene segments and subtypes (e.g. H5, N1, and M gene segment for typing) can be conducted in parallel or multiplexed, but this adds extra complexity and cost to the overall process that may be prohibitive. A presumptive test would certainly be useful during a pandemic where rapid screening for a particular subtype is the highest priority, but thorough tracking of all subtypes circulating in the population in a pre-pandemic phase is vital (Department of Health and Human Services, 2006).

Microarray-based diagnostics

Microarray-based approaches have great potential in the field of molecular diagnostics for influenza, and include bead-based array systems with optical detection, traditional flat substrate microarrays (glass, nylon, silicon, etc.), and a variety of other unique platforms (Bourinbaiar *et al.*, 2006). Examples of the bead-based approach are the Luminex xTAG™ Respiratory Virus Panel (Luminex Corp., Austin, TX, USA). This assay is FDA-approved in the United States for simultaneous detection of 12 respiratory pathogens, and is also certified in Europe to detect an additional 8 pathogens, including influenza H5. The assay utilizes the Luminex proprietary universal tagging system coupled with multiplex bead-based detection technology. Also in development (but not yet commercially available) is the ResPlex

III assay (Qiagen, Valencia, CA, USA) which utilizes the proprietary target-enriched multiplex PCR (tem-PCR) coupled again with the Luminex bead-based detection technology for the simultaneous detection of influenza A subtypes H1, H2, H3, H5, H7, H9, N1, N2, as well as typing of influenza A and B (via the NS gene segment). These bead-based approaches clearly have great utility in a high volume clinical diagnostic setting where detection of a broad range of respiratory pathogens is desired, but the start-up costs of the required instrumentation would probably be prohibitive in any resource-limited setting.

CombiMatrix Corporation (Mukilteo, WA, USA) has developed a commercial influenza microarray and detection system (currently for research use only) based on RT-PCR and a semiconductor-based microarray with electrochemical detection. The microarray is fairly high density (~12,000 sequences), provides high-resolution information about all known HA and NA subtypes, and can additionally provide information about reassortant viruses. The relatively high cost per chip (~$550), however, may preclude its use for routine clinical and global strain surveillance. Alternatively, resequencing arrays have also been utilized for influenza identification (Wang *et al.*, 2006; Lin *et al.*, 2007). Resequencing arrays contain ~300,000 25-mers that are designed to 'read' the genetic sequence of a target based on hybridization patterns. Each base in the amplified genetic target is queried using a set of four probes that are identical except for a central mismatch so that each of the 4 bases is present in one of the probes. The resulting pattern of relative intensities after hybridization of the target allows the sequence to be read at each position. By tiling across an entire genetic region with overlapping probe sequences, the full sequence of that region can be determined. Resequencing arrays clearly have high information content with potential benefits over traditional sequencing, but the disadvantages of this approach with respect to surveillance include high cost per specimen, complicated statistical analysis of the resulting data, and limited sensitivity (Hudson, 2008). In addition, there have also been developments of several traditional flat substrate microarrays that focus on a panel of respiratory pathogens, including influenza (Wang *et al.*, 2002; Quan *et al.*, 2007). Wang *et al.* (2002) utilized the full genomes of 140 viral species to generate an array ~6000 70-mers (ViroChip) that was targeted towards highly parallel virus screening. Quan *et al.* (2007) targeted 21 different respiratory viruses using a microarray (GreeneChipResp) consisting of ~15,000 60-mers. These examples of relatively high-density microarrays clearly have high information content, but require more complicated data analysis and have relatively high cost per assay.

Low-density diagnostic microarrays

The use of lower density microarrays for routine diagnosis and surveillance has been reviewed, highlighting the benefits over more expensive, complex higher density arrays (Zammatteo *et al.*, 2002; Mikhailovich *et al.*, 2008). Several instances of flat substrate low-density DNA microarrays specifically for influenza detection and subtyping using multiple gene targets have been described (Li *et al.*, 2001; Kessler *et al.*, 2004; Sengupta *et al.*, 2003; Mehlmann *et al.*, 2006; Townsend *et al.*, 2006). Li *et al.* (2001) utilized immobilized capture sequences ~500 nt in length and identified 3 isolates from pre-1977 viruses. Kessler *et al.* (2004) used shorter (45–65 nt) capture oligonucleotides designed from pre-1998 viruses to identify four influenza A subtypes and influenza B on a total of 9 specimens. Sengupta *et al.* (2003) utilized an online primer database to choose ~450 influenza specific sequences for use as capture sequences and evaluated the microarray using five pre-1997 viruses. These studies provided proof of concept for microarray subtyping of influenza, but the primary limitation was the amplification of multiple genes, including HA and NA, with specific primer sets (Li *et al.*, 2001; Kessler *et al.*, 2004) or with universal influenza primers that give variable amplification efficiency for the different gene segments (Sengupta *et al.*, 2003). Our group previously demonstrated an influenza diagnostic microarray that utilized DNA capture probes designed to be complementary to portions of the M gene segment for typing and to HA and NA genes for subtyping (Mehlmann *et al.*, 2006; Townsend *et al.*, 2006). With a multiple gene approach, the method sensitivity was limited due to a relatively high failure rate in the multiplex RT-PCR used to amplify the genes (Markoulatos *et al.*, 2002).

MChip

The failure of the HA and/or NA genes to amplify consistently during the testing of our previous influenza microarray (Mehlmann *et al.*, 2006; Townsend *et al.*, 2006) led us to examine the potential of using the M gene segment for subtyping. When both HA and NA failed to amplify for a specific specimen, the hybridization pattern in the capture sequences corresponding to the M gene segment was very distinct depending on the subtype of the virus. This result was unexpected, as the M gene segment is typically used as a universal target due to its relatively high conservation among all influenza A viral subtypes. Through additional targeted sequence design, we focused on the differences in the M gene segment between influenza A subtypes in order to further explore the relationship between the M gene segment patterns and subtype specificity. With fifteen sequences on the microarray, the correlation between the hybridization pattern and subtype was surprisingly strong. Although the 'pattern recognition' was initially just a simple visual analysis, we recognized the importance of removing the human subjectivity from the assay. This led to the development of a more automated pattern recognition procedure in the form of an artificial neural network (ANN).

Outlined in the remainder of this chapter is the development of the 'MChip', a single gene microarray for influenza A for both typing and subtyping based on the matrix (M) gene segment. The M gene segment is 1027 nt in length and codes for two proteins, M1 and M2. M1 is the most abundant protein in the virion and lies underneath the lipid envelope, whereas M2 is a more minor component of the virion that has ion channel activity (Lamb *et al.*, 1996). From a diagnostics standpoint, influenza is often described as 'moving target' due to the high mutation rate of HA and NA (Steinhauer *et al.*, 2002). Utilizing the M segment for diagnostic purposes offers a significant advantage due to its much higher nucleotide level conservation (1×10^{-3} nucleotide substitutions per nucleotide per year) (Steinhauer *et al.*, 2002), alleviating the need to continually design new primer and probe sequences. Even with its relatively high conservation, however, the M gene segment does contain sufficient genetic diversity between subtypes to enable reliable identification and inference of influenza A viral subtype in a low-density microarray format. The benefits of this approach are simplicity, rapidity (~7 hour time to result), low cost, improved sensitivity, and improved reliability due to the need for amplification of only one short, relatively well-conserved gene segment. Here we detail the overall microarray design process, and several recent studies demonstrating influenza A/H1N1, H3N2, and H5N1 subtyping utilizing an artificial neural network-based pattern recognition approach for subtyping.

MChip sequence selection process

Overview/general requirements

Most current microarray sequence selection tools are targeted towards the more widely used gene expression microarrays (Tomiuk *et al.*, 2001; Russell, 2003). The overall strategy for probe design depends on the particular application. In general, gene expression probes are intended to be completely specific for the intended gene target and to have adequate 'coverage' of the gene target. Criteria to be considered in terms of physical properties of the probes are similar melting temperatures and hybridization conditions, probe length, and inability of probes to cross-hybridize. For a low-density microarray targeting a highly mutable virus like influenza where subtype discrimination is desired, the design process can be more complicated. For example, the microarray probe sequences should not only target a single gene of a specific virus strain, but should have similar hybridization efficiencies to a wide variety of strains all belonging to the same subtype.

The microarray sequence design process used for MChip was adapted from an approach utilized with our first generation FluChip that targeted the HA, NA, and M gene segments (Mehlmann *et al.*, 2005). In fact, seven microarray sequence sets designed from the M gene segment used on the original FluChip were carried through and utilized on the MChip. Targeting only a single gene segment to identify and subtype added complexity over our initial approach, however, as the goal was to maximize differences in hybridization for different subtypes using a very limited amount of sequence space. Therefore, further sequence selection was undertaken in an effort to choose additional sequences that would either maximize discrimination between the targeted subtypes or act as universal influenza A probes.

Database compilation

M gene segment sequences for influenza A subtypes H1N1 (43 sequences), H1N2 (67 sequences), H3N2 (306 sequences), H5N1 (249 sequences), H3N8 (8 sequences), and H9N2 (92 sequences) were compiled using the publicly available online database from Los Alamos National Laboratory (LANL) (Macken *et al.*, 2001) and a smaller collection from the Centres for Disease Control (CDC). All sequences were from collection years 2000 and forward. As the goal was to find probe sequences that were unique to a particular subtype and/or host species, these subtype-specific databases were separated by host species.

Sequence selection

The host-specific subtype databases (e.g. human H1N1) described above were mined for conserved regions using our automated ConFind algorithm (Smagala *et al.*, 2005). The ConFind algorithm was designed to identify conserved regions from aligned sequence files, and to provide robust handling for alignments containing missing or incomplete sequence information. ConFind also allows user-defined limits for length, positional variability, exceptions to the positional variability, and number of sequences in an alignment containing non-ambiguous characters at each position. The conserved regions identified were then used to design appropriate 'capture' and 'label' sequence pairs 16–25 nucleotides (nt) each in length (one to be used as an immobilized capture, and the other to be used as a label sequence in a sandwich assay as shown in Fig. 9.1A). The number of mismatches between the designed sequence pairs and each host-specific subtype database was determined, and sequences were chosen that were anticipated to be broadly reactive with all influenza subtypes or with viruses of a specific host species or subtype (e.g. all avian viruses, only H3N2 viruses). This overall process resulted in numerous additional sequence pairs that were experimentally tested as described next.

Cross-reactivity testing

All capture and label pairs were experimentally examined for cross-reactivity by conducting six replicate hybridizations of fluorophore-conjugated label sequences in the absence of any target influenza under otherwise identical conditions. Where measurable signals on the microarray

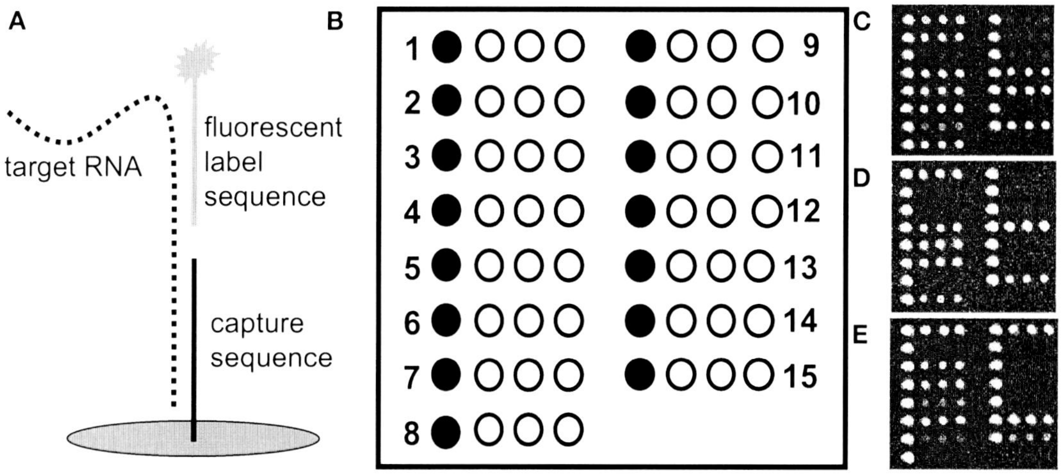

Figure 9.1 (A) Schematic of two-step hybridization detection utilized. (B) MChip layout where filled circles are hybridization positive control and empty circles are triplicate spots for each capture sequence. Typical patterns in the 15 capture sequences for H3N2, H1N1, and H5N1 viruses in (C), (D) and (E), respectively. Brighter white spots indicate higher fluorescence intensity. Reproduced in part with permission from Anal. Chem. 2006, 78, 7610–7615. Copyright 2006 American Chemical Society.

occurred (signal-to-noise >3) on a majority of hybridized slides, the capture and corresponding label sequence were removed and not used further. Fifteen capture and label pairs were found to not cross-react with one another and were utilized for the final MChip layout. Sequence information can be obtained by not-for-profit enterprises for research use only from the University of Colorado Technology Transfer Office.

Experimental

The full experimental details of each study have been given elsewhere (Dawson *et al.*, 2006, 2007; Mehlmann *et al.*, 2007; Moore *et al.*, 2007). A brief description is included here for an understanding of the overall experimental process. Since each study discussed utilized a unique set of influenza specimens, the details concerning clinical specimens and viral isolates are discussed alongside each individual study.

Nucleic acid extraction and amplification

All handling of H5N1 viruses was performed at CDC by authorized personnel under BSL 3+ conditions. Briefly, viral RNA was extracted using commercially-available column-based kits and subsequently amplified using reverse-transcription (RT) followed by PCR (or alternatively, single-step RT-PCR). PCR was performed with the full length M gene segment primers (Zou, 1997), with the 5′ primer containing a T7 promoter. *In vitro* transcription using the PCR product as a template was then performed utilizing T7 RNA polymerase.

Array fabrication

All capture oligonucleotides were 5′ amino-C6 modified for covalent attachment to aldehyde-modified glass (CEL Associates, Inc., Pearland, TX, USA). Capture oligonucleotides were spotted at 10 μM from a solution of 3× SSC, 50 mM phosphate (pH 7.5), and 0.005% sarkosyl. A MicroGrid arrayer (Genomic Solutions Inc., Ann Arbor, MI, USA) with two 100-μm-diameter solid core pins and 70% relative humidity was utilized, washing the pins between solutions to prevent carryover. Slides were placed in a 100% relative humidity environment for 18 h after spotting and washed according the following procedure prior to hybridization: (1) 5 min in 4× SSPE/0.1% SDS, (2) 5 min in 4× SSPE, (3) 5 min in 18 MΩ water, and (4) 5 min in 90 – 100°C in 18 MΩ water.

Hybridization

Amplified RNA (4 μl) was fragmented by adding 1 μl of 5× fragmentation buffer (200 mM Tris-acetate, 500 mM potassium acetate, and 150 mM magnesium acetate, pH 8.4) for 25 min at 75°C as described by Mehlmann *et al.* (2005). Hybridization to the microarray was performed immediately after fragmentation. Briefly, 15 μl of quenching/hybridization buffer was added to the fragmented RNA solution (final concentration was 4× SSPE, 30 mM EDTA, 2.5× Denhardt's solution, 30% deionized formamide, 200 nM of each label (Quasar® 570-modified, Biosearch Technologies, Novato, CA, USA), and 5 nM of a positive control label oligonucleotide). Microarray hybridizations were carried out for 2 h at room temperature under saturated humidity.

Microarray imaging and data extraction

Hybridized slides were scanned using a VersArray ChipReader scanner (Bio-Rad Laboratories, Hercules, CA, USA) with 532 nm detection, laser power of 60%, PMT sensitivity of 700 V, and 5 μm resolution. Fluorescence images were analysed using VersArray Analyzer software, version 4.5 (Bio-Rad Laboratories). The positive control spots (i.e. capture and label sequences complementary to one another) were used to monitor the hybridization process and as spatial markers. Intensities of the 15 capture sequences were extracted for each spot in the triplicate set, and the mean signal was calculated. Relative signal values were used for analysis to account for different hybridized influenza concentrations. For a given microarray image, the sequence with highest mean intensity was used to defined the maximum signal, which was scaled to 100. The mean intensities for all other sequences were scaled appropriately relative to 100.

Artificial neural network image interpretation

Artificial neural networks (ANNs) are computerized algorithms modelled after the structure and function of the biological neural network of the central nervous system. In practice, ANNs are commonly used to determine complex interactions between inputs and outputs for problems like pattern recognition and classification. The network 'learns' how input information such as numeric data or images are related to a series of desired output categories through a process of error minimization (Khan *et al.*, 2001). ANNs are considered a supervised learning algorithm, as the network must first determine the mathematical relationships that connect a set of paired inputs and output categories known as a 'training set', and then use these relationships to predict the output category for sets of input data for which the output category is unknown. ANNs have been previously used in microarray analysis (Linder *et al.*, 2007; Valafar, 2002), as well as in cancer prediction (Khan *et al.*, 2001; Chatterjee *et al.*, 2006; Berrar *et al.*, 2003). The reader is directed to the concise introduction to artificial neural networks recently written by Krogh (2008).

Either EasyNN-Plus (www. easynn.com) or the WEKA software package (Waikato Environment for Knowledge Analysis, version 3.4.8), were used to develop the artificial neural networks used for MChip. The ANNs utilized 16 inputs and 4 outputs with a hidden layer that consisted of 11 nodes. For each dataset, the relative fluorescence intensities of the 15 capture sequences and the maximum mean intensity (MAX) of the array (i.e. the mean absolute intensity of the most intense capture sequence) were used as input. The MAX input was added in order to help differentiate negative samples from influenza A positive samples, as the MAX value for negative samples should be near background levels. The four output categories were H3N2, H1N1, H5N1 and negative, and the known outputs were entered as either 1 or 0, designating true or false. Microarray results of known subtype (H3N2, H1N1, and H5N1), as well as samples known to be negative for influenza A were selected to train the ANN. Of the training samples utilized in each study, 20–25% were used to validate the network during learning. The complete details of each neural network used can be found in the original publications (Dawson *et al.*, 2006, Moore *et al.*, 2007).

MChip studies

Initial testing and proof of principle for pattern recognition

Fig. 9.1A shows the detection strategy in which short oligonucleotides (~16–25-mer) immobilized in an array format on a solid support were used to capture amplified, fragmented RNA. The hybridization events were then labelled with complementary fluorescently labelled oligonucleotides in a sandwich assay. The MChip layout is shown in Fig. 9.1B, where the open circles represent triplicate spots for each of the 15 capture sequences. Initial characterization of the MChip was performed with a set of 58 known influenza A viral isolates representing a wide variety of subtypes including: H1N1, H3N2, H5N1, H3N8, H9N2, H3N7, H7N7. The M gene segment was successfully amplified for all 58 samples, and all resulted in positive fluorescent signals on the microarray. This high success rate for amplification highlights the reliability of the M gene segment as a diagnostic target. Representative images of human-adapted H3N2 and H1N1 amplified RNA hybridized to the MChip are shown in Figures 1C and 1D, and a typical image of hybridized H5N1 amplified RNA is shown in Fig. 9.1E. From these images, it is obvious that the three different influenza A viral subtypes generate different visual patterns on the microarray. Although not shown, all viruses within a subtype exhibited a similar microarray pattern. While certain sequences (such as 1, 4, 5, 6, and 15) were designed to be broadly reactive, others were designed and observed to hybridize only to a certain subtype. For example, sequences 3, 9, and 14 were selective for the H5N1 viruses, while sequences 2 and 13 were selective for the H3N2 subtype. In addition, differences in overall patterns were a result of differences in relative fluorescence intensities for some of the capture sequences.

An additional set of 53 clinical samples was utilized to test the hypothesis that the patterns of relative signal intensity from the fluorescence images of the hybridized MChip could be used to identify unknown influenza A viruses. Specimens from throat or nasopharyngeal swabs known to be positive for influenza A (but whose subtypes were blinded to us) were provided by the Colorado Department of Public Health and Environment (CDPHE) and by Air Force Institute for Operational Health (AFIOH) Virology and Bacteriology Laboratories (Brooks City-Base, TX, USA). Several negative control samples were provided by CDC. These negative specimens included true negatives, samples positive for influenza B, and samples positive for other respiratory pathogens such the SARS CoA causing severe acute respiratory syndrome, human parainfluenza virus type 3 (hPIV3), respiratory syncytial virus (RSV), and human metapneumovirus (hMPV).

As described in the Experimental section, an artificial neural network (ANN) was trained to recognize the patterns associated with each subtype using influenza A virus samples of known subtype. Microarray results from 50 samples of known subtype (H3N2, H1N1, and H5N1), and 10 samples known to be negative for influenza A were selected to train the ANN. The trained ANN was then used to determine the subtypes for 53 unknown samples in a blind study, with a threshold output score of 0.75 used for subtype assignment. Table 9.1 summarizes the ANN assignments for the 53 patient samples analysed. In some cases the 'known' consisted of only partial subtype (i.e. samples were identified using immunofluorescence assay designed to probe only the antigenicity of the HA protein). In this study, 50 of 53 samples were correctly identified and fully subtyped (for influenza A) or identified as negative. There was one false positive result and two false negative results. Negative samples (either true negatives or other pathogens that cause influenza-like illnesses, e.g. RSV) were defined as 'influenza A negative' and were not further characterized.

Identification of H5N1 and other non-human adapted subtypes

Highly pathogenic avian A/H5N1 influenza was first isolated from a human in Hong Kong in 1997 (Subbarao *et al.*, 1998). As of June 19, 2008, 385 human infections of this avian virus have been reported resulting in 243 deaths (World Health Organization, 2008) with the majority of cases clearly arising from direct avian-to-human transmission. No widespread human-to-human transmission of the H5N1 virus has been reported to date, but evidence of limited human-to-human transmission of the H5N1 virus in a family cluster in Indonesia in May of 2006 was reported (Normile, 2006). It is known that the A/H5N1 virus has undergone significant reassortment with other avian viruses since its occurrence in humans in 1997, with a number of different reassortants being identified as early as 2001 (Guan *et al.*, 2002; Li *et al.*, 2004). Antigenic analyses of A/H5N1 viruses from 2002–2003 have also shown significant antigenic drift compared to viruses from 1997–2001 (Guan *et al.*, 2004). In addition, the evolution of the HA gene of A/H5N1 since 2004 has resulted in several different lineages, and, according to phylogenetic analyses, there is evidence that the M gene has coevolved with the HA gene (Aubin *et al.*, 2005). As this particular virus continues to mutate it is important to have a diagnostic tool capable of detecting as many of the A/H5N1 variants as possible for reliable global strain surveillance efforts.

In order to test the hypothesis that the M gene is a useful diagnostic target even for rapidly evolving influenza viruses like A/H5N1, MChip was evaluated with a diverse set of A/H5N1 influenza viruses in combination with a probabilistic artificial neural network. H5N1 samples were originally received by the CDC (Atlanta, GA), and subsequently propagated in the allantoic cavities of 11-day-old embryonated chicken eggs or in Madin-Darby canine kidney (MDCK) cells. All handling of H5N1 samples was performed at the CDC under biohazard safety level 3-enhanced (BSL-3+) conditions. The samples were initially tested for haemagglutination, and the HA subtypes of all haemagglutination-positive samples were identified by HI assay.

The training set consisted of processed microarray data (including replicate experiments) from influenza A positive samples: (57 A/H3N2, 23 A/H1N1, 54 A/H5N1) and 29 negative samples (influenza B positive samples and other respiratory pathogens). A list of the A/H5N1 viral isolates, and replicates, used in training is given in Table 9.2. The trained ANN was used to identify a set of unknowns composed of 24 distinct A/H5N1 samples. In a second analysis, H1N1 and H3N2 viruses were added to the set of 'unknowns' in order to assess how well it would perform with a broader range of subtypes. While the study was not conducted 'blind', (i.e. the viral subtype was known to the users), the samples to be identified were 'unknown' to the trained neural network (i.e. they were not used for training or validation).

As observed in Table 9.2, 21 of the 24 A/H5N1 samples were correctly assigned as A/H5N1 with ≥95% probability. Only one virus was assigned as A/H5N1 with less than 90% certainty (#17 = 0.83). Furthermore, 6 of the 7 negative controls were correctly assigned with a probability ≥98%. It is interesting to note that the most recent examples of A/H5N1 from the family cluster in Indonesia were all correctly identified despite controversial reports of significant genetic mutation (Butler, 2006). If the probability threshold is set to 95%, 4 samples out of the 31 were incorrectly identified and the corresponding clinical sensitivity and specificity are 88% and 100%, respectively.

The ability to identify H5N1 viruses in a more realistic situation was also investigated. In this case, datasets representing H3N2, H1N1, H5N1, and negative specimens were all utilized as unknowns and subsequently identified by the ANN. Only 2 of the 47 samples were not definitively identified. One of these was A/Indonesia/CDC624/2006 (H5N1) for which the probabilities for H1N1, H3N2, H5N1, and neg were 0%, 27%, 73%, and 0%, respectively. This 73% probability was considered below an acceptable threshold. The second was an influenza negative specimen (containing RSV) for which the probabilities were 0, 53%, 0%, and 47% for H1N1, H3N2, H5N1, and negative, respectively. Again, this negative specimen was not included in the sensitivity calculation as the results did not meet the definition of a false negative. All of the other samples were correctly identified with >95% certainty, resulting in a clinical sensitivity of 97% and 100% clinical specificity.

Samples representing a variety of subtypes other than H1N1, H3N2, and H5N1 have also been examined. Fig. 9.2 shows hybridized MChip images for a variety of avian viruses and an equine H3N8 virus, all of which gave different and unique relative fluorescence patterns. For example, sequence 9 is positive in Fig. 9.2B, C, and F, but negative for A, D, and E. These results indicate that the differences in the microarray relative fluorescence intensity patterns observed for the H3N2, H1N1, and H5N1 samples may be extended to other subtypes that represent atypical human infections.

Application of MChip to historic H1N1 viruses, including the 1918 pandemic strain

Given the ability of influenza viruses to rapidly mutate and reassort, there is concern about novel influenza viruses. Since the MChip sequences were designed from M gene sequences of recently circulating viruses, it was desirable to test the MChip response to 'novel' viruses. In this case, the novel viruses were represented by historic viruses that are genetically dissimilar to the recent sequences from which the microarray capture and label sequences were designed. Sixteen H1N1 viruses of human, swine, and avian origins were chosen for evaluation based on the availability of sample and sequence data and are shown in Table 9.3. The viruses were obtained by reconstitution of lyophilized egg passaged virus stored at the Centres for Disease Control and Prevention (CDC) Influenza Branch. A/Brevig Mission/1/1918 vRNA was a generous gift from Dr. Terrence Tumpey of the CDC. A/Puerto Rico/8/1934 was purchased from American Type Culture Collection (ATCC, Bethesda, MD.)

Fig. 9.3A and B shows microarray fluorescence images for a modern H1N1 virus and for A/Brevig Mission/1/1918, respectively. The sequences previously identified as broadly reactive to influenza A (numbers 1, 4, 5, 6, and 15) exhibited good signal intensities for the historic viruses. Additionally, the sequence that was previously identified as H1N1 specific (sequence 12) was observed for all of the samples except for the WSN33 lab strain. Interestingly, the three sequences designed to be sensitive to H5N1 avian influenza viruses (sequences 3, 9, and 14) were also sensitive to several older H1N1 viruses. We have found sequence 14 to be very important for distinguishing avian H5N1 results, and several in this sample set showed positive signals on sequence 14: A/Swine/Iowa/1930 and A/Turkey/Kansas/1980 (both isolated from non-human hosts), the lab strain WSN33, A/Brevig Mission/1/1918, and A/Missouri/301/1952.

Table 9.1 Artificial neural network performance for influenza A subtyping

	ANN output				Known subtype	
	H1N1	H3N2	H5N1	Negative		
1	0.01	0.99	0.00	0.00	A/H3	✓
2	0.01	0.98	0.00	0.00	A/H3	✓
3	0.03	0.91	0.00	0.00	A/H3	✓
4	0.03	0.96	0.00	0.00	A/H3	✓
5	0.01	0.99	0.00	0.00	A/H3	✓
6	0.01	0.99	0.00	0.00	A/H3	✓
7	0.21	0.79	0.00	0.00	A/H3	✓
8	0.05	0.92	0.00	0.00	A/H3	✓
9	0.00	1.00	0.00	0.00	A/H3	✓
10	0.00	1.00	0.00	0.00	A/H3	✓
11	0.00	1.00	0.01	0.00	A/H3	✓
12	0.00	1.00	0.00	0.00	A/H3	✓
13	0.00	1.00	0.00	0.00	A/H3	✓
14	0.00	1.00	0.00	0.00	A/H3	✓
15	0.00	1.00	0.01	0.00	A/H3	✓
16	0.01	1.00	0.01	0.00	A/H3	✓
17	0.00	1.00	0.01	0.00	A/H3	✓
18	0.00	1.00	0.01	0.00	A/H3	✓
19	0.00	1.00	0.00	0.00	A/H3	✓
20	0.00	1.00	0.01	0.00	A/H3	✓
21	0.00	1.00	0.00	0.00	A/H3N2	✓
22	0.00	1.00	0.00	0.00	A/H3N2	✓
23	0.00	1.00	0.00	0.00	A/H3N2	✓
24	0.00	1.00	0.00	0.00	A/H3N2	✓
25	0.00	1.00	0.00	0.00	A/H3N2	✓
26	0.00	0.99	0.00	0.00	A/H3N2	✓
27	0.00	1.00	0.00	0.00	A/H3N2	✓
28	0.00	0.98	0.00	0.01	A/H3N2	4
29	0.00	1.00	0.00	0.00	A/H3N2	4
30	0.00	1.00	0.00	0.00	A/H3N2	4
31	0.00	0.99	0.00	0.00	A/H3N2	4
32	0.00	1.00	0.00	0.00	A/H3N2	4
33	0.01	0.99	0.00	0.00	A/H3N2	4
34	0.84	0.14	0.01	0.00	A/H1	4
35	1.00	0.01	0.00	0.00	A/H1	4

Table 9.1 Continued

	ANN output					
	H1N1	H3N2	H5N1	Negative	Known subtype	
36	1.00	0.01	0.00	0.00	A/H1	4
37	1.00	0.00	0.00	0.00	A/H1N1	4
38	1.00	0.00	0.00	0.00	A/H1N1	4
39	0.63	0.27	0.00	0.01	A/H1	7
40	0.00	0.00	0.00	1.00	B	4
41	0.01	0.00	0.00	0.99	B	4
42	0.00	0.35	0.00	0.58	B	7
43	0.00	0.01	0.00	1.00	B	4
44	0.00	0.00	0.00	1.00	B	4
45	0.00	0.16	0.00	0.91	B	4
46	0.00	0.86	0.00	0.21	B	X
47	0.01	0.00	0.00	1.00	SARS	✓
48	0.00	0.00	0.00	1.00	SARS	✓
49	0.05	0.00	0.00	1.00	hMPV	✓
50	0.02	0.00	0.00	1.00	hMPV	✓
51	0.11	0.00	0.00	0.98	RSV	✓
52	0.01	0.03	0.00	0.97	hPIV3	✓
53	0.00	0.00	0.00	1.00	hPIV3	✓

HA partial subtype identified by immunofluorescence assay by the Colorado Department of Public Health and Environment (CDPHE); Full antigenic characterization provided by Centers for Disease Control (CDC); Samples 47–53 are influenza-like illnesses included as negative controls: SARS (severe acute respiratory syndrome), hMPV (human metapneumovirus), RSV (respiratory syncytial virus), hPIV3 (human parainfluenza virus type 3). Output scores >0.75 are highlighted. Checkmarks indicate correct determination by the ANN and 'X' indicates either incorrect assignment or no output score >0.75.

Positive signals on sequences that were designed to be sensitive to avian influenza viruses may be evidence of the avian ancestry of H1N1 viruses.

When the ANN was trained using only modern viruses (collected after 2000), the historic H1N1 viruses were not accurately identified. However, when the ANN was trained with temporally related viruses (by including some of the historic viruses in the training set), the ANN was able to accurately assign subtype a set of historic viruses. Specifically, 12 of the historic viruses were used in training, and 3 were tested as unknowns (see Table 9.3). The ANN assigned the 1918 virus as an A/H1N1 with a 94% probability. Interestingly, the improved training provided an accurate subtype assignment for both A/Swine/Iowa/1930 (albeit with only 70% probability) and A/Turkey/Kansas/1980 (97% probability). Although this was a limited study, the MChip robustly detected a range of historic viruses, and visual inspection indicated that the overall signal patterns were distinct from those of currently circulating strains. Such an indication from a rapid diagnostic would be cause for more extensive analysis of the virus.

Table 9.2 A/H5N1 training set and artificial neural network probability distribution for 'unknown' A/H5N1 viruses and negative controls

Training set
Specimen ID[1]
A/chicken/Nigeria/42/2006 (4)
A/chicken/Vietnam/NCVD15/2003 (2)
A/chicken/Vietnam/NCVD30/2003 (2)
A/chicken/Vietnam/NCVD-CDC22/2005 (2)
A/chicken/Vietnam/NCVD-CDC33/2005 (4)
A/chicken/Vietnam/NCVD-CDC36/2005 (2)
A/chicken/Vietnam/NCVD-CDC37/2005 (2)
A/chicken/Vietnam/NCVD-CDC42/2005 (2)
A/chicken/Vietnam/NCVD-CDC51/2005 (2)
A/chicken/Vietnam/NCVD-CDC52/2005 (2)
A/duck/Vietnam/NCVD25/2003 (2)
A/duck/Vietnam/NCVD-CDC17/2005 (2)
A/duck/Vietnam/NCVD-CDC41/2005 (2)
A/duck/Vietnam/NCVD-CDC50/2005 (2)
A/goose/Kazakhstan/464/2005 (6)
A/Indonesia/CDC357/2006 (2)
A/Indonesia/CDC370T/2006 (2)
A/Prachi Nburi/6231/2004 (2)
A/Vietnam/JP20–2/2005 (2)
A/Vietnam/JP36–2/2005
A/chicken/Vietnam/01/2004
A/swine/Vietnam/6774/2004
A/Vietnam/HN30408/05
A/Vietnam/JP4207/05
A/muscovy duck/Vietnam/04/2004
A/muscovy duck/Vietnam/17/2004
A/chicken/Vietnam/07/2004

1 Number in parentheses represents the number of replicate experimental results utilized for training (total of 54 datasets).

2 Specimens 1–24 are all influenza A/H5N1, and probabilities >0.80 are highlighted.

3 'Neg' indicates the sample is known to be influenza A negative. Samples 29–31 are influenza-like illnesses included as negative controls: SARS CoA (Coronavirus that causes severe acute respiratory syndrome), RSV (respiratory syncytial virus), hPIV3 (human parainfluenza virus type 3).

4 Checkmarks indicate correct identification with at least 80% probability.

Table 9.2 Continued

'Unknown' samples to be identified by artificial neural network						
	Probability distribution[2]					
	H1N1	H3N2	H5N1	Negative	Specimen ID[3]	Correct ID?[4]
1	0.01	0.00	1.00	0.00	A/Iraq/207-NAMRU3/2006	✓
2	0.00	0.00	1.00	0.00	A/chicken/Vietnam/NCVD-CDC3/2005	✓
3	0.00	0.00	1.00	0.00	A/environment/Vietnam/NCVD-CDC54/2005	✓
4	0.00	0.00	1.00	0.00	A/duck/Kulon Progo/BBVET/IX/2004	✓
5	0.00	0.00	1.00	0.00	A/Anhui/2/2005	✓
6	0.00	0.00	1.00	0.00	A/Indonesia/CDC390/2006	✓
7	0.00	0.00	0.99	0.01	A/environment/Vietnam/NCVD-CDC60/2005	✓
8	0.00	0.00	1.00	0.00	A/Muscovy duck/Vietnam/NCVD28/200	✓
9	0.00	0.00	1.00	0.00	A/environment/Vietnam/NCVD-CDC53/2005	✓
10	0.00	0.00	1.00	0.00	A/Anhui/1/2005	✓
11	0.00	0.00	1.00	0.00	A/chicken/Vietnam/NCVD-CDC37/2005	✓
12	0.00	0.00	1.00	0.00	A/chicken/Vietnam/NCVD-CDC23/2005	✓
13	0.00	0.00	1.00	0.00	A/duck/Kazakhstan/467/2005	✓
14	0.00	0.00	1.00	0.00	A/Indonesia/CDC326T/2006	✓
15	0.00	0.00	1.00	0.00	A/chicken/Vietnam/NCVD-CDC49/2005	✓
16	0.01	0.00	0.99	0.00	A/feline/Indonesia/CDC1/2006	✓
17	0.01	0.16	0.83	0.00	A/Indonesia/CDC624/2006	✓
18	0.01	0.00	0.99	0.00	A/Indonesia/CDC623/2006	✓
19	0.10	0.00	0.90	0.00	A/Indonesia/CDC594/2006	✓
20	0.08	0.01	0.91	0.00	A/Indonesia/CDC595/2006	✓
21	0.04	0.00	0.96	0.00	A/Indonesia/CDC597/2006	✓
22	0.02	0.00	0.98	0.00	no strain designation	✓
23	0.05	0.00	0.95	0.00	A/Indonesia/CDC596/2006	✓
24	0.05	0.00	0.96	0.00	A/Indonesia/CDC599/2006	✓
25	0.00	0.00	0.00	1.00	Negative	✓
26	0.02	0.00	0.00	0.98	Negative	✓
27	0.00	0.00	0.00	1.00	Negative	✓
28	0.00	0.00	0.00	1.00	Negative	✓
29	0.00	0.00	0.00	1.00	SARS	✓
30	0.00	0.20	0.00	0.80	RSV	✓
31	0.00	0.00	0.00	1.00	hPIV3	✓

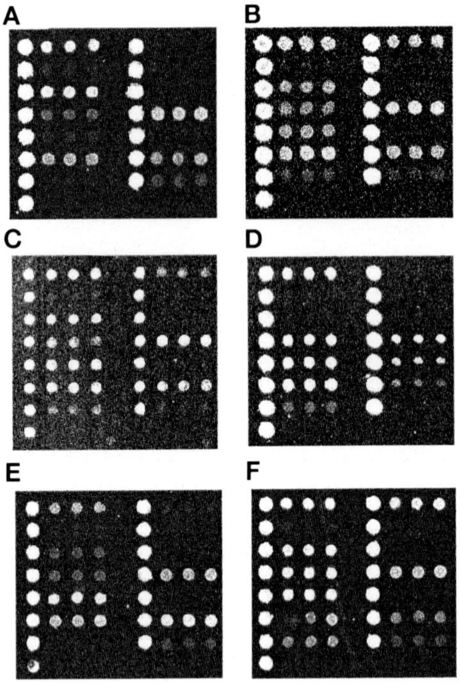

Figure 9.2 MChip images for other subtypes: (A) equine H3N8, (B) avian H10, (C) avian H7N3, (D) avian H7N2, (E) avian H10N7, (F) avian H9N2. Microarray layout is given in Fig. 9.1B. Reproduced with permission from Anal. Chem. 2007, *79*, 378–384. Copyright 2007 American Chemical Society.

Table 9.3 Historic A/H1N1 viruses tested

1	A/Brevig Mission/1/1918
2	A/Swine/Iowa/1930
3	A/WSN/1933
4	A/Puerto Rico/8/1934*
5	A/Alaska/1935*
6	A/Galenby/1937*
7	A/Bellamy/1942*
8	A/Iowa/1943*
9	A/Fort Monmouth/1/1947*
10	A/Fort Warren/1/1950*
11	A/Missouri/301/1952*
12	A/Tasmania/30/1954*
13	A/England/19/1955*
14	A/Scotland/38/1956*
15	A/Prussia/1957*
16	A/Turkey/Kansas/4880/1980

* Indicates inclusion in the artificial neural network training set.

Reproduced with permission from J. Clin. Microbiol. 2007, *45*, 3807–3810. Copyright 2007 American Society for Microbiology.

Comparison of MChip performance to standard methods

It is important to compare any newly developed diagnostic test to the current gold standard methods. Here, a comparison of the ability of several techniques to identify the presence of influenza A between is summarized. The MChip and associated assay is compared a number of well-established influenza diagnostic methods including viral culture, QuickVue® Influenza A+B rapid immunoassay (Quidel Corporation, San Diego, CA, USA), and real-time RT-PCR.

For the comparative study, one hundred two (102) respiratory specimens that had been collected from patients with influenza-like illness and tested by QuickVue Influenza A+B and viral culture were selected for subsequent analysis by real-time RT-PCR and MChip. The median age of the patients from whom the specimens were collected was 7.5 years (range = 3 months to 86 years). A full description of the clinical specimen collection, real-time RT-PCR, and shell vial experiments is given elsewhere (Mehlmann *et al.*, 2007). Of the 102 samples, 57 (56%) were positive for influenza A, 9 (9%) were positive for respiratory syncytial virus (RSV), and 36 (35%) were negative. RT-PCR detected influenza A in 61 (60%) samples and obtained negatives for 41 (40%). QuickVue® detected influenza A in 53 (52%) samples and gave negative results for 49 (48%) samples. MChip detected influenza A in 57 (56%) samples and gave negative results for 45 (44%) samples.

Table 9.4 shows a quantitative comparison of the methods investigated with respect to their ability to identify influenza A virus. With viral culture as the reference method, MChip exhibited clinical sensitivity and specificity of 98%. Within the calculated confidence intervals, the MChip sensitivity and positive and negative predictive values were similar to those of both QuickVue® and real-time RT-PCR. The clinical specificity of MChip (98%) was similar to that

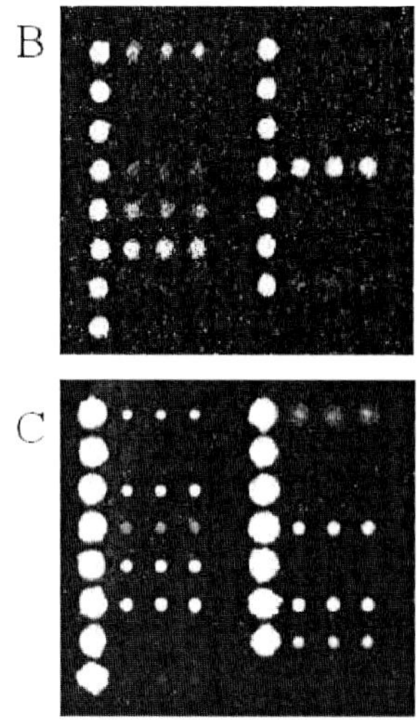

Figure 9.3 Comparison of modern H1N1 virus and the 1918 'Spanish Flu.' (A) Fluorescence image from a representative modern H1N1 virus. (B) Fluorescence image from A/Brevig Mission/1/1918. Array layout is in Fig. 9.1B. Brighter spots indicate higher signal intensity. Reproduced with permission from J. Clin. Microbiol. 2007, *45*, 3807–3810. Copyright 2007 American Society for Microbiology.

of the QuickVue® test (100%), but substantially greater than for RT-PCR (89%) due to four specimens that were only positive by real time RT-PCR. The two most likely explanations for relatively poor performance by RT-PCR is the increased sensitivity of RT-PCR compared to viral culture and/or a higher rate of false positive results for RT-PCR. Previous comparative studies for influenza detection have indicated higher sensitivity of RT-PCR compared to culture (Boivin *et al.*, 2001; Ellis *et al.*, 1997; Herrmann *et al.*, 2001; Kehl *et al.*, 2001), and arguments have been made that RT-PCR would be a more reliable reference standard than viral culture (Ruest *et al.*, 2003). However, its inherent sensitivity makes RT-PCR more susceptible to false positives (Kwok *et al.*, 1989).

For comparison, the MChip results referenced to RT-PCR, rather than culture, are summarized in Table 9.4. In summary, the MChip performance is comparable to other standard diagnostic methods when referenced to either viral culture or RT-PCR. In fact, the clinical sensitivity and specificity determined for the MChip in this blinded study are better than the average values reported for most rapid tests (Dominguez *et al.*, 1993; Herrmann *et al.*, 2001; Kaiser *et al.*, 1999; Leonardi *et al.*, 1994; Liolios *et al.*, 2001; Reina *et al.*, 1996; Schultze *et al.*, 2001).

A significant advantage of the MChip is the ability to simultaneously detect and subtype influenza positive specimens in a single experiment. None of the other methods evaluated in this comparative study provided subtype information

Table 9.4 Comparison of influenza diagnostic methods

	MChip	QuickVue®	RT-PCR
Referenced to viral culture			
Sensitivity (%) *	98 (91–100)	93 (83–97)	98 (90–100)
Specificity (%) *	98 (89–100)	100 (92–100)	89 (77–95)
PPV (%)*	98 (91–100)	100 (93–100)	92 (82–96)
NPV (%)*	98 (88–100)	92 (81–97)	98 (87–100)
Referenced to RT-PCR			
Sensitivity (%)*	92 (82–96)	85 (74–92)	92 (82–96)
Specificity (%)*	98 (87–100)	97 (87–100)	98 (87–100)
PPV (%)*	98 (91–100)	98 (90–100)	98 (91–10)
NPV (%)*	89 (77–95)	82 (69–90)	89 (77–95)

*Numbers in parenthesis are the 95% confidence intervals.

Reproduced from J. Clin. Microbiol. 2007, *45*, 1234–1237. Copyright 2007 American Society for Microbiology.

without additional testing. For viral culture, an additional test involving reference antibodies must be conducted to determine influenza subtype. For RT-PCR several additional experiments targeting the H1, H3 and H5 genes are necessary to determine the subtype. So although MChip performed comparably to other standard methods for the identification of influenza A, MChip demonstrated a significant advantage over the other test methods by providing simultaneous detection as well as subtype identification for influenza positive samples. This simultaneous identification and subtyping capability could provide a significant advantage to public health officials, vaccine manufacturers, and physicians involved in direct patient care, for all of whom early detection and rapid subtype information can be critically important.

Concluding remarks

While it is unclear which influenza A viral subtype will ultimately cause the next pandemic (Horimoto *et al.*, 2005), it is clear that the recent emergence of H5N1 in the human population has caused global concern and a demonstrated need for increased surveillance. In areas of such as Southeast Asia where H1N1 and H3N2 and H5N1 viruses can co-circulate and the threat of human infection with H5N1 is very real, improved diagnostic tests capable of rapidly providing subtype information on human clinical specimens are imperative. We envisage that MChip could be to address the following critical questions: (1) is a patient infected with influenza A virus, and (2) if so, is the virus a typical human-adapted subtype or is it atypical? In a real-world application, any patient that tests positive for influenza A that cannot be definitively identified as a currently circulating subtype (H3N2 or H1N1) would be triaged in an appropriate manner, and health officials would be directed to a more detailed investigation.

MChip provides a maximum amount of information without the need to amplify several gene segments, and can provide a result in seven hours. Although the analysis utilizes only the M gene segment, we have so far not circumvented the need to RT-PCR amplify it prior to microarray hybridization and analysis. The RT-PCR amplification in its commonly used format requires a certain level of laboratory infrastructure and a careful workflow to perform well, but future advances in integrated 'lab on a chip' devices (Zhang *et al.*, 2006; Zhang *et al.*, 2007) may enable RT-PCR-based amplification to be performed in a more closed 'hands-off' format. In addition, the proof-of-principle studies presented here, a relatively expensive fluorescence reader was utilized for MChip detection and imaging. The typical microarray confocal fluorescence detection methodology is not desirable for widespread adoption of low-density microarray technology, particularly in resource limited settings. Improvements in microarray detection technology are being developed that could significantly reduce both the start-up cost and individual assay cost associated with processing and imaging low-density microarrays (Kuck and Taylor, 2008). It is our hope that combined with technological advancements such as those mentioned above, MChip and other low-density diagnostic microarrays will assist in enabling improved worldwide surveillance and diagnosis of influenza and other viral infections.

Acknowledgements

This work was conducted while both authors were affiliated with the Department of Chemistry and Biochemistry, UCB 215, University of Colorado at Boulder, Boulder, CO 80309. The authors gratefully acknowledge funding from NIH/NIAID (U01AI056528) and the cooperation of the Centers for Disease Control and Prevention in Atlanta. We would like to thank the Colorado Department of Public Health and Environment (CDPHE) Division of Laboratory Services (Denver, CO), the Air Force Institute for Operational Health (San Antonio, TX), University of Iowa College of Public Health Centre for Emerging Infectious Diseases (Coralville, IA), and Quidel Corporation (San Diego, CA) for providing influenza clinical specimens.

References

Aubin, J.-T., Azebi, S., Balish, A., Banks, J., Bhat, N., Bright, R.A., Brown, I., Buchy, P., Burguiere, A.-M., Chen, H.-I., *et al.* (2005). Evolution of H5N1 avian influenza viruses in Asia. Emerg. Infect. Dis. *11*, 1515–1521.

Beladi, S.P., Ghorashi, S.A., and Morshedi, D. (2005). Using nested-PCR for detection of avian influenza virus. Acta Vet. Brno. *74*, 581–584.

Berrar, D.P., Downes, C.S., and Dubitzky, W. (2003). Multiclass cancer classification using gene expression profiling and probabilistic neural networks. In Pacific Symposium on Biocomputing (Lihue, Hawaii), pp. 5–16.

Boivin, G., Hardy, I., Kress. A. (2001). Evaluation of a rapid optical immunoassay for influenza viruses (FLU OIA test) in comparison with cell culture and reverse transcription-PCR. J. Clin. Microbiol. *39*, 730–732.

Bourinbaiar, A.S., Timofeev, I.V., and Agwale, S.M. (2006). Recent advances in development of avian flu and influenza diagnostics. Expert Rev. Mol. Diagn. *6*, 783–795.

Butler, D. (2006). Family tragedy spotlights flu mutations. Nature *442*, 114–115.

Chatterjee, M., Mohapatra, S., Ionan, A., Bawa, G., Ali-Fehmi, R., Wang, X., Nowak, J., Ye, B., Nahhas, F.A., Lu, K., *et al.* (2006). Diagnostic markers of ovarian cancer by high-throughput antigen cloning and detection on arrays. Cancer Res. *66*, 1181–1190.

Chomel, J.J., Pardon, D., Thouvenot, D.,Allard, J.P., and Aymard, M. (1991). Comparison between three rapid methods for direct diagnosis of influenza and the conventional isolation procedure. Biologicals *19*, 287–292.

Daum, L.T., Canas, L.C., Arulanandam, B.P., Niemeyer, D., Valdes, J.J., and Chambers, J.P. (2007). Real-time RT-PCR assays for type and subtype detection of influenza A and B viruses. Influenza and Other Respir. Viruses, *1*, 167–175.

Dawson, E.D., Moore, C.L., Smagala, J.A., Dankbar, D.M., Mehlmann, M., Townsend, M.B., Smith, C.B., Cox, N.J., Kuchta, R.D., Rowlen, K.L., and Stears, R.L. (2006). MChip: A tool for influenza surveillance. Anal. Chem. *78*, 7610–7615.

Dawson, E.D., Moore, C.L., Smagala, J.A., Dankbar, D.M., Mehlmann, M., Townsend, M.B., Smith, C.B., Cox, N.J., Kuchta, R.D., and Rowlen, K.L. (2007). Identification of A/H5N1 influenza viruses using a single gene diagnostic microarray. Anal. Chem. *79*, 378–384.

Department of Health and Human Services (2006). HHS Pandemic Influenza Plan. Available from: http://www.hhs.gov/pandemicflu/plan/pdf/HHSPandemicInfluenzaPlan.pdf

DiTrani, L., Bedini, B., Donatelli, I., Campitelli, L., Chiappini, B., DeMarco, M.A., Delogu, M., Buonavoglia, C., and Vaccari, G. (2006). A sensitive one-step real-time PCR for detection of avian influenza viruses using a MGB probe and an internal positive control. BMC Infect. Dis. *6*, 87.

Dominguez, E.A., Taber, L.H., and Couch, R.B. (1993). Comparison of rapid diagnostic techniques for respiratory syncytial and influenza A virus respiratory infections in young children. J. Clin. Microbiol. *31*, 2286–2290.

Ellis, J.S., Fleming, D.M., and Zambon., M.C. (1997). Multiplex reverse transcription-PCR for surveillance of influenza A and B viruses in England and Wales in 1995 and 1996. J. Clin. Microbiol. 35, 2076–2082.

Fouchier, R.A.M., Bestebroer, T.M., Herfst, S., Van Der Kemp, L., Rimmelzwaan, G.F., and Osterhaus, A.D.M.E. (2000). Detection of influenza A viruses from different species by PCR amplification of conserved sequences in the matrix gene. J. Clin. Microbiol. *38*, 4096–4101.

Guan, Y., Poon, L.L.M., Cheung, C.Y., Ellis, T.M., Lim, W., Lipatov, A.S., Chan, K.H., Sturm-Ramirez, K.M., Cheung, C.L., Leung, Y.H.C., *et al.* (2004). H5N1 influenza: A protean pandemic threat. Proc. Natl. Acad. Sci. USA *101*, 8156–8161.

Guan, Y., Peiris, J.S.M., Lipatov, A.S., Ellis, T.M., Dyrting, K.C., Krauss, S., Zhang, L.J., Webster, R.G., and Shortridge, K.F. (2002). Emergence of multiple genotypes of H5N1 Avian influenza viruses in Hong Kong SAR. Proc. Natl. Acad. Sci. U.S.A. *99*, 8950–8955.

Herrmann, B., Larsson, C., and Zweygberg., B.W. (2001). Simultaneous detection and typing of influenza viruses A and B by a nested reverse transcription-PCR: comparison to virus isolation and antigen detection by immunofluorescence and optical immunoassay (FLU OIA). J. Clin. Microbiol. 39, 134–138.

Horimoto, T., and Kawaoka, Y. (2005). Influenza: Lessons from past pandemics, warnings from current incidents. Nat. Rev. Microbiol. 3, 591–600.

Hudson, M.E. (2008). Sequencing breakthroughs for genomic ecology and evolutionary biology. Mol. Ecol. Resour. *8*, 3–17.

Kaiser, L., Briones, M.S., and Hayden, F.G. (1999). Performance of virus isolation and Directigen Flu A to detect influenza A virus in experimental human infection. J. Clin. Virol. *14*, 191–197.

Kehl, S.C., Henrickson, K.J., Hua, W., and Fan., J. (2001). Evaluation of the Hexaplex assay for detection of respiratory viruses in children. J. Clin. Microbiol. *39*, 1696–1701.

Kessler, N., Ferraris, O., Palmer, K., Marsh, W., and Steel, A. (2004). Use of the DNA Flow-Thru Chip, a three-dimensional biochip, for typing and subtyping of influenza viruses. J. Clin. Microbiol. *42*, 2173–2185.

Khan, J., Wei, J.S., Ringner, M., Saal, L.H., Ladanyi, M., Westermann, F., Berthold, F., Schwab, M., Antonescu, C.R., Peterson, C., *et al.* (2001). Classification and diagnostic prediction of cancers using gene expression profiling and artificial neural networks. Nature Med. 7, 673–679.

Krogh, A. (2008). What are artificial neural networks? Nature Biotech. *26*, 195–197.

Kuck, L.R., and Taylor, A.W. (2008). Photopolymerization as an innovative detection technique for low-density microarrays. Biotechniques, in press (Please update if possible).

Kwok, S., and Higuchi, R. (1989). Avoiding false positives with PCR. Nature, *339*, 237–238.

Lamb, R.A., and Krug, R.M. (1996). Orthomyxovyridae: The viruses and their replication. In Field's Virology, Knipe, D.M., ed. (Philadelphia: Lippincott, Williams and Wilkins), pp. 605–647.

Leonardi, G.P., Leib, H., Birkhead, G.S., Smith, C., Costello, P., Conron. W. (1994). Comparison of rapid detection methods for influenza A virus and their value in health-care management of institutionalized geriatric patients. J. Clin. Microbiol. 32, 70–74.

Li, J., Chen, S., and Evans, D.H. (2001). Typing and subtyping influenza virus using DNA microarrays and multiplex reverse transcriptase PCR. J. Clin. Microbiol. 39, 696–704.

Li, K.S., Guan, Y., Wang, J., Smith, G.J.D., Xu, K.M., Duan, L., Rahardjo, A.P., Puthavathana, P., Buranathai, C., Nguyen, T.D., *et al.* (2004). Genesis of a highly pathogenic and potentially pandemic H5N1 influenza virus in eastern Asia. Nature *430*, 209–213.

Lin, B.-C., Blaney, K.M., Malanosky, A.P., Ligler, A.G., Schnur, J.M., Metzgar, D., Russell, K.L., and Stenger, D.A. (2007). Using a resequencing microarray as a multiple respiratory pathogen detection assay. J. Clin. Microbiol. *45*, 443–452.

Linder, R., Richards, T., and Wagner, M. (2007). Microarray data classified by artificial neural networks. In Methods in Molecular Biology, Rampal, J.B., ed. (Humana Press), pp. 345–372.

Liolios, L., Jenney, A., Spelman, D., Kotsimbos, T., Catton, M., Wesselingh, S. (2001). Comparison of a multiplex reverse transcription-PCR-enzyme hybridization assay with conventional viral culture and immunofluorescence techniques for the detection of seven viral respiratory pathogens. J. Clin. Microbiol. *39*, 2779–2783.

Lu, P.S. (2006). Early Diagnosis of Avian Influenza. Science, *312*, 337.

MacKay, W.G., van Loon, A.M., Niedrig, M., Meijer, A., Lina, B., and Niesters, H.G.M. (2008). Molecular detection and typing of influenza viruses: are we ready for an influenza pandemic? J. Clin. Virol. *42*, 194–197.

Macken, C., Lu, H., Goodman, J., and Boykin, L. (2001). The value of a database in surveillance and vaccine selection. In Options for the Control of Influenza IV, A.D.M.E. Osterhaus, N. Cox, and A.W. Hampson, eds. (Amsterdam, Elsevier Science), pp. 103–106.

Markoulatos, P., Siafakas, N., and Moncany, M. (2002). Multiplex polymerase chain reaction: A practical approach. J. Clin. Lab. Anal. *16*, 47–51.

Matthey, S., Nicholson, D., Ruhs, S., Alden, B., Knock, M., Schultz, K., and Schmuecker, A. (1992). Rapid detection of respiratory viruses by shell vial culture and direct staining by using pooled and individual monoclonal antibodies. J. Clin. Microbiol. 30, 540–544.

Mehlmann, M., Townsend, M.B., Stears, R.L., Kuchta, R.D., and Rowlen, K.L. (2005). Optimization of fragmentation conditions for microarray analysis of viral RNA. Anal. Biochem. 347, 316–323.

Mehlmann, M., Bonner, A.B., Williams, J.V., Dankbar, D.M., Moore, C.L., Kuchta, R.D., Posdiad, A.B., Tamerius, J.D., Dawson, E.D., and Rowlen, K.L. (2007). Comparison of the MChip to viral culture, reverse transcription-PCR, and the QuickVue Influenza A+B test for rapid diagnosis of influenza. J. Clin. Microbiol. 45, 1234–1237.

Mehlmann, M., Dawson, E.D., Townsend, M.B., Smagala, J.A., Moore, C.L., Smith, C.B., Cox, N.J., Kuchta, R.D., and Rowlen, K.L. (2006). FluChip: Robust sequence selection method for a diagnostic microarray. J. Clin. Microbiol. 44, 2857–2862.

Mikhailovich, V., Gryadunov, D., Kolchinsky, A., Makarov, A.A., and Zasedatelev, A. (2008). DNA microarrays in the clinic: Infectious diseases. BioEssays, *30*, 673–682.

Moore, C.L., Smagala, J.A., Smith, C.B., Dawson, E.D., Cox, N.J., Kuchta, R.D., and Rowlen, K.L. (2007). Evaluation of MChip with historic subtype H1N1 influenza A viruses, including the 1918 'Spanish flu' strain. J. Clin. Microbiol. *45*, 3807–3810.

Ng, L.F.P., Barr, I., Nguyen, T., Noor, S.M., Tan, R.S.-P., Agathe, L.V., Gupta, S., Khalil, H., To, T.L., Hassan, S.S., and Ren, E.-C. (2006). Specific detection of H5N1 avian influenza A virus in field specimens by a one-step RT-PCR assay. BMC Infect. Dis. *6*, 40.

Normile, D. (2006). Human transmission but no pandemic in Indonesia. Science *312*, 1855.

Payungporn, S., Phakdeewirot, P., Chutinimitkul, S., Theamboonlers, A., Keawcharoen, J., Oraveerakul, K., Amonsin, A., and Poovorawan, Y. (2004). Single-step multiplex reverse transcription-polymerase chain reaction (RT-PCR). for influenza A virus subtype H5N1 detection. Viral Immunol. *17*, 588–593.

Payungporn, S., Chutinimitkul, S., Chaisingh, A., Damrongwantanapokin, S., Buranathai, C., Amonsin, A., Theamboonlers, A., and Poovorawan, Y. (2006). Single step multiplex real-time RT-PCR for H5N1 influenza A virus detection. J. Virol. Meth. *131*, 143–147.

Poddar, S.K. (2002). Influenza virus types and subtypes detection by single step single tube multiplex reverse transcription-polymerase chain reaction (RT-PCR) and agarose gel electrophoresis. J. Virol. Meth. *99*, 63–70.

Quan, P.-L., Palacios, G., Jabado, O.J., Conlan, S., Hirschberg, D.L., Pozo, F., Jack, P.J.M., Cisterna, D., Renwick, N., Hui, J., *et al.* (2007). Detection of respiratory viruses and subtype identification of influenza a viruses by GreenChipResp oligonucleotide microarray. J. Clin. Microbiol. *45*, 2359–2364.

Reina, J., Munar, M., and Blanco, I. (1996). Evaluation of a direct immunofluorescence assay, dot-blot enzyme immunoassay, and shell vial culture in the diagnosis of lower respiratory tract infections caused by influenza A virus. Diagn. Microbiol. Inf. Dis. 25, 143–145.

Ruest, A., Michaud, S., Deslandes, S., and Frost., E.H. (2003). Comparison of the directigen flu A+B test, the QuickVue influenza test, and clinical case definition to viral culture and reverse transcription-PCR for rapid diagnosis of influenza virus infection. J. Clin. Microbiol. *41*, 3487–3493.

Russell, R. (2003). Designing microarray oligonucleotide probes. Brief. Bioinf. *4*, 361–367.

Schultze, D., Thomas, Y., and Wunderli., W. (2001). Evaluation of an optical immunoassay for the rapid detection of influenza A and B viral antigens. Eur. J. Clin. Microbiol. Inf. Dis. *20*, 280–283.

Sengupta, S., Onodera, K., Lai, A., and Melcher, U. (2003). Molecular detection and identification of influenza viruses by oligonucleotide microarray hybridization. J. Clin. Microbiol. *41*, 4542–4550.

Smagala, J.A., Dawson, E.D., Mehlmann, M., Townsend, M.B., Kuchta, R.D., and Rowlen, K.L. (2005).

ConFind: A robust tool for conserved sequence identification. Bioinform. *21*, 4420–4422.

Smith, A.B., Mock, V., Melear, R., Colarusso, P., and Willis, D.E. (2003). Rapid detection of influenza A and B viruses in clinical specimens by Light Cycler real time RT-PCR. J. Clin. Virol. *28*, 51–58.

Spackman, E., Senne, D.A., Myers, T.J., Bulaga, L.L., Garber, L.P., Perdue, M.L., Lohman, K., Daum, L.T., and Suarez, D.L. (2002). Development of a real-time reverse transcriptase PCR assay for type A influenza virus and the avian H5 and H7 hemagglutinin subtypes. J. Clin. Microbiol. *40*, 3256–3260.

Steinhauer, D.A., and Skehel, J.J. (2002). Genetics of influenza viruses. Ann. Rev. Genet. *36*, 305–332.

Stone, B., Burrows, J., Schepetiuk, S., Higgins, G., Hampson, A., Shaw, R., and Kok, T. (2004). Rapid detection and simultaneous subtype differentiation of influenza A viruses by real time PCR. J. Virol. Meth. *117*, 103–112.

Storch, G.A. (2003). Rapid diagnostic tests for influenza. Curr. Opin. Pediatr. 15(1), 77–84.

Suarez, D.L., Das, A., and Ellis, E. (2007). Review of rapid molecular diagnostic tools for avian influenza virus. Avian Dis. *51*, 201–208.

Subbarao, K., Klimov, A., Katz, J., Regnery, H., Lim, W., Hall, H., Perdue, M., Swayne, D., Bender, C., Huang, J., *et al.* (1998). Characterization of an avian influenza A (H5N1). virus isolated from a child with a fatal respiratory illness. Science *279*, 393–396.

Taubenberger, J.K.; Layne, S.P. (2001). Diagnosis of influenza virus: Coming to grips with the molecular era. Molec. Diag., *6*, 291–305.

Templeton, K.E., Scheltinga, S.A., Beersma, M.F.C., Kroes, A.C.M., Claas, E.C.J. (2004). Rapid and sensitive method using multiplex real-time PCR for diagnosis of infections by influenza A and influenza B viruses, respiratory syncytial virus, and parainfluenza viruses *1, 2, 3*, and 4. J. Clin. Microbiol. *42*, 1564–1569.

Thontiravong, A., Payungporn, S., Keawcharoen, J., Chutinimitkul, S., Wattanodorn, S., Damrongwatanapokin, S., Chaisingh, A., Theamboonlers, A., Poovorawan, Y., Oraveerakul. (2007). The single-step multiplex reverse transcription-polymerase chain reaction assay for detecting H5 and H7 avian influenza A viruses. Tohoku J. Exp. Med. *211*, 75–79.

Townsend, M.B., Dawson, E.D., Mehlmann, M., Smagala, J.A., Dankbar, D.M., Moore, C.L., Smith, C.B., Cox Nancy, J., Kuchta, R.D., and Rowlen, K.L. (2006). FluChip: Experimental evaluation of a diagnostic influenza microarray. J. Clin. Microbiol. *44*, 2863–2871.

Tomiuk, S., and Hofmann, K. (2001). Microarray probe selection strategies. Brief. Bioinf. *2*, 329–340.

Valafar, F. (2002). Pattern recognition techniques in microarray data analysis: A survey. Ann. NY Acad. Sci. *980*, 41–64.

Wang, D., Coscoy, L., Zylberberg, M., Avila, P.C., Boushey, H.A., Ganem, D., and DeRisi, J.L. (2002). Microarray-based detection and genotyping of viral pathogens. Proc. Natl. Acad. Sci. U.S.A. *99*, 15687–15692.

Wang, Z., Daum, L.T., Vora, G.J., Metzgar, D., Walter, E.A., Canas, L.C., Malanosky, A.P., Lin, B., and Stenger, D.A. (2006). Identifying influenza viruses with resequencing arrays. Emerg. Inf. Dis. *12*, 638–646.

Wei, H.-L., Bai, G.-R., Mweene, A.S., Zhou, Y.-C., Cong, Y.-L., Pu, J., Wang, S., Kida, H., Liu, J.-H. (2006). Rapid detection of avian influenza virus A and subtype H5N1 by single step reverse transcription-polymerase chain reaction. Virus Genes, *32*, 261–267.

Whiley, D.M., and Sloots, T.P. (2005). A 5′-nuclease real-time reverse transcription-polymerase chain reaction assay for the detection of a broad range of influenza A subtypes, including H5N1. Diagn. Microbiol. Infect. Dis. *53*, 335–337.

World Health Organization (2005). Recommended laboratory tests to identify avian influenza A virus in specimens from humans. Available from: http://www.who.int/csr/disease/avian_ influenza/guidelines/avian_labtests2.pdf

World Health Organization (2008). Cumulative number of confirmed human cases of avian influenza A/(H5N1). reported to WHO. Available from: http://www.who.int/csr/disease/avian_ influenza/country/cases_table_2008_06_19/en/index.html

Wu, C., Cheng, X., He, J., Lv, X., Wang, J., Deng, R., Long, Q., and Wang, X. (2008). A multiplex real-time RT-PCR for detection and identification of influenza virus types A and B and subtypes H5 and N1. J. Virol. Meth. *148*, 81–88.

Xie, Z., Pang, Y.-S., Liu, J., Deng, X., Tang, X., Sun, J., and Khan, M.I. (2006). A multiplex RT-PCR for detection of type A influenza virus and differentiation of avian H5, H7, and H9 hemagglutinin subtypes. Mol. Cell. Probes *20*, 245–249.

Zammatteo, N., Hamels, S., de Longueville, F., Alexandre, I., Gala, J.-l., Brasseur, F., and Remacle, J. (2002). New chips for molecular biology and diagnostics. Biotech. Ann. Rev. *8*, 85–101.

Zhang, C., and Xing, D. (2007). Miniaturized PCR chips for nucleic acid amplification and analysis: latest advances and future trends. Nucl. Acids Res. *35*, 4223–4237.

Zhang, C., Xu, J., Ma, W., and Zheng, W. (2006). PCR microfluidic devices for DNA amplification. Biotech. Adv. *24*, 243–284.

Zou, S. (1997). A practical approach to genetic screening for influenza virus variants. J. Clin. Microbiol. *35*, 2623–2627.

Computer-assisted Vaccine Design

10

Hao Zhou, Ramdas S. Pophale and Michael W. Deem

Abstract

We define a new parameter to quantify the antigenic distance between two H3N2 influenza strains. We use this parameter to measure antigenic distance between circulating H3N2 strains and the closest vaccine component of the influenza vaccine. For the data between 1971 and 2004, the measure of antigenic distance correlates better with efficacy in humans of the H3N2 influenza A annual vaccine than do current state-of-the-art measures of antigenic distance such as phylogenetic sequence analysis or ferret antisera inhibition assays. We suggest that this measure of antigenic distance could be used to guide the design of the annual flu vaccine. We combine the measure of antigenic distance with a multiple-strain avian influenza transmission model to study the threat of simultaneous introduction of multiple avian influenza strains. For H3N2 influenza, the model is validated against observed viral fixation rates and epidemic progression rates from the World Health Organization FluNet – Global Influenza Surveillance Network. We find that a multiple-component avian influenza vaccine is helpful to control a simultaneous multiple introduction of bird-flu strains. We introduce Population at Risk (PaR) to quantify the risk of a flu pandemic, and calculate by this metric the improvement that a multiple vaccine offers.

Introduction

Circulating influenza virus uses the ability to change its surface proteins, along with its high transmission rate, to flummox the adaptive immune response of the host. The random accumulation of mutations in the haemagglutinin (HA) and neuraminidase (NA) epitopes, the regions on surface of the viral proteins that are recognized by host antibodies, poses a formidable challenge to the design of an effective annual flu vaccine. Under current practice, the World Health Organization (WHO) and government health agencies rely on historical experience and phylogenetic analysis of HA and NA protein sequences from the circulating human strains to decide upon the components in the annual influenza vaccine (Munoz *et al.*, 2005). Every year, the authorities concerned make a projection about the circulating influenza strains for the coming flu season. The flu vaccine currently contains three strains that are as similar as possible to those strains that are predicted to be the most prominent. At present, the three strains in the vaccine are one H3N2 A, one H1N1 A, and one influenza B component. Historically, the vaccine efficacy has seldom reached the 100% mark. Over the years, it has hovered between 30% and 60% against influenza-like illnesses. In fact, owing to a phenomenon known as the 'original antigenic sin' (Davenport *et al.*, 1953; Fazekas *et al.*, 1966; Deem *et al.*, 2003), the vaccine efficacy has even been negative at times. If original antigenic sin is operative, the host antibodies that are produced in response to a viral strain tend to suppress creation of new and different antibodies in response to a different viral strain. Whether such a phenomenon takes place or whether the vaccine is effective depends to a great extent on how similar the vaccine component strains are to the circulating viral strains. The methods employed to calculate such an antigenic distance draw heavily from the ferret antisera haemagglutinin inhibition assays. It has been assumed that the antigenic distance thus obtained from ferrets correlates well with the efficacy of the influenza vaccine in humans. However, to our knowledge there is limited evidence in the literature of such correlations.

We introduce the reader to a new and effective way of measuring the antigenic distance between different strains. We define a quantity, $p_{epitope}$, that measures the difference between the dominant epitope regions of the haemagglutinin proteins of any two H3N2 influenza viruses in general. The term $p_{epitope}$ can be used to measure difference between the dominant epitope of the H3N2 vaccine component and the dominant circulating viral strain. A dominant epitope is one that elicits the most significant response from the adaptive immune system for a particular strain in a particular year (Fitch *et al.*, 1991, 2000; Bush *et al.*, 1999; Plotkin *et al.*, 2003). We show that for data spanning last 35 years, the $p_{epitope}$ measure of antigenic distance correlates better with the efficacy studies in humans of the influenza vaccine than do the current measures of antigenic distance, even those derived from ferret animal model studies. The $p_{epitope}$ measurement can also be used to measure difference between two circulating viral strains. The $p_{epitope}$ measurement can give us an idea of antigenic variability of circulating influenza strains.

Methods

We have developed a statistical mechanics-based theory to model the response of an immune system, free of immunosenescence, to disease and vaccination. We compare this theory to experimental studies of vaccine efficacies for 18- to 64-year- old subjects over the past 35 years, when the H3N2 subtype of influenza A was the dominant strain. H3N2 is the most common stain and has caused significant morbidity and mortality (Macken *et al.*, 2001). As is the norm, we focused on the five epitopes of the haemagglutinin protein. We used the generalized NK model to calculate the affinity constants quantifying the immune response following exposure to an antigen after vaccination. The model takes into account three types of interactions within an antibody. These include interactions within subdomains, interactions between different subdomains, and interaction between an antibody and an antigen (Jun *et al.*, 2005). The binding constant is given as $K = \exp(a - b\langle U \rangle)$, where U is the energy function of an antibody (Deem *et al.* 2003), $a = -18.56$, and $b = 1.67$. The values of the constants are determined after comparing the dynamics of the model with experimental results (Deem *et al.*, 2003). In order to capture the antigenic distance between the vaccine strain and the circulating strain, we use $p_{epitope}$ as an order parameter in our model. It represents the fraction of amino acids that differ between the dominant epitope regions of the two strains.

$$p_{epitope} = \frac{\text{number of amino acid differences in the dominant epitope}}{\text{total number of amino acids in the dominant epitope}} \qquad 13.1$$

In order to build the model we needed the identity and sequence of the dominant epitope in both the candidate vaccine and the circulating strain. This definition of the five epitopes in the H3N2 haemagglutinin protein was taken from reference (Macken *et al.*, 2001). Our model assumes the following correlation between the vaccine efficacy, E, and the binding constants. $E = \alpha \ln(K_{secondary}\ (p_{epitope})/K_{primary})$, where the constant α is selected such that a perfect match between the vaccine and the circulating strain corresponds to the average historical vaccine efficacy of 45% (Gupta *et al.*, 2006). $K_{primary}$ is the binding constant corresponding to the primary immune response, and $K_{secondary}$ is the binding constant corresponding to the secondary immune response that follows vaccination. Other than the constant, α, no parameter was fit to data, making the model predictive. For example, the point where vaccine efficacy becomes zero is independent of the value of α.

Results

Table 10.1 compares vaccine efficacy values from experimental studies (Smith *et al.*, 1979; Clements *et al.*, 1986; Keitel *et al.*, 1988, 1997; Edwards *et al.*, 1994; Nichol *et al.*, 1995; Campbell *et al.*, 1997; Grotto *et al.*, 1998; Bridges *et al.*, 2000; Mixeu *et al.*, 2002; Millot *et al.*, 2002; Kawai *et al.*,

2003; Lester *et al.*, 2003; Dolan *et al.*, 2004) and predictions from our theory as a function of the $p_{epitope}$. In literature, vaccine efficacy is defined as

$$\text{Efficiency} = \frac{u - v}{u} \qquad 13.2$$

where u and v are the influenza-like illness rates in unvaccinated and vaccinated individuals, respectively. The epidemiological estimates of u and v contain some noise; however, these are the best estimates there are of influenza vaccine efficacy in humans. Our model demonstrates the effectiveness of using $p_{epitope}$ as a measure of antigenic drift between the circulating strain and the vaccine strain. Crystallographic data and immunoassays have shown that only the epitope regions of the viral proteins are significantly involved in the recognition of the antigen via the antibodies (Air *et al.*, 1985). Our definition of $p_{epitope}$ follows from this observation. When $p_{epitope}$ value in the dominant epitope is greater than 0.19 according to experimental records or is greater than 0.22 according to our theory, the efficacy of the influenza vaccine drops to the negative territory (Table 10.1 and Fig. 10.1A) indicating that while designing the vaccine, this regime needs to be avoided. As an example, during the 1997/98 influenza season in the Northern hemisphere, when the Sydney/5/97 strain was widespread, the value of $p_{epitope}$ was 0.238, resulting in an efficacy of −17% (Bridges *et al.*, 2000). Only one data point falls outside our theoretical predications, that for the 1989/90 epidemic (Nguyen-Van-Tran *et al.*, 2003). During that year there were probably multiple strains circulating, including the strains of influenza B (Edwards *et al.*, 1994; Ikonen *et al.*, 1996).

The WHO uses as a first approximation an alternative definition of the antigenic drift. It considers the sequence difference of the entire haemagglutinin protein of the circulating and viral strains.

$$p_{sequence} = \frac{\text{number of amino acid differences in the sequence}}{\text{total number of amino acids in the sequence}} \qquad 13.3$$

The correlation with the experimentally observed efficacy is not as positive when using the $p_{sequence}$ definition for the antigenic distance. This conclusion is seen from Table 10.1 and Fig. 10.1B, and it follows from the fact that many amino acids in the protein are either inaccessible to the human antibodies or not recognizable with high probability by them. Immune recognition of the epitope regions results in the vaccine efficacy being more correlated with the $p_{epitope}$ measure of antigenic distance in equation 10.1 than with the approximate measure of distance in equation 10.3.

The gold standard measure of antigenic drift is that derived from ferret antisera (Smith *et al.*, 1999; Lee *et al.*, 2004). Comparison of the vaccine efficacy to this measure of antigenic drift also shows less than ideal success. From Table 10.1 and Fig. 10.1C, the haemagglutinin assays in ferret antisera are unable to capture a significant amount of human efficacy information. For example, an antigenic distance of zero according to ferret antisera experiments does not mean that the two strains are identical. As an example, for the 1996/97 season, the antigenic distance derived from ferret antisera between the vaccine strain of A/Nanchang/933/95 and the circulating strain of A/Wuhan/359/95 was zero, whereas the $p_{epitope}$ value was 0.095. The corresponding vaccine efficacy in the Northern and Southern hemispheres was 28% (Millot *et al.*, 2002) and 11% (Mixeu *et al.*, 2002) respectively. When compared with the average efficacy corresponding to a perfect match between the vaccine strain and the circulating strain, which is 45%, these efficacy values are clearly much lower. Thus, ferret-derived distances will occasionally misidentify antigenically distant strains as antigenically identical.

Discussion

Design of the influenza vaccine is always a race against time. Currently, under the supervision of the WHO and the national health agencies, in the Northern hemisphere, the components of the annual influenza vaccine are determined between February and April. The mass production of the vaccine is then carried out by growing the virus in hens' eggs. After regulatory tests in mid-July, the vaccine is distributed in September. Data are collected to determine the vaccine efficacy starting in October and continuing through the winter flu season. By January a good measure of the effectiveness of the season's vaccine is obtained.

Table 10.1 Summary of results for $p_{epitope}$ analysis (Gupta *et al.*, 2006); individual references also provided

Year	Vaccine strain	Circulating strain	Vaccine efficacy (%)
1971–1972	Aichi/2/68(V01085)	HongKong/1/68(AF201874)	7 (Smith *et al.*, 1979)
1972–1973	Aichi/2/68(V01085)	England/42/72(AF201875)	15 (Smith *et al.*, 1979)
1973–1974	England/42/72(ISDNENG72)	PortChalmers/1/73(AF092062)	11 (Smith *et al.*, 1979)
1975–1976	PortChalmers/1/73(AF092062)	Victoria/3/75(ISDNVIC75)	–3 (Smith *et al.*, 1979)
1984–1985	Philippines/2/82(AF233691)	Mississippi/1/85(AF008893)	–6 (Keitel *et al.*, 1988)
1985–1986	Philippines/2/82(AF233691)	Mississippi/1/85(AF008893)	–2 (Keitel et. al, 1997; Demicheli *et al.*, 2004)
1987–1988	Leningrad/360/86(AF008903)	Shanghai/11/87(AF008886)	17 (Edwards *et al.*, 1994; Keitel *et al.*, 1997)
1989–1990	Shanghai/11/87(AF008886)	England/427/88(AF204238)	–5 (Edwards *et al.*, 1994)
1992–1993	Beining/32/92(Af008812)	Beining/32/92(Af008812)	59 (Campbell *et al.*, 1997)
1993–1994	Beining/32/92(Af008812)	Beining/32/92(Af008812)	38 (Demicheli *et al.*, 2004)
1994–1995	Shangdong/9/93(Z46417)	Johannesburg/33/94(AF008774)	25 (Nichol *et al.*, 1995)
1995–1996	Johannesburg/33/94(AF008774)	Johannesburg/33/94(AF008774)	45 (Grotto *et al.*, 1998)
1996–1997	Nanchang/933/95(AF008725)	Wuhan/359/95(AF008722)	28 (Millot *et al.*, 2002)
1997	Nanchang/933/95(AF008725)	Wuhan/359/95(AF008722)	11 (Mixeu *et al.*, 2002)
1997–1998	Nanchang/933/95(AF008725)	Sydney/5/97(AJ311466)	–17 (Bridges *et al.*, 2000)
1998–1999	Sydney/5/97(AJ311466)	Sydney/5/97(AJ311466)	34 (Bridges *et al.*, 2000)
1999–2000	Sydney/5/97(AJ311466)	Sydney/5/97(AJ311466)	43 (Lester *et al.*, 2003)
2001–2003	Panama/2007/99(ISDNCDA001)	Panama/2007/99(ISDNCDA001)	55 (Kawai et. al, 2003)
2003–2004	Panama/2007/99(ISDNCDA001)	Fujian/411/2003(ISDN38157)	12 (Dolan *et al.*, 2004)

The definitions of $p_{epitope}$ and $p_{sequence}$ have been given previously (fractional change in the immunodominant epitope and fractional change in the whole sequence respectively). The terms d_1 and d_2 are the two available measures of distance based on ferret antiserum. Figures below provide the analysis of the data presented in the table. References used are shown. When more than one antisera assay has been performed, the

Dominant epitope	$p_{epitope}$	$p_{sequence}$	d_1	d_2	N_u	N_v
A	0.158	0.033			25,202	26,317
B	0.190	0.055			26,130	26,778
B	0.143	0.018	5 (Smith *et al.*, 1999)	4 (Smith *et al.*, 1999)	26,536	28,158
B	0.190	0.055	4 (Kendal *et al.*, 1983)	16 (Kendal *et al.*, 1983)	25,591	29,247
B	0.190	0.033	2 (WHO, 1988)	2 (WHO, 1988)	241	171
B	0.190	0.033	2 (WHO, 1988)	2 (WHO, 1988)	25,391	15,388
B	0.143	0.024	2 (WHO, 1988)	1 (WHO, 1988)	1,451,064	1,211,060
A	0.105	0.021			1016	1016
	0.0	0.0	0 (Ellis *et al.*, 1995)	0 (Ellis *et al.*, 1995)	131	131
	0.0	0.0	0 (Ellis *et al.*, 1995)	0 (Ellis *et al.*, 1995)	12	26
A	0.108	0.021			424	422
	0.0	0.0	0 (CDC, 1997; Corias *et al.*, 2001)	0 (CDC, 1997; Corias *et al.*, 2001)	652	684
B	0.095	0.006	0 (CDC, 1997; Corias *et al.*, 2001)	0 (CDC, 1997; Corias *et al.*, 2001)	2978	273
B	0.095	0.006	0 (CDC, 1997; Corias *et al.*, 2001)	0 (CDC, 1997; Corias *et al.*, 2001)	299	294
B	0.238	0.040	4.5 (Corias *et al.*, 2001; Pontoriero *et al.*, 2001)	27.3 (Corias *et al.*, 2001; Pontoriero *et al.*, 2001)	554	576
	0.0	0.0	0 (Corias *et al.*, 2001; Cox wt. al., 2003)	0 (Corias *et al.*, 2001; Cox *et al.*, 2003)	596	582
	0.0	0.0	0 (Corias *et al.*, 2001; Cox wt. al., 2003)	0 (Corias *et al.*, 2001; Cox wt. al., 2003)	324	342
	0.0	0.0	0 (Corias *et al.*, 2001; Cox wt. al., 2003)	0 (Corias *et al.*, 2001; Cox *et al.*, 2003)	982	968
B	0.143	0.040	2 (Cox *et al.*, 2003)	8 (Cox *et al.*, 2003)	402	1000

calculated distances are averaged. Error bars are calculated assuming binomial statistics for each data set: $\varepsilon^2 = (\sigma_v^2/u^2/N_v + (v/u^2)^2\ \sigma_u^2/N_u)$, where $\sigma_v^2 = v(1-v)$ and $\sigma_u^2 = u(1-u)$. If two sets of data are averaged in 1 year, then $\varepsilon^2 = \varepsilon_1^2/4 + \varepsilon_2^2/4$.

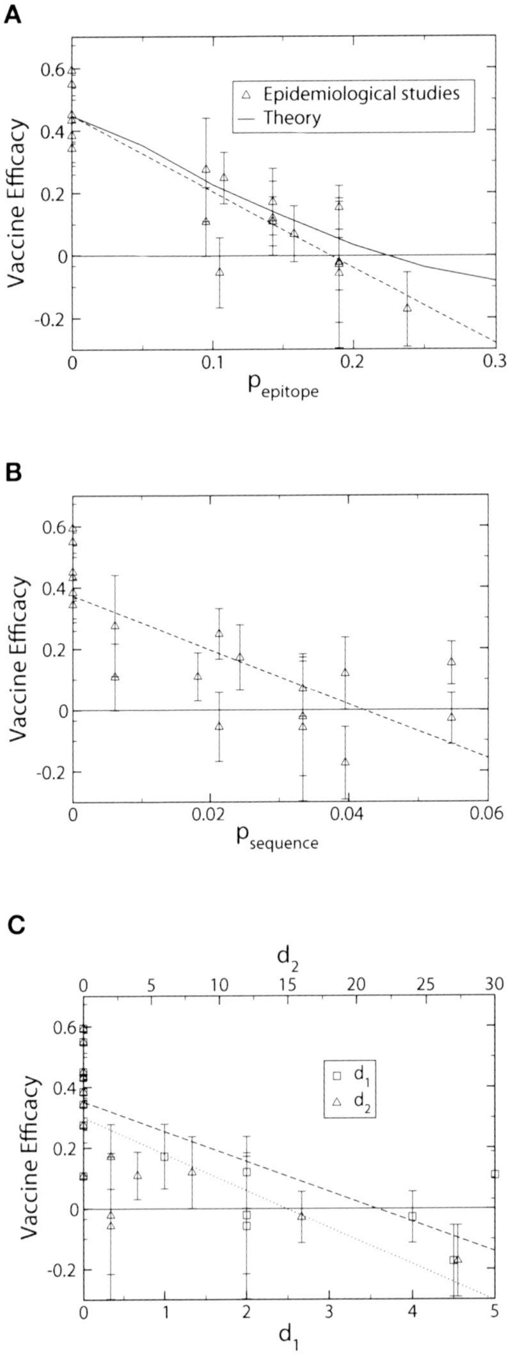

Choice of the vaccine strain is contingent upon various biological and manufacturing constraints. For example, among the egg-cultured strains, the availability of high growth strains is an additional criterion. The organizations in charge have to weigh in all these constraints before they decide on the best match for the anticipated circulating strain for the following flu season.

The p_{epitope} measure of antigenic distance can influence and contribute to the vaccine design process in two ways. First, p_{epitope} can help to identify the strains to be included in the annual flu vaccine. For every season, given a list of strains and their probabilities of outbreak, the value of p_{epitope} can help define the weighted distance of a particular vaccine strain from the circulating strains. This procedure will enable us to identify the vaccine strain closest to the strains causing a possible outbreak. Alternatively, p_{epitope} can also help identify the 'like' strains and to quantify 'likeness.' Various manufacturing constraints in the vaccine production process mean it is often not feasible to grow large quantities of the exact strain that is desired for the annual vaccine. Under such a scenario, it is required to choose several strains that are similar to the chosen one. The value of p_{epitope} can help quantify the 'likenesses' of a given strain to the desired strain. We can extend p_{epitope} measure of antigenic distance to other strains of influenza as well. Using information about the epitope regions of other H*x*N*y* strains of influenza A or influenza B, we can calculate the value of p_{epitope} to quantify the antigenic distance for these strains of influenza.

It is clear that the immune system response is neither linear nor monotonic in the antigenic distance. As a result, the original antigenic sin leading to a negative vaccine efficacy exists only in an intermediate regime of the antigenic distance. If the vaccine falls in this regime, however, the vaccinated individual appears to be more

Figure 10.1 Vaccine efficacy for influenza-like illness as observed in epidemiological studies as a function of three different measures of antigenic distance (Gupta *et al.*, 2006). A linear least squares fit to the data is also shown. Error bars are one standard error ε, calculated as described in Table 10.1. (A) p_{epitope}: (long dashed, R^2= 0.81). (B) p_{sequence}: (long dashed, R^2= 0.59). Figure shows the same epidemiological data as in (A). Only the definition of the x-axis is different. (C) d_1 (long dashed, $R^2=0.57$) and d^2 (short dashed, $R^2=0.43$), derived from ferret antisera experiments. Results were averaged when multiple haemagglutinin inhibition (HI) studies had been performed for a given year. These HI binding arrays measure the ability of ferret antisera to block the agglutination of red blood cells by influenza viruses. The epidemiological data are the same as in (A). Only the definition of the x-axis is different.

susceptible to influenza-like illnesses compared to an unvaccinated individual. Over the past 33 years, this phenomenon seems to have taken place 26% of the time according to the H3N2 epidemiological data as indicated by 5 of the total 19 data points that are negative in Table 10.1 and Fig. 10.1B. The negative points are not necessarily a result of experimental errors; rather they may have their roots in the original antigenic sin phenomenon where the immune system is unable to distinguish between the new and the old viral strains. It is desired that the regime corresponding to the original antigenic sin is avoided, not only for the obvious immunological consequences, but also for the negative impact it creates against the acceptance of public health policy. Our theory rightfully captures the underlying physics of immune response, and corroborates the experimental findings. Our theory and the p_{epitope} measure of antigenic distance can be used to ensure that vaccines chosen would not fall in the original antigenic sin region against expected strains. Saying this more conservatively, our theory can be used to ensure that the vaccine strains chosen are antigenically close enough to the expected strains so that the vaccine efficacy is expected to be positive.

It appears that p_{epitope} is a measure of the antigenic distance between viral strains that correlates more strongly with vaccine efficacy than does the sequence analysis or the ferret antisera. The population at large may benefit if the health authorities incorporate p_{epitope} in the prediction and design of the influenza vaccine in addition to (or even instead of) the current practices. We now present an example to show how our theory can assist shaping public health policies. We examine the 2004/2005 flu season in the Northern hemisphere. With the aid of p_{epitope} we can stipulate a priori the extent of protection that a particular vaccine strain provides against the circulating strain for a particular season. The comparison among various candidates would provide us selection criteria for their inclusion in the annual flu vaccine. During the 2003/2004 flu epidemic, A/Fujian/411/2002 strain was predominant, and it was expected that this strain would dominant in 2004/2005 as well. In order to counter this strain, the Advisory Council of the FDA recommended using A/Wyoming/2003 as the H3N2 component of the 2004/2005 vaccine, since according to the prevalent measures, this strain and A/Fujian/411/2002 were found to be 'antigenically equivalent' (Harper *et al.*, 2004). Our calculations, however, yielded a p_{epitope} value of 0.095 for the pair in question, suggesting that the two strains were not antigenically equivalent, and the predicted efficacy is 20% (Fig. 10.1A). Also, in 2004/2005, A/California/7/2004 strain was in circulation along with significant amount of influenza B. The p_{epitope} value for the pair of A/California/7/2004 and the A/Wyoming/2003 strain was 0.286. According to our theory the vaccine would not have provided a positive protection from the A/California/7/2004 strain. The vaccine efficacy that year was 9.2% (Chan *et al.*, 2008), which is roughly the average of the efficacies (0% and 20%) predicted by p_{epitope} theory. We note that another of the WHO approved candidates for that year was A/Kumamoto/102/02 (ISDN38180) (WHO, 2004), and we found p_{epitope} value to be zero versus A/Fujian/411/2002. According to our theory, a vaccine containing the Kumamoto strain would have provided a better protection against the Fujian strain than did the Wyoming vaccine strain.

Suggestions to improve predictability of vaccine efficacy

At the heart of our approach lies the identification of the dominant haemagglutinin epitope recognized in humans. The identity of the dominant epitope for humans is currently not measured. We, thus, define and postulate the dominant epitope for humans for a certain strain and for a certain season as the epitope that undergoes the largest fractional change in the sequence of amino acids in comparison to the vaccine strain. A measurement of the epitope that is dominant in humans for various circulating strains and vaccines should help to improve the predictive power of our approach even more, as we can then use the measured dominant epitope to calculate p_{epitope} instead of using our postulated dominant epitope. It is important to continue measurement and sequence analysis of the prominent circulating strains in the flu season. These data will help to validate and perhaps better calibrate the p_{epitope} measure of antigenic distance. Another important piece of the puzzle is epidemiological study to relate vaccine efficacy to the antigenic drift. We believe that use of the p_{epitope} measure of antigenic distance will enable the health authorities to predict the severity of the yearly flu season, to design better vaccines for the same, and help to manage the health resources for this period.

In more general terms, the results presented here have implications for the fight against diseases that stem from rapidly mutating viruses and that are managed through antibody responses. The p_{epitope} measure of antigenic distance can predict efficacies for vaccine strains against multiple circulating strains, and it could then be used towards redesigning of the vaccine in terms of frequency and composition. One such disease about which there is concern for pandemic potential is the avian influenza, or bird flu disease. In the next section, we present a multiple-strain transmission model for avian influenza that would enable one to manage the risk from such an outbreak. We use p_{epitope} to evaluate the effectiveness of a multiple-component vaccine to counter such threat.

Vaccination strategies and risk management of flu pandemic

Since its first appearance (Hong Kong, 1997), H5N1 avian influenza is known to have spread to various parts of the world, including South-East Asian countries, parts of central and Middle East Asia, Africa, and Europe. This rapid spread has prompted the WHO to suggest that 'we are closer to a pandemic than at any time since 1968' (WHO report, 2005; Mills *et al.*, 2006). Bird flu has been observed in pigs (Cyranoski, 2005), and occurrences of person-to-person transmission have probably been detected (Ungchusak *et al.*, 2005; Normile, 2006; Yang *et al.*, 2006). Efficacy of the vaccine produced against the original Hong Kong strain of H5N1 has been poor against some of the new strains. In addition, the high mutation rates observed in the avian influenza have raised the prospect of simultaneous introduction of multiple strains (Mills *et al.*, 2006; Ducatez *et al.*, 2006; Chen *et al.*, 2006; Capua *et al.*, 2007), putting a question mark against the efficacy of a single-component vaccine. Although in the event of appearance of multiple strains, eventual emergence of a single dominant strain is likely; the lack of a priori knowledge of which strain will be dominant makes it worthwhile to look into the alternative of a multiple-component vaccine. Here, we use an epidemiological model to study the efficacy and cross-protection of a multiple-component bird flu vaccine. We further extend the concept of p_{epitope} through a stochastic model to manage the risk of a flu pandemic. The strategy is inspired by similar risk management studies in finance, and the population at risk for infection in an epidemic plays the role of the risk variable. Here p_{epitope} represents the difference between circulating strains.

Multiple-strain introduction transmission model

As mentioned above, there is evidence, both theoretical and otherwise, that deems introduction of multiple-strains of avian influenza is a distinct possibility. There have been no mathematical models, however, that explore such a scenario to evaluate efficacy of practicable vaccine strategies. Our model uses the concepts of hierarchical scale-free network, and virus transmission, and viral evolution to evaluate possible, and worst case, epidemic spreading scenarios and to evaluate efficacy of single or multiple-component vaccine strategies.

Hierarchical scale-free network

We divide the total human population (6.7×10^9) into (6.7×10^6) groups. Each group consists of 10^3 people that exhibit similar health status and social behaviour. These groups are distributed over N_{city} cities. The distribution of cities with i groups ($N_{\text{city}}(i)$) goes as $N_{\text{city}}(i) \propto i^{-2.1}$ (Zipf, 1949; Newman, 2005). The largest and the smallest city areas in our simulation have 6.2×10^4 and 400 groups respectively (Newman, 2005). The total number of cities fluctuates around 4000 (Guimera *et al.*, 2005).

The cities are connected through a network that extends globally and is defined by the network spawned by the airlines. The distribution of cities with i contacts is given by $N_{\text{city\_contact}}(i) \propto i^{-2.0}$ (Guimera *et al.*, 2005). The network connecting groups within a city is defined by the ground transportation network. In each city, the distribution of groups with i contacts is given by $N_{\text{group\_contact}}(i) \propto i^{-2.8}$ (Eubank *et al.*, 2004). Population size differences between cities mean that movement from one city to another leads to a final averaged distribution over cities that is different for different contact numbers. Any hierarchical structure present in the actual distributions of within- or between-city contacts, which is not detected by the scale-free analyses performed to date, would generally be expected to slow down the spread of an epidemic through the population.

The scale-free network is generated in two steps. The first step is to randomly assign a degree to each node based on the given power law distribution. This is achieved by the classical transformation method (Press *et al.*, 2002). The second step is to connect nodes. This is achieved by connecting nodes with a probability that is proportional to their degree. This scale-free network generation method has been used in other scenarios (Albert *et al.*, 2002).

Virus transmission and evolution

In our simulation, the viruses are introduced in groups that are chosen randomly. The virus has a latency period of 2 days, which is followed by the infectious period. During the infectious period, the virus can either be killed by the host immune system, or it can be transmitted between different groups globally. During both these periods, the virus can mutate as well as be killed. The term $R_{\text{mutation}} = 1.6 \times 10^{-5}$/amino acid/day (Sato *et al.*, 2006) gives the mutation rate, whereas, the virus killing probability is given by $\text{Kill\_prob} = \text{cKill} + \text{cAlpha} \times \max((\text{cEpiMax} - \text{epi}),\ 0)$. Here cKill is the intrinsic probability of killing virus, and its baseline value is fitted in our model to be 0.36. A value of 1.0 proved to be optimal for cAlpha. cEpiMax is the largest possible epitope-based distance between the vaccine strain and the current viral strain before the vaccine efficacy goes to zero, and its value is 0.19 (Gupta *et al.*, 2006). The term epi is a function of the host immune history, and circulating virus strain. It is defined as the smallest distance among the epitope-based distances between the current viral strain and the viral strains (50) in the immune history of the infected group. The transmission rate takes into account the seasonal effect (Ferguson *et al.*, 2003), and is different for transmission within a city ($\text{Ffac1} \times \tau_0 \times (1 + 0.25 \times \text{Sin}(2\pi T/360))$, $\text{Ffac1} = 1.0$) and transmission between cities ($\text{Ffac} \times \tau_0 \times (1 + 0.25)$, $\text{Ffac} = 7.0$). FCap is the number of different groups that people travelling on a flight come from, and it is set to 100 for our model. The term τ_0 is the intrinsic transmission ability of the virus, and equals 0.07 for the virus introduced on day 1. The mutants, once created, can transmit according to probability cProb. For a baseline parameter, cProb equals 0.26.

Vaccination strategies

As an example, for a very rapid public health policy response, we consider that after 40 days, individuals are vaccinated. We consider single-component as well as multiple-component vaccines. In single-component case, the most dominant strain at day 10 is taken as the vaccine strain, whereas, for multiple-component case, the top 10 strains that are most dominant on day 10 are incorporated into the vaccine. Day 10 is taken as the data collection day by public health authorities. We calculated the efficacy and the cumulative attack rates for both cases. We calculate the effect of vaccine on a population as a whole. Thus we compare the rate of infection, u, in a population that is vaccinated, at some point at which some percentage of the population has been vaccinated with the rate of infection, v, in an unvaccinated population. Then we use Eqn. 10.2 to calculate population-level efficacy. Thus, efficacy represents the percentile reduction in disease occurrences in people who were vaccinated compared to those who were not (Nichol *et al.*, 1994; Chow, 2003). Note that this is a non-standard definition of efficacy. Efficacy in Fig. 10.1 is on a per-person basis rather than a per-population basis as in Fig. 10.4.

The model's ability to predict an influenza pandemic was tested against the existing epidemiological data (Fig. 10.2A). The model successfully predicted the average trend of H3N2 isolates data in the WHO FluNet database from 1995 to 2006 (Fig. 10.2B). We also successfully predicted the fixation rate for dominant and non-dominant epitopes of influenza (Fig. 10.3) versus those measured (Ferguson *et al.*, 2003).

WHO FluNet data analysis

We acquired the WHO FluNet data for the period of 1995–2006 (WHO FluNet, 2006). We separately averaged the isolates data for countries

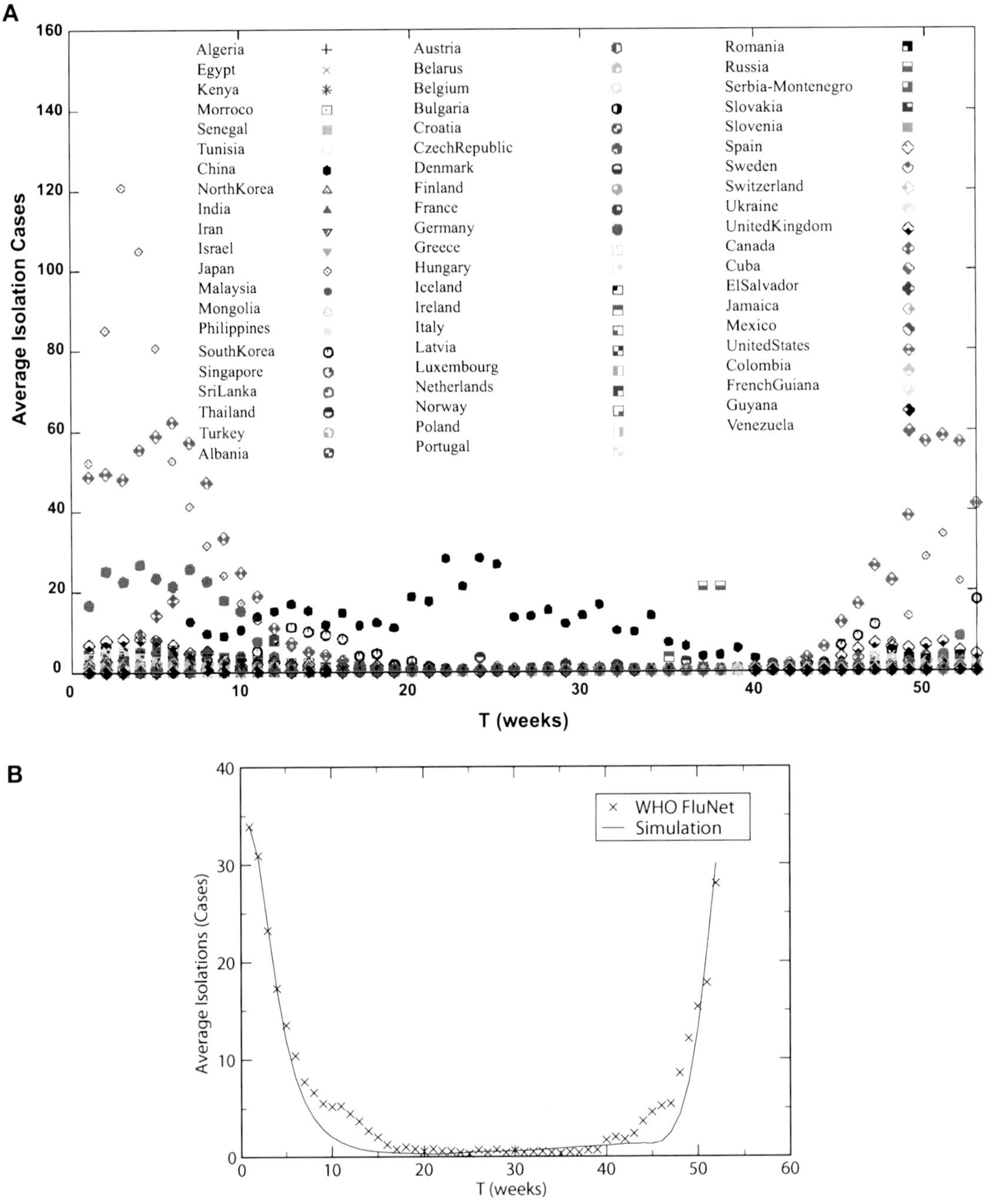

Figure 10.2 Average isolates cases of human H3N2 influenza. (A) Cases reported during last 10 years for various countries around the world. (B) Comparison between the WHO FluNet database and as predicted by our immunological and epidemiological model.

from the Northern and the Southern hemispheres over the 1997–2006 seasons (Fig. 10.2A). For this period, the annual flu epidemic seems to have several peaks. It is interesting to note the different epidemic dynamics in China, with an incidence peak in the summer, which is very different from other Northern hemisphere countries. The highest peak occurs in summer. China probably acts as a reservoir for the flu virus and also as a resource for transmission of the virus from pigs to humans in the summer. Thus, the summer peak in China is likely to be related to the role played

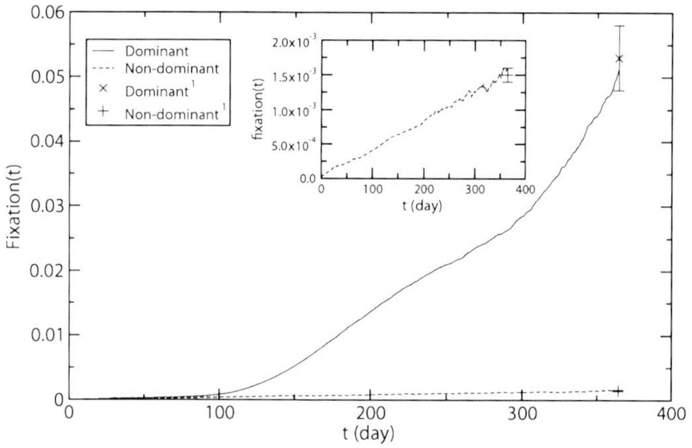

Figure 10.3 Influenza fixation rate for dominant and non-dominant epitopes. Dominant: fixation rate for dominant epitope from simulation; non-dominant: fixation rate for non-dominant epitopes from simulation; Dominant[1]: fixation rate and error bar for dominant epitopes from sequence analysis (Ferguson *et al.*, 2003); and non-dominant[1]: fixation rate and error bar for non-dominant epitope from sequence analysis (Ferguson *et al.*, 2003). Here $p_{\text{epitope}} = 0$, and 40% of the population is vaccinated on day 40.

by this virus reservoir. For each country for each year, we aligned the week with the highest peak, $\text{week}_{\text{peak}}$, to week_1 so that we could match the burst times for different years. This results in all data from a week preceding the peak week being shifted to $(\text{week}_{52} - \text{week}_{\text{peak}})$, and all data from a week following the peak week being shifted to $(\text{week}_i - \text{week}_{\text{peak}})$. After rearranging the data in such manner, we compared the results for average isolates from our simulation with those from the WHO FluNet data.

Multiple-component vaccine

We evaluated the viability of using a multiple-component vaccine against simultaneous introduction of multiple viral strains. In our model, we considered the possible scenario where two initial viruses, differing by p_{epitope}, cause a pandemic. A p_{epitope} value of zero indicates that viruses are the same, whereas a p_{epitope} value of 1 indicates that the viruses are completely different. The model then predicted the efficacy of both single and multiple-component vaccines as a function of the p_{epitope} (Fig. 10.4A). The cumulative attack rates are also predicted (Fig. 10.4B). A multiple-component vaccine was found to exhibit greater efficacy for the case where multiple-strains are introduced. There always exists a lag between viral outbreaks and administration of pertinent vaccine. We considered the effect such a lag has on single and multiple-component vaccines. It was found that the lag affects single-component vaccine more than it affects multiple-component vaccine (Fig. 10.4C and D).

Population at risk

In order to manage the risk of possible introduction of multiple strains of avian influenza, we combined together our stochastic model and a bioinformatics study. Such risk management methods are in common use in financial institutions (Hull, 2005). In economics, Value at Risk (VaR) is used as a metric in risk management studies. The term stands for the maximum loss over a given period of time with a certain confidence level. We use an analogous term, Population at Risk (PaR) in our study. It defines the maximum percentage of the world's population that is at risk from viral infection in the event of an influenza pandemic over the period of 1 year with a confidence level. As an example, a PaR value of 0.1 at the 95% confidence level means that if there are 100 pandemics, the maximum percentage of the global population that would be infected by the circulating viral strains in 95 of these pandemics stands at 0.1. The method could be applied to the risk management of infectious diseases in general. In the case of avian influenza, we analysed all H5N1 strains in the National Centre for Biotechnology Information (NCBI)

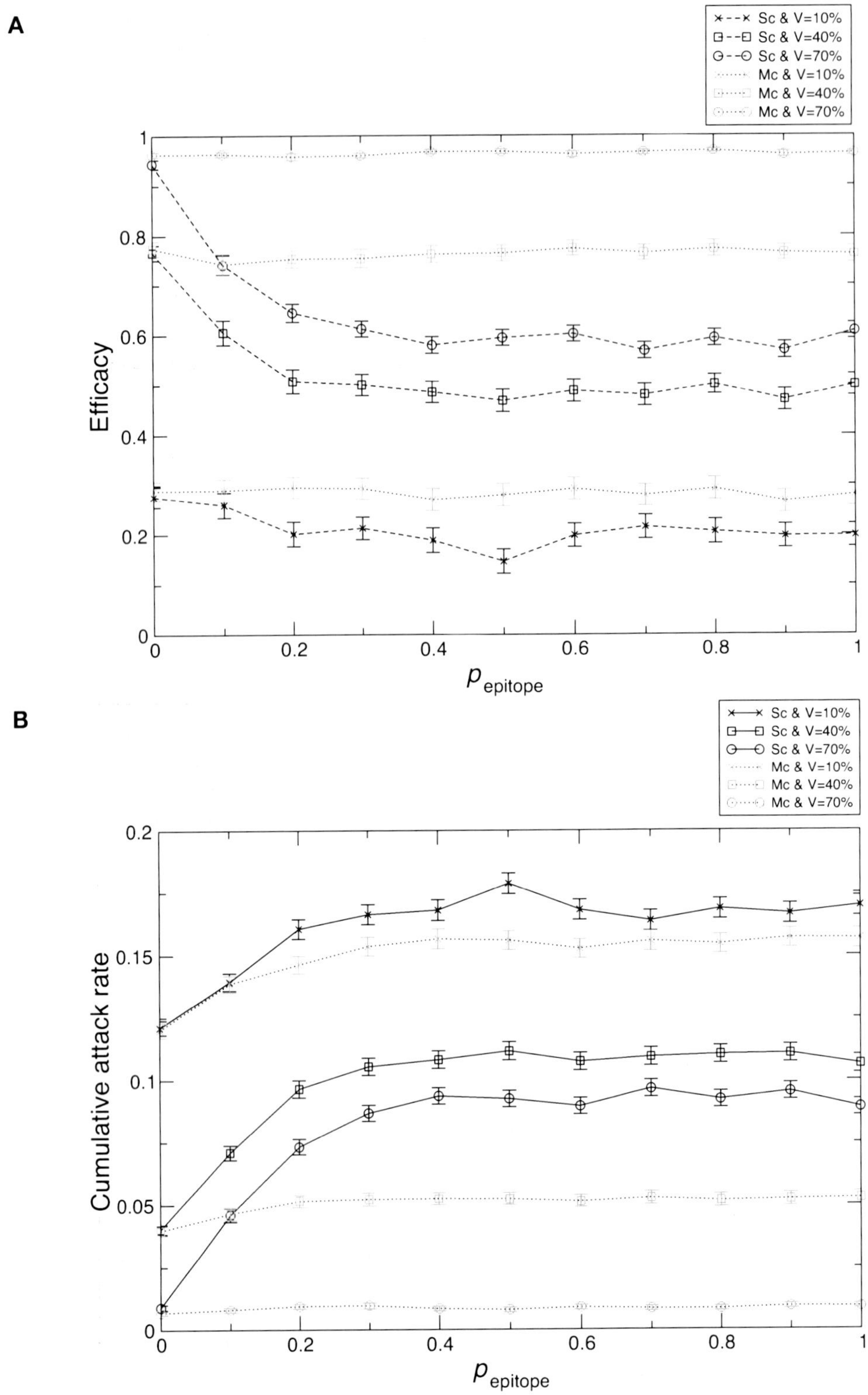

Figure 10.4 The effect of a vaccine for the initial introduction of two-strains. This indicates the multiple-component vaccine is more efficacious than single-component vaccine for different vaccination day and vaccination populations. Sc, single-component vaccine; Mc, multiple-component vaccine; *V*, vaccination population over total population; and Vac Day, the day when vaccine is administered. Vaccination population

C

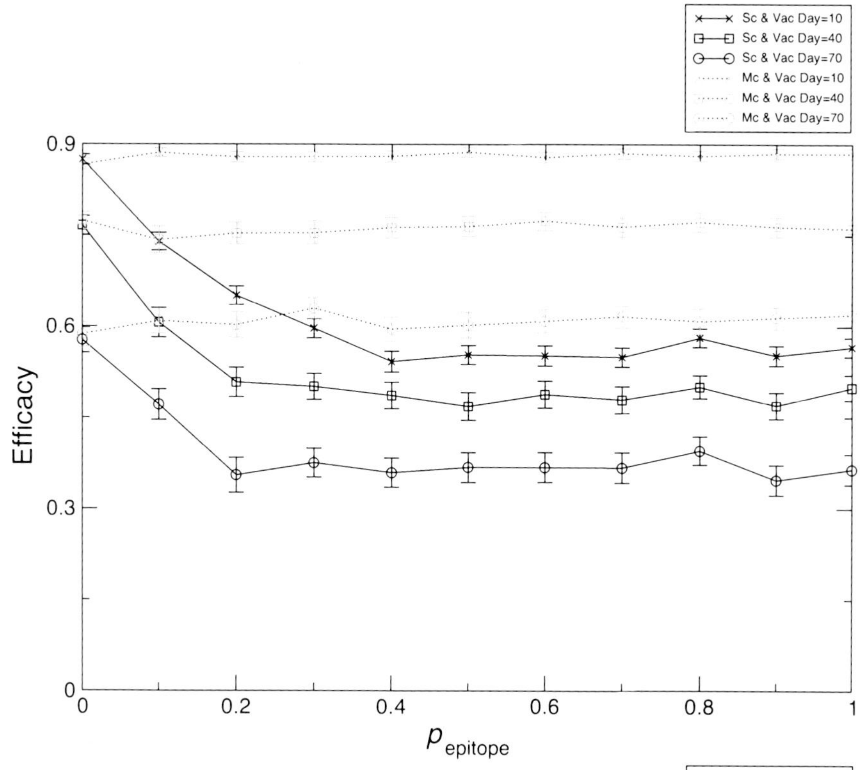

D

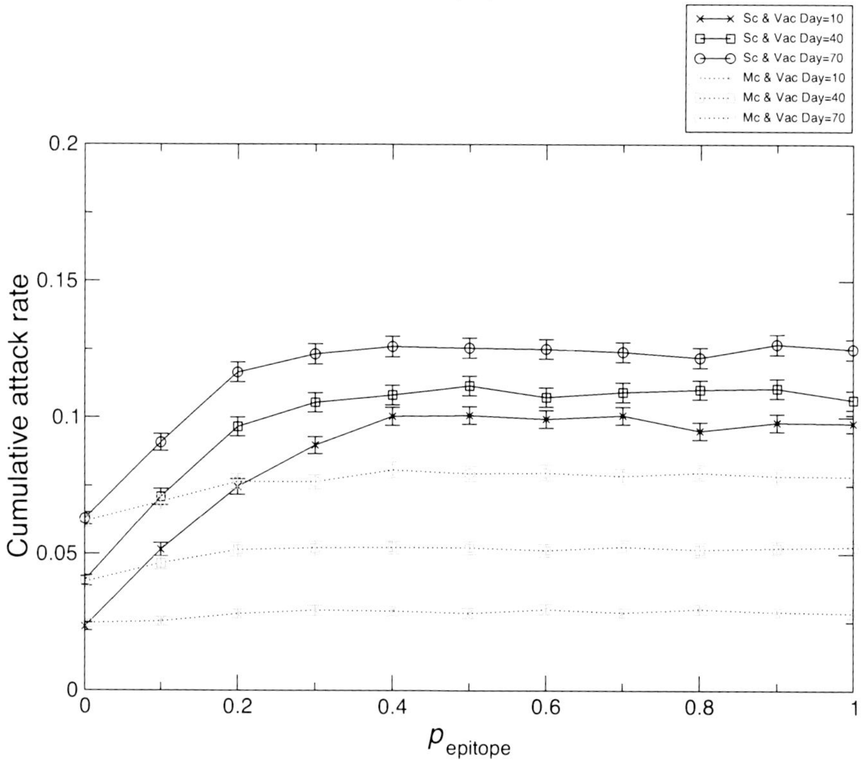

dependency at Vac Day=40. (A) Efficacy as a function of p_{epitope} for single and multiple component vaccines with different vaccination populations. (B) Cumulative attack rate as a function of p_{epitope}. Vaccination day dependence at V=40%. (C) Efficacy as a function of p_{epitope}. (D) Cumulative attack rate as a function of p_{epitope}.

database (NCBI database, 2007), and calculated the $p_{epitope}$ for all pairs of the viral strains. The average value turned out to be $p_{epitope}=0.118$. We generated two initial viruses using the $p_{epitope}$ value, and carried out simulations for 2000 yearly pandemics for three cases: with no vaccination, vaccination with single-component vaccine, and vaccination with multiple-component vaccine. The PaR was calculated for confidence interval between 0.9 and 0.99. As seen from Fig. 10.5, our model is able to predict the virus transmission (Fig. 10.5A) and basic production number (Fig. 10.5B). It also shows that the fewer the vaccine components, the more the number of people that get infected in these pandemics (Fig. 10.5C). This clearly shows that a multiple-component vaccine not only improves the efficacy of a vaccine, but also allows one to manage worst-case scenarios.

Summary

We have developed a pandemic model for avian influenza. Virus evolution, inter-personal transmission, vaccination, and immune history are taken into account. The model is able to reproduce the observed data for H3N2 isolates. Vaccination against multiple-components is not a foregone conclusion in case of simultaneous introduction of multiple strains. The parameter $p_{epitope}$, helps decide if use of multiple-component vaccine is likely to be beneficial. The $p_{epitope}$ measure of antigenic distance also helps decide what strains to incorporate in the vaccine under such scheme. The model shows that a vaccine with multiple-components outperforms a single-component vaccine when there are greater than one circulating strains that are antigenically different from each other. The model also shows that compared to the traditional single-component vaccine, a multiple-component vaccine is affected to a lesser degree by the lag between viral outbreak and vaccine administration. This result has important implications if there is a sudden attack of avian influenza pandemic, and the process of designing, producing, delivering, and administrating a vaccine is delayed significantly. The model proves that the spread of the virus is contingent upon number of emergent strains, fraction of the population vaccinated, and the time of administration of the vaccine following the outbreak. Lastly, the model provides a new parameter, PaR, which foretells how severe a potential epidemic could be. Combined with multiple-component vaccine, PaR would be valuable to quantify the benefit from actual vaccination strategies, beyond the broad notion of addressing multiple isolates. One can calculate virus differences through WHO FluNet to get an average $p_{epitope}$ value that could be used in our model. The model can yield PaR values for different areas of the world. The results of PaR calculations could alert policy makers and the general public to the current pandemic risk level, and along with $p_{epitope}$ calculations also help pharmaceutical companies to prepare optimal vaccine components. An improved predictive ability regarding the extent of impending epidemic and appropriate vaccine design would enhance our ability to better manage precious public health resources.

Appendix: sensitivity analysis

We carried out a sensitivity analysis of the model for all parameters for Fig. 10.5. A sensitivity analysis is important for model verification. It also indicates to which public health policy measures an influenza outbreak may be most sensitive. First, we did sensitivity analysis for the network structure. The variable parameters were the exponents in the power law relation of population distribution (N_{city}), group distribution ($N_{group\_contact}$) and flight distribution ($N_{city\_contact}$). For the population distribution, when the exponent was decreased by 5%, the cumulative attack rate was reduced by 41%, and when the exponent was increased by 5%, the cumulative attack rate was increased by 62%. For the group distribution, when the exponent was decreased by 4%, the cumulative attack rate was increased by 27%, and it decreased by 19% for an increase of the same magnitude in the variable. The epidemic curve remained practically unaffected for changes in both these variables. For the flight distribution, a 2.5% decrease in the exponent caused a jump of nearly 85%, and an increase of 2.5% in the same variable caused a decrease of 50% in the cumulative attack rate. In the epidemic trend, one can see the average isolations decrease almost 50% at week 40, for an increase of 2.5% in the exponent value.

We checked the model's sensitivity against the flight transportation parameters. A change of

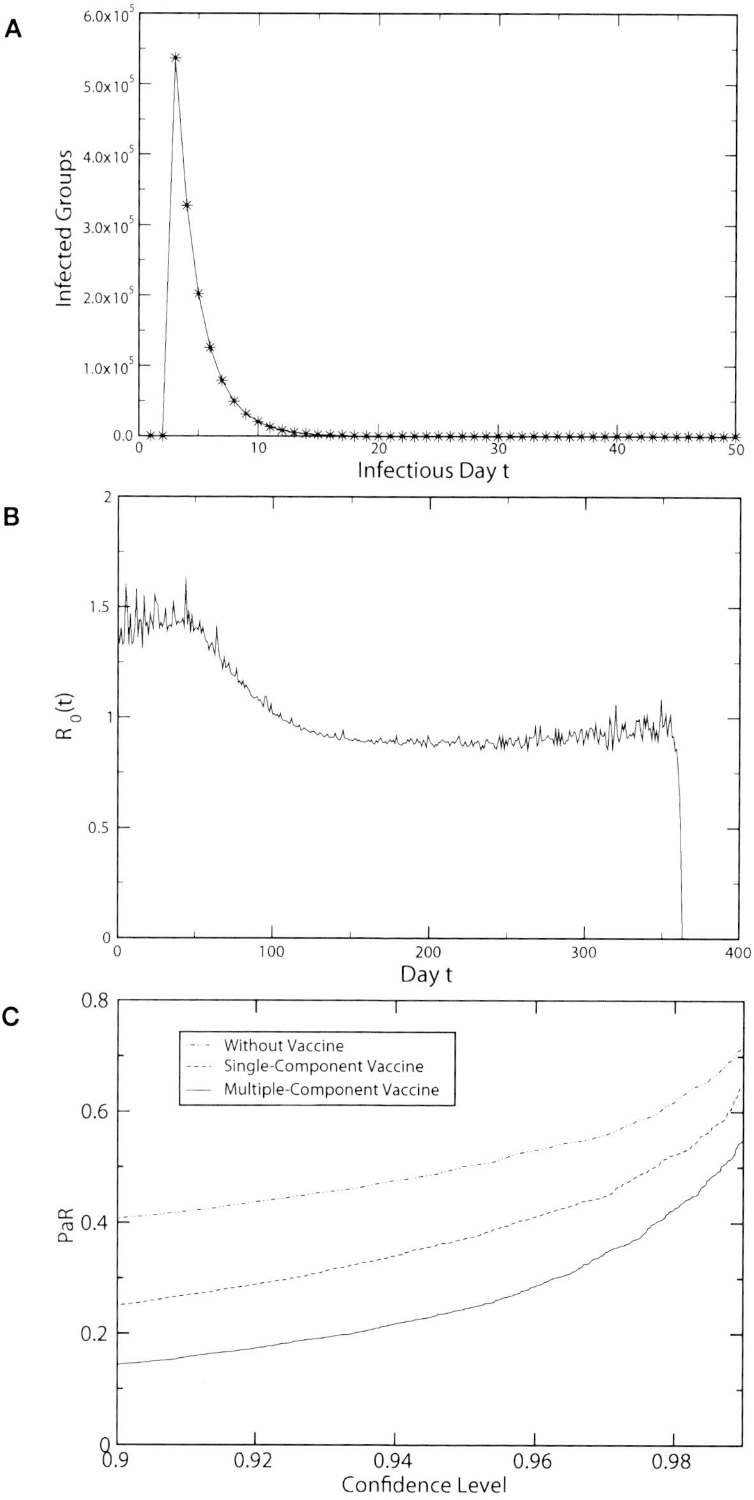

Figure 10.5 Virus transmission, basic production number and Population at Risk (PaR) from our immunological and epidemiological model. (A) Averaged infectious cases. (B) The basic production number (R_0). Here $p_{epitope} = 0.1182$ and 40% of the population is vaccinated at day 40. (C) Population at Risk (PaR) over a year for three cases: without vaccination, single-component vaccination, and multiple-component vaccination.

7% in Ffac in either direction resulted in a proportional change of 28% in the cumulative attack rate. The epidemic trend curve was practically unchanged, with only a small jump in average isolations around week 40 with increasing Ffac. A change of 8% in either direction for ffactor1 also resulted in proportional change of 7% for the cumulative attack rate. The epidemic curve remained unaltered for these changes. When FCap was decreased by 15%, the cumulative attack rate was decreased by 43%, and when FCap was increased by 15%, the rate jumped by 65%. The cumulative attack rate has a non-linear relationship with the FCap. The epidemic curve was affected by these changes in FCap. Around week 10, and week 40, the average cases increased up to two fold, when FCap increased from 100 to 115. Reducing air travel would seem to be effective for mitigating an influenza epidemic.

Sensitivity analysis was carried out for the initial virus transmission probability, τ_0, the immune killing, cKill, and for the immune history capacity. For a 7% change in τ_0, the cumulative attack rate changed by nearly 40%. The function increased linearly with the variable. The epidemic trend saw almost twofold jump at week 40 in the average cases. The cumulative attack rate increased nearly 39% when cKill decreased by 6%. It decreased by 32% for an increase of same magnitude in cKill. The epidemic trend showed nearly 50% reductions in average cases at week 40, for a 6% increase in cKill. The cumulative attack rate remained independent of the immune history capacity. A two fold increase in the parameter brought about 1% decrease in the attack rate. The epidemic trend remained unaltered. Measures such as face masks would seem to be effective.

Next, we checked the dependence of the cumulative attack rate on the mutation rate, R_{mutation}, and the transmission probability of the mutants, cProb. A twofold increase in R_{mutation} rendered the cumulative attack rate the same (<1% change). A twofold decrease in the variable yielded the same result. The epidemic trend saw nearly threefold increase in the average isolations around week 40 when R_{mutation} was doubled. The cProb parameter, when decreased by 50%, saw a decrease of 1% in the cumulative attack rate, but when the parameter was increased twofold, the cumulative attack rate jumped by nearly 10%. This indicated a non-linear relationship between the function and the variable. The epidemic trend showed nearly three times increase in the average cases at week 40 when the parameter was doubled.

We changed the initially infected group number. The sensitivity analysis showed that doubling the number of initially infected groups pushed the cumulative attack rate up by 22%, whereas halving the value of the parameter decreased the attack rate by nearly 13%. The epidemic trend curve remained unchanged for these variations.

The fixation rate depends upon the mutation rate, R_{mutation}, and the mutation transmission probability, cProb, the most. We carried out a sensitivity analysis against these two variables. For both the parameters for non-dominant epitopes, the fixation rate stayed relatively flat. However, for the dominant epitope, the fixation rate nearly doubled with a twofold increase in the variable.

We carried out sensitivity tests for a single-component vaccine at $p_{\text{epitope}}=0.0$ and $p_{\text{epitope}}=1.0$, and for multiple-component vaccine at $p_{\text{epitope}}=1.0$, both at day 40 with 40% of the population vaccinated.

The network structure parameter (exponent for population distribution) affects cumulative attack rate and efficacy. The cumulative attack rate increases while efficacy decreases with increasing value of the parameter. For the case of $p_{\text{epitope}}=1.0$, a change of 5% in the parameter results in 2 to 4 fold change in the cumulative attack rate for both single and multiple-component vaccines. A similar trend is seen for the case of $p_{\text{epitope}}=0.0$ for single-component vaccine. The efficacy drops by 6–7% for multiple-component vaccine at $p_{\text{epitope}}=1.0$ as well as for single-component vaccine at $p_{\text{epitope}}=0.0$, and it drops by around 10% for the single-component vaccine at $p_{\text{epitope}}=1.0$ for every 5% increase in the value of the network structure parameter.

We checked the sensitivity of the cumulative attack rate and efficacy against the exponent of flight distribution. Once again, in terms of absolute values, a multiple-component vaccine outperforms a single-component vaccine over the tested range of the parameter. The cumulative attack rate decreases with increasing value of the variable. For the case of $p_{\text{epitope}}=1.0$, a change of 4% in the parameter results in 40% to 50% change in the cumulative attack rate for both single and multiple-component vaccines. A similar trend is observed for the case of $p_{epitope} = 0.0$ for a single-component vaccine. For $p_{epitope} = 1.0$, and for the same change in the parameter, the efficacy of single-component vaccine increases around 4%, and that of multiple-component vaccine increases by about 5%, whereas, for $p_{epitope} = 0.0$, the efficacy for the single-component vaccine increases by 5%.

Sensitivity analysis of the flight transportation parameter gives the following results. In the case that $p_{epitope} = 1.0$, when we change Ffac by 7%, the cumulative attack rate changes by around 40% for the multiple-component vaccine, and it changes by 45% for the single-component vaccine. The efficacy drops by 6% for the multiple-component case, and it drops by nearly 10% for the single-component vaccine. When $p_{epitope} = 0.0$, the cumulative attack rate of the single-component vaccine changes by nearly 50%, and its efficacy changes by nearly 5% for a 7% change in Ffac. The attack rate increases, whereas, efficacy decreases with increase in the value of Ffac.

With FCap as the variable, the sensitivity analysis shows that for the case of $p_{epitope} = 1.0$, with a 15% increase, the cumulative attack rate more than doubles, and efficacy drops by 15% for the multiple-component vaccine, and the cumulative attack rate doubles, and efficacy drops by nearly 13% for the single-component vaccine. For the case of $p_{epitope} = 0.0$, there is a twofold increase, and a 15% decrease for cumulative attack rate and efficacy, respectively.

We analysed the sensitivity of our model to the virus transmission probability, τ_0. For the case of $p_{epitope} = 1.0$, and both for multiple as well as single-epitope vaccine, the cumulative attack rate changes nearly 50% for a 7% change in the value of τ_0. The efficacy varies 6% and 12% respectively for the two cases. For the case of $p_{epitope} = 0.0$, the single-component vaccine sees cumulative attack rate change 50% and efficacy change 6% for a 7% change in the value of τ_0. The attack rate increases, and efficacy decreases with increasing τ_0.

For the case of $p_{epitope} = 1.0$, a change of 6% in intrinsic killing probability, cKill, changes the cumulative attack rate by threefold, and 2.5-fold respectively, for multiple and single-component vaccines. Efficacy changes by 14% in multiple-component vaccine, and by 18% in single-component vaccine. When $p_{epitope} = 0.0$, the cumulative attack rate changes 3 fold, and efficacy changes 14% for single-component vaccine for a similar change in cKill. Once again, attack rate increases, and efficiency decreases with increase in cKill.

Sensitivity analysis was also applied to the mutation rate, $R_{mutation}$. With $p_{epitope} = 1.0$, cumulative attack rate changes by 15% in multiple-component, and by 10% in single-component vaccine. For multiple-component vaccine, a drop in $R_{mutation}$ by 50% increased the efficacy by 7%, but a twofold increase in $R_{mutation}$ decreased the efficacy by 13%. A similar trend was observed for a single-component vaccine. For the case of $p_{epitope} = 0.0$, a 50% decrease in $R_{mutation}$ raised the cumulative attack rate by nearly 25%, and reduced efficacy by nearly 3%. A twofold increase in $R_{mutation}$ on the other hand raised the attack rate by 25%, but reduced efficacy by nearly 16%.

Finally, sensitivity analysis of the mutant transmission probability, cProb, yielded the following results. When $p_{epitope} = 1.0$, a twofold increase in cProb brings 17% and 10% increase in the cumulative attack rate of multiple and single-component vaccines respectively. For the same change in cProb, efficacies drop by 7%, and 4% for multiple and single-component vaccines respectively. When $p_{epitope} = 0.0$, for a twofold change in cProb, the cumulative attack rate of the single-component vaccine shows a jump of around 25%. The efficacy drops by around 7% for the same change in cProb.

Acknowledgement

This research was supported in part by the FunBio program of DARPA.

References

Air, G.M., Els, M.C., Brown, L.E., Laver, W.G., and Webster, R.G. (1985). Location of antigenic sites on the three-dimensional structure of the influenza N2 virus neuraminidase. Virology. *145*, 237–248.

Both, G.W., Sleigh, M.J., Cox, N.J., and Kendal, A.P. (1983). Antigenic drift in influenza virus H3 hemagglutinin from 1968 to 1980: Multiple evolutionary pathways and sequential amino acid changes at key antigenic sites. J. Virol. *48*, 52–60.

Bush, R.M., Bender, C.A., Subbarao, K., Cox, N.J., and Fitch, W.M. (1999). Predicting the evolution of human influenza A. Science. *286*, 1921–1925.

Bridges, C.B., Thompson, W.W., Meltzer, M.I., Reeve, G.R., Talamonti, W.J., and Cox, N.J. *et al.* (2000). Effectiveness and cost-benefit of influenza vaccination of healthy working adults: A randomized controlled trial. J. Am. Med. Assoc. *284*, 1655–1663.

Campbell, D.S., and Rumley, M.H. (1997). Cost-effectiveness of the influenza vaccine in a healthy, working-age population. J. Occup. Environ. Med. *39*, 408–414.

Capua, I. and Marangon, S. (2007). The use of vaccination to combat multiple introductions of noticeable avian influenza viruses of the H5 and H7 subtypes between 2000 and 2006 in Italy. Vaccine. *25*, 4987–4995.

Chan, A.L. F., Shie, H.J., Lee, Y.J., and in, S.J. (2008). The evaluation of free influenza vaccination in health care workers in a medical center in Taiwan. *30*, 39–43.

Chen, H., Smith, G.J.D., Li, K.S., Wang, J., Fan, X.H., Rayner, J.M., Vijaykrishna, D., Zhang, J.X., Zhang, L.J., Guo, C.T., Cheung, C.L., Xu, K.M., Duan, L., Huang, K., Qin, K., Leung, Y.H.C., Wu, W.L., Lu, H.R., Chen, Y., Xia, N.S., Naipospos, T.S.P., Yuen, K.Y., Hassan, S.S., Bahri, S., Nguyen, T.D., Webster, R.G., Peiris, J.S.M., and Guan, Y. (2006). Establishment of multiple sublineages of H5N1 influenza virus in Asia: Implications for pandemic control. Proc. Natl. Acad. Sci. U.S.A. *103*, 2845–2850.

Chow, S.C. (2003). Encyclopedia of biopharmaceutical statistics. (London: Informa Healthcare, 2nd edn).

Clements, M.L., Betts, R.F., Tierney, E.L., and Murphy, B.R. (1986). Resistance of adults to challenge with influenza A wild-type virus after receiving live or inactivated virus vaccine. J. Clin. Microbiol. 23, 73–76.

Corias, M.T., Aguilar, J.C., Galiano, M., Carlos, S., Gregory, V., Lin, Y.P., Hay, A., and Perez-Biena, P. (2001). Rapid molecular analysis of hemagglutinin gene of human influenza A H3N2 viruses isolated in Spain from 1996 to 2000. Arch. Virol. *146*, 2133–2147.

Cox, N., Balish, A., Brammer, L., Fukuda, K., Hall, H., Klimov, A., Lindstrom, S., Marby, J., Perez-Oronoz, G., Postema, A., *et al.* (2003). Information for the vaccines and related biological products advisory committee, CBER, FDA, WHO Collaborating Center for surveillance epidemiology and control of influenza. Centers for Disease Control.

Cyranoski, D. (2005). Bird flu spreads among Java's pigs. Nature *435*, 390–391.

Davenport, F.M., Hennessy, A.V., and Francis, T. (1953). Epidemiologic and immunologic significance of age distribution of antibody to antigenic variants of influenza virus. J. Exp. Med. *98*, 641–656.

Deem, M.W., and Lee, H.-Y. (2003). Sequence space localization in the immune system response to vaccination and disease. Phys. Rev. Lett. *91*, 68–101.

Demicheli, V., Rivetti, D., Deeks, J., and Jefferson, T. (2004). Vaccines for preventing influenza in healthy adults. Cochrane Database Syst. Rev. 3, CD001269.

Dolan, S., Nyquist, A.C., Ondrejka, D., Todd, J., Gershman, K., Alexander, J., Bridges, C., Copeland, J., David, F., Fuler, G., *et al.* (2004). Preliminary assessment of the effectiveness of the 2003 4-inactivated influenza vaccine Colorado Denver. Centers for Disease Control and Prevention Morbidity and Mortality Weekly Report. *53*, 8–11.

Ducatez, M.F., Olinger, C.M., Owoade, A.A., De Landtsheer, S., Ammerlaan, W., Niesters, H.G.M., Osterhaus, A.D.M.E., Fouchier, R.A.M., and Muller, C.P. (2006). Multiple introductions of H5N1 in Nigeria. Nature. *442*, 37–37.

Edwards, K.M., Dupont, W.D., Westrich, M.K., Plummer, Jr. W.D., Palmer, P.S., and Wright, P.F. (1994). A randomized controlled trial of cold-adapted and inactivated vaccines for the prevention of influenza A disease. J. Infect. Dis. *169*, 68–76.

Ellis, J.S., Chakravarty, P., and Clewly, J.P. (1995). Genetic and antigenic variation in the hemagglutinin of recently circulating human influenza A H3N2 viruses in the United Kingdom. Arch. Virol. *140*, 1889–1904.

Eubank, S., Guclu, H., Anil Kumar, V.S., Marathe, M.V., Srinivasan, A., Toroczkai, Z., and Wang, N. (2004). Modeling disease outbreaks in realistic urban social networks. Nature. *429*, 180–184.

Fazekas, de S., Groth, St., and Webster, R.G. (1966). Disquisition on original antigenic sin: Evidence in man. J. Exp. Med. *124*, 331–345.

Ferguson, N.M., Galvani, A.P., and Bush, R.M. (2003). Ecological and immunological determinants of influenza evolution. Nature. *422*, 428–433.

Fitch, W.M., Butch, R.M., Bender, C.A., Subbarao, K., and Cox, N.J. (2000). Predicting the evolution of human influenza A. J. Hered. *91*, 183–185.

Fitch, W.M., Leiter, J.M., Li, X., and Palese, P. (1991). Positive Darwinian evolution in human influenza A viruses. Proc. Natl. Acad. Sci. U.S.A. *88*, 4270–4274.

Gupta, V., Earl, D.J., and Deem, M.W. (2006). Quantifying influenza vaccine efficacy and antigenic distance. Vaccine. *24*, 3881–3888.

Grotto, I., Mandel, Y., Green, M.S., Varsano, N., Gdalevich, M., Ashkenazi, I., and Shemer, J. (1998). Influenza vaccine efficacy in young, healthy adults. Clin. Infect. Dis. *26*, 913–917.

Guimera, R., Mossa, S., Turschi, A., and Amaral, L.A.N. (2005). The worldwide air transportation network: Anomalous centrality, community structure, and cites' global roles. Proc. Natl. Acad. Sci. U.S.A. *102*, 7794–7799.

Harper, S.A., Fukuda, K., Uyeki, T.M., Cox, N.J., and Bridges, C.B. (2004). Recommendations of the advisory committee in immunization practices. Centers for Disease Control and Prevention Morbidity and Mortality Weekly Report. 53(RR06), 1–40.

Hull, J.C. (2005). Options, futures and other derivatives. (New Jersey: Prentice Hall).

Ikonen, N., Kinnunen, L., Forsten, T., and Pyhala, R. (1996). Recent influenza B viruses in Europe: A phylogenetic analysis. Clin. Diag. Virol. *6*, 63–71.

Influenza virus resource at national center for biotechnology information NCBI (2007). http://www.ncbi.nlm.nih.gov/genomes/FLU/Database/multiple.cgi

Information for FDA vaccine advisory panel meeting (1997). Atlanta, GA: Centers for Disease Control.

Kawai, N., Ikematsu, H., Iwaki, N., Satoh, I., Kawashima, T., and Tsuchimoto, T., *et al.* (2003). A prospective, internet-based study of the effectiveness and safety of influenza vaccination in the 2002 influenza season. Vaccine. *21*, 4507–4513.

Keitel, W.A., Cate, T.R., and Couch, R.B. (1988). Efficacy of sequential annual vaccination with inactivated influenza virus vaccine. Am. J. Epidemiol. *127*, 353–364.

Keitel, W.A., Cate, T.R., Couch, R.B., Huggins, L.L., and Hess, K.R. (1997). Efficacy of repeated annual immunization with inactivated influenza virus vaccines over a five year period. Vaccine. *15*, 1114–1122.

Lee, M.S., and Chen, J.S.E. (2004). Predicting antigenic variants of influenza A/H3N2 viruses. Emerg. Infect. Dis. *10*, 1385–1390.

Lester, R.T., McGeer, A., Tomlinson, G., and Detsky, A.S. (2003). Use of, effectiveness of, attitudes regarding influenza vaccine among house staff. Infect. Control. Hosp. Epidemiol. *24*, 839–844.

Macken, C., Lu, H., Goodman, J., and Boykin, L. (2001). The value of a database in surveillance and vaccine selection. In: Osterhaus ADME, Cox N, Hampson AW, editors. Options for the control of influenza IV. Elsevier Science; (hemagglutinin H3 epitope structural mapping, http://www.flu.lanl.gov/).

Millot, J.L., Aymard, M., and Bardol, A. (2002). Reduced efficiency of influenza vaccine in prevention of influenza-like illness in working adults: A 7 month prospective survey in EDF Gaz de France employees, in Rhone-Alpes, 1996–1997. Occup. Med. Oxford. *52*, 281–292.

Mills, C.E., Robins, J.M., Bergstrom, C.T., and Lipsitch, M. (2006). Pandemic influenza: Risk of multiple introductions and the need to prepare for them. PLoS. Med. *3*, e135.

Mixeu, M.S.A.G., Vespa, C.N.R., Forleo-Neto, E., Toniolo-Neto J., and Alves, P.M. (2002). Impact of influenza vaccination on civilian aircrew illness and absenteeism. Aviat. Space. Environ. Med. *73*, 876–880.

Munoz, E., and Deem, M.W., (2005). Epitope analysis for influenza vaccine design. Vaccine. *23*, 1144–1148.

Newman, M.E.J. (2005). Power laws, Pareto distributions and Zipf's law. Contemporary Physics. *46*, 323–351.

Nguyen-Van-Tarn, J.S., and Hampson, A.W. (2003). The epidemiology and clinical impact of pandemic influenza. Vaccine. *21*, 1762–1768.

Nichol, K.L., Lind, A., Margolis, K.L., Murdoch, M., McFadden, R., Hauge, M., Magnan, S., and Deake, M. (1995). The effectiveness of vaccination against influenza in healthy, working adults. N. Engl. J. Med. *333*, 889–893.

Nichol, K.L., Margolis, K.L., Wuorenma, J., and Von. Sternberg, T. (1994). The efficacy and cost effectiveness of vaccination against influenza among elderly persons living in the community. New Engl. J. Med. *331*, 778–784.

Normile, D. (2006). Human transmission but no pandemic in Indonesia. Science. *312*, 1855–1855.

Plotkin, J.B., and Dushoff, J. (2003). Codon bias and frequency-dependent selection on the hemagglutinin epitopes of influenza A virus. Proc. Natl. Acad. Sci. U.S.A. *100*, 7152–7157.

Pontoriero, A.V., Baumeister, E.G. Campos, A.M., Savy, V.L., Lin, Y.P., and Hay, A.J. (2001). Pan. Am. J. Public Health. *9*, 246–253.

Press, W.H., Teukolsky, S.A., Vetterling, W.T., and Flannery, B.P. (2002). Numerical recipes in c++. The Art of Scientific Computing, 2n edn (Cambridge, UK: Cambridge University Press).

Smith, D.J., Forrest, S., Ackley, D.H., and Perelson, A.S. (1999). Variable efficacy of repeated annual influenza vaccination. Proc. Natl. Acad. Sci. U.S.A. *96*, 14001–14006.

Smith, J.W., and Pollard, R. (1979). Vaccination against influenza: A five-year study in the post office. J. Hyg. *83*, 157–170.

Sun, J., Earl, D.J., and Deem, M.W. (2006). Localization and glassy dynamics in the immune system. Mod. Phys. Lett. B. *20*, 63–95.

Ungchusak, K., Auewarakul, P., Dowell, S.F., Kitphati, R., Auwanit, W., Puthavathana, P., Uiprasertkul, M., Boonnak, K., Pittayawonganon, C., Cox, N.J., Zaki, S.R., Thawatsupha, P., Chittaganpitch, M., Khontong, R., Simmerman, J.M., and Chunsutthiwat, S. (2005). Probable person-to-person transmission of avian influenza A (H5N1). New Engl. J. Med. *352*, 333–340.

WHO FluNet (2006). http://gamapserver.who.int/GlobalAtlas/home.asp

WHO Weekly Epidemiological Record (1988). *63*, 57–59.

WHO Weekly Epidemiological Record (2004). *79*, 88–92.

World Health Organization (2005). H5N1 avian influenza: First steps towards development of a human vaccine. Weekly Epidemiological Record. *80*, 277–278.

Yang. Y., Halloran, M.E., Sugimoto, J.D., and Longini, I.M., Jr. (2006). Detecting human-to-human transmission of avian influenza A (H5N1). Emerg. Infect. Dis. *312*, 1855–1855.

Zipf, G.K. (1949). Human Behavior and the Principle of Least Effort (Cambridge, MA: Addison-Wesley Press, Inc.).

Index

N

O

P

R

Other Books of Interest

Caliciviruses: Molecular and Cellular Virology	2010
Epstein-Barr Virus: Latency and Transformation	2010
Lentiviruses and Macrophages: Molecular and Cellular Interactions	2010
RNA Interference and Viruses: Current Innovations and Future Trends	2010
Retroviruses: Molecular Biology, Genomics and Pathogenesis	2010
Frontiers in Dengue Virus Research	2010
Metagenomics: Theory, Methods and Applications	2010
Microbial Population Genetics	2010
Lab-on-a-Chip Technology: Biomolecular Separation and Analysis	2009
Lab-on-a-Chip Technology: Fabrication and Microfluidics	2009
Real-Time PCR: Current Technology and Applications	2009
RNA and the Regulation of Gene Expression	2008
Animal Viruses: Molecular Biology	2008
Segmented Double-Stranded RNA Viruses:	2008
Coronaviruses: Molecular and Cellular Biology	2008
Bacteriophage: Genetics and Molecular Biology	2007
Alpha Herpesviruses: Molecular and Cellular Biology	2007
AIDS Vaccine Development: Challenges and Opportunities	2007

Nanotechnology in Water Treatment Applications	2010
Iron Uptake and Homeostasis in Microorganisms	2010
Anaerobic Parasitic Protozoa: Genomics and Molecular Biology	2010
Borrelia: Molecular Biology, Host Interaction and Pathogenesis	2010
Aspergillus: Molecular Biology and Genomics	2010
Environmental Molecular Microbiology	2010
Neisseria: Molecular Mechanisms of Pathogenesis	2010
Frontiers in Dengue Virus Research	2010
ABC Transporters in Microorganisms	2009
Pili and Flagella: Current Research and Future Trends	2009
Bacterial Polysaccharides: Current Innovations and Future Trends	2009
Microbial Toxins: Current Research and Future Trends	2009
Acanthamoeba: Biology and Pathogenesis	2009
Bacterial Secreted Proteins: Secretory Mechanisms and Role in Pathogenesis	2009
Lactobacillus Molecular Biology: From Genomics to Probiotics	2009
Mycobacterium: Genomics and Molecular Biology	2009
Clostridia: Molecular Biology in the Post-genomic Era	2009
Plant Pathogenic Bacteria: Genomics and Molecular Biology	2009
Microbial Production of Biopolymers and Polymer Precursors	2009
Epigenetics	2008